era 代	*period 紀*			
	Permian 二疊紀	Upper		
		Lower		280
UPPER PALAEOZOIC 上古生代	**Carboniferous** 石炭紀	Upper 上	Stephanian 斯蒂芬 Westphalian 威斯法 Namurian 納繆爾	
		Lower 下	Visean 維憲 Tournaisian 杜內	345
	Devonain 泥盆紀	Upper 上	Famennian 法門 Frasnian 弗拉斯	
		Middle 中	Givetian 吉維特 Couvinian 艾菲爾	
		Upper 下	Emsian 艾姆斯 Siegenian 西根 Gedinnian 吉丁	400
	Silurian 志留紀	Upper 上	Přidolian 普列多里 Ludlow 羅德洛 Wenlock 温洛克	
		Lower 下	Llandovery 蘭多維列	435
LOWER PALAEOZOIC 下古生代	**Ordovician** 奧陶紀	Upper 上	Ashgill 阿什極 Caradoc 卡拉道克	
		Lower 下	Llandeilo 蘭代洛 Llanvirn 蘭維恩 Arenig 阿倫尼 Tremadoc 特馬豆克*	500
	Cambrian 寒武紀	Upper 上		
		Middle 中		
		Lower 下		570
	Pre-Cambrian 前寒武紀			

* included with the Upper Cambrian in some classifications.
某些劃分方案中劃入上寒武紀

NOTES The age and stage names given here for the various gelological sysgems are in common sue in western Europe, but many other classifications are also used.
備　註 上述各地質系統中的期和階的名稱在西歐是通用的，此外還有其他劃分方案。

The meanings of the words era, sub-era, period,epoch, age, and stage willbe found on page 113.
代、亞代、紀、世、期和階等詞的含義見第 113 頁。

Approximate ages for the boundaries between the geological sysgems are given in millions fo years (Ma).
地質系統之間的分界年齡以百萬年 (Ma) 為單位。

前言

朗文科學系列圖解詞典包括生物、化學、科學、植物和地質五冊。這是一套內容既有聯繫而又各自獨立成冊的系列詞典，《朗文英漢地質圖解詞典》為其中的一冊。

本書收詞1500多個，包括地貌學、地質物理學、地質化學、結晶學、岩石學、構造地質學、板塊構造、地層學、侵蝕、沉積物和礦物等全部地質學基本原理的詞彙。

這些詞按詞義分科目編排，英漢雙解對照，釋義簡明，概念準確。每一科目的上、下各個詞條，內容互有關聯。釋義深入淺出，易於理解；其中又標出頁碼和箭嘴號，引導讀者查找相關詞條，提供更多資料作比較，以加深理解，掌握更多詞彙。

詞典中收入近350幅印刷精美的彩色插圖和圖表，直觀地顯示所闡釋的題目和原理，以助讀者更好理解釋義，但釋義並不依賴這些插圖。

此外，本詞典還收入國際單位制、地質學常用縮寫詞、地質學用的詞頭和詞尾四個實用附錄，並標注 K.K. 音標，自成一個英漢地質學詞彙表，方便讀者檢索。

本詞典適合具高中至大學一、二年級程度的學生以及需要深入了解地質學術語的非地質學專業的讀者查閱使用。

朗文出版（遠東）有限公司
一九九二年三月

朗文英漢地質圖解詞典

LONGMAN ENGLISH-CHINESE ILLUSTRATED DICTIONARY OF GEOLOGY

Longman 朗文 YORK PRESS

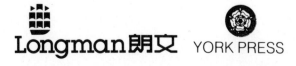

Longman Group (Far East) Ltd
18/F., Cornwall House
Tong Chong Street
Quarry Bay
Hong Kong
Tel: 811 8168
Fax: 565 7440
Telex: 73051 LGHK HX

First published 1992

ISBN 962 359 362 7 (Hong Kong Edition)
ISBN 962 359 724 X(Taiwan Edition)

Produced by Longman Group (Far East) Ltd.
Printed in Hong Kong

朗文出版（遠東）有限公司
香港鰂魚涌糖廠街
康和大廈十八樓
電話：811 8168
圖文傳真：565 7440
電傳：73051 LGHK HX

一九九二年初版

國際書號 962 359 362 7（香港版）
國際書號 962 359 724 X（台灣版）

出版：朗文出版（遠東）有限公司
印刷：香港

Contents 目錄

How to use the dictionary 本詞典的用法

This dictionary contains over 1500 words used in the geological sciences. These are arranged in groups under the main headings listed on pp.3–4. The entries are grouped according to the meaning of the words to help the reader to obtain a broad understanding of the subject.

At the top of each page the subject is shown in bold type and the part of the subject in lighter type. For example, on pp.12 and 13:

本詞典共收地質科學用詞約 1,500 多條。這些詞按第 3 至 4 頁所列的主要標題分類編排。所有詞條均按詞義歸類以幫助讀者對所查找之科目獲得一個概括的瞭解。

在每頁的上方以黑體字印出有關科目名稱，並以秀麗體印出該科目下的分段名稱。例如，第 12 頁和第 13 頁：

12 · GEOPHYSICS/EARTHQUAKES AND SEISMOLOGY

GEOPHYSICS/THE EARTH'S MAGNETISM · **13**

12 · 地球物理學/地震與地震學

地理物理學/地震與地震學 · 13

In the definitions the words used have been limited so far as possible to about 1500 words in common use. These words are those listed in the 'defining vocabulary' in the *New Method English Dictionary* (fifth edition) by M. West and J.G. Endicott (Longman 1976). Words closely related to these words are also used: for example, *characteristic*, defined under *character* in West's *Dictionary*. For some definitions other words have been needed. Some of these are everyday words that will be familiar to British readers; others are scientific words that are not central to geology. These are listed in Appendix 1 (pp.151–61). If therefore you find a word in a definition that is not familiar to you, you should turn to Appendix 1.

釋義部分所用的詞盡可能限於常用的 1,500 個詞左右。這些詞列於 M · 韋斯特和 J · G · 恩迪科特合編的《新法英語詞典》(第五版，朗文公司 1976 年版)中的釋義詞彙表內。釋義時也使用一些與這些詞密切相關的詞，例如：使用韋斯特詞典中在 "character (特性)" 條下解釋的 "characteristic (特徵)" 這個詞。對於某些釋義還要用另外一些詞。其中有些是英國讀者熟悉的日常用詞彙，有些則是非側重於地質學方面的科學用詞彙。這些詞彙都收列於附錄一(151 至 161 頁)中，因此，如您在釋義文句中發現不熟悉的詞時應該查閱附錄一。

1. To find the meaning of a word

Look for the word in the alphabetical index at the end of the book, then turn to the page number listed.

The description of the word may contain some words with arrows in brackets (parentheses) after them. This shows that the words with arrows are defined near by.

(↑) means that the related word appears above or on the facing page;

(↓) means that the related word appears below or on the facing page.

A word with a page number in brackets (parentheses) after it is defined elsewhere in the dictionary on the page indicated. Looking up the words referred to in either of these two ways may help in understanding the meaning of the word that is being defined.

The explanation of each word usually depends on knowing the meaning of a word or words above it. For example, on p.123 the meaning of *axial plane, fold-axis,* and the words that follow depends on the meaning of the word *fold*, which appears above them. Once the earlier words are understood those that follow become easier to understand.

1. 查明詞的意義

在詞典末尾的字母順序索引中找出欲查的詞，然後翻到該詞旁註明的頁碼。

詞的釋義中遇有一些詞後面帶箭號括在括弧(圓括弧)內，表示其解釋在附近。

(↑)表示和這個相關的詞出現在本詞條之前或前一頁上；

(↓)表示和這個相關的詞出現在本詞條之後或後一頁上。

遇到某個詞後面帶頁碼括在括弧(圓括弧)內，表示該詞的解釋在所註明的頁碼上。參照這兩種方式查出這些詞並閱讀其解釋，可幫助您更好地理解原先所查那個詞的詞義。

對每個詞的解釋通常都依賴於理解該詞前面出現的一個或幾個詞的意義。例如第 123 頁，"軸面"、"褶皺軸"及接着出現的幾個詞的詞義，都依賴於在上述這些詞前面出現的"褶皺"這個詞的意義。理解了在前面出現的那些詞的詞義之後，就比較容易理解接着出現的那些詞的意義。

2. To find related words

2. 查找相關的詞

Look in the index for the word you are starting from and turn to the page number shown. Because this dictionary is arranged by ideas, related words will be found in a set on that page or one near by. The illustrations will also help here.

For example, words relating to volcanic eruptions are on pp.68–70. On p.68 *volcano* is followed by words used to describe various kinds of volcanic eruption and types of volcano; p.61 gives words for solid materials thrown out from volcanoes; p.62 lists words for liquid and gaseous materials from volcanoes.

從索引中查出作為起頭的詞，翻到標明的頁碼。由於本詞典是按照概念編排詞條，所以在同一頁或其前、後的一頁上可找出一組相關的詞，插圖也有助於理解。

例如，在第 68 至 70 頁上列出有關火山噴發有關的詞。第 68 頁，在"火山"這詞後面編排的各詞條都是闡釋不同類型火山噴發和火山類型；在第 69 頁列舉有關噴出的各種固體物質的詞彙；在第 70 頁列出有關火山噴出的各種液體和氣體物質的詞彙。

3. As an aid to studying or revising

3. 學習和複習的輔助工具

There are two methods of using the dictionary in studying or revising a topic. You may wish to see if you know the words used in that topic or you may wish to revise your knowledge of a topic.

(*a*) To find the words used in crystallography, you would look up *crystallography* in the alphabetical index. Turning to the page indicated, p.40, you would find *crystal, crystallize, crystallography, crystal lattice*, and so on. Turning over to p.41 you would find *crystallographic axis, intercept,* and so on; and on p.42 *symmetry* etc.

(*b*) Suppose that you wished to revise your knowledge of a topic; e.g. *sediments*. If, say, the only word you could remember was *deposit* you could look in the alphabetical index and find *deposit*. The page reference is to p.80. There you would find the words *sediment, sedimentation, deposit, deposition, bed, stratum, bedding-plane,* etc. If you next looked at p.81, you would find words relating to the various environments in which sediments are deposited; then on p.82 you would find words relating to the characters of sediments; on p.83 words describing sedimentary structures; and so on.

本詞典可用於學習和複習某一個課題，有兩種用法。您可以利用本詞典查核自己是否認識該課題中所用的詞彙，也可以用本詞典複習某一項課題的知識。

(a) 查找結晶體方面用的詞時先在字母順序索引查出"結晶學"(crystallography) 這個詞。翻到標註的第 40 頁，可找到"晶體"、"使結晶"、"結晶學"、"晶格"等詞條。在第 41 頁可查到"結晶軸"、"截距"等詞；在第 42 頁可查到"對稱"等詞。

(b) 複習某項課題的知識，例如有關沉積物的知識。如您只記得"沉積"這個詞，則只需在字母順序索引中查出這個詞，翻到標註的第 80 頁，可查到"沉積物"、"沉積形成作用"、"沉積"、"沉積作用"、"層"、"地層"、"層面"等詞。再翻到第 81 頁，可查到和沉積物沉積的各種環境有關的一些詞；再翻到第 82 頁，可查到與沉積物特性有關的一些詞；在第 83 頁上可查到闡釋沉積構造的詞等等。

4. To find a word to fit a required meaning

4. 查找適當的詞，以表達確切的意義

It is almost impossible to find a word to fit a meaning in most dictionaries, but it is easy with this book. For example, if you had forgotten the word for the angle between the axis of a fold and the horizontal all you would have to do would be to look up *fold axis* in the alphabetical index and turn to the page indicated, p.123. There you would find the word *pitch* with a diagram to illustrate its meaning.

在大多數詞典中，您想查找一個適當的詞來確切表達某一意義，這幾乎是不可能的，但用本詞典却可輕易做到這一點。例如，您忘了表達褶皺與水平面的交角這個含意的詞，您祇需從字母順序索引中查出"褶皺軸"(fold axis)這個詞，並翻到所標示的第 123 頁。即可查到您所要的"側伏角"(pitch)這一個詞以及解釋其意義的插圖。

THE DICTIONARY
詞典正文

section through the earth
地球的斷面圖

geology (*n*) the science of the Earth: how it was formed, what it is made of, its history and the changes that take place on it and in it. Geology includes parts of geophysics (p.11), mineralogy (p.44), petrology (p.62), stratigraphy (p.112), palaeontology (p.98), and structural geology (p.122). **geological, geologic** (*adj*).

Earth sciences a group of sciences that includes geology, geophysics (p.11), geochemistry (p.15), oceanography (p.34), meteorology (the study of the weather), and astronomy (the study of the heavenly bodies) so far as it concerns the Earth.

crust (*n*) the part of the Earth above the Mohorovičić discontinuity (p.10). It is less dense than the mantle (↓). The *continental crust* of the great land areas is thicker, less dense, and older than the *oceanic crust*. **crustal** (*adj*).

mantle (*n*) the part of the Earth between the crust (↑) and the core (↓), i.e. between the Moho (p.10) and the Gutenberg discontinuity (p.10). It probably consists largely of MgO and SiO_2 with sodium, calcium, and aluminium.

core (*n*) the central part of the Earth, below the Gutenberg discontinuity (p.10) at a depth of about 2900 km below the Earth's surface. The core is thought to consist almost entirely of iron. It can be divided into the *outer core*, which is probably liquid, and the *inner core*, which is probably solid, at a depth of 5100 km. The density (p.154) of the core is more than twice the density of the mantle (↑).

lithosphere (*n*) the outer, solid part of the Earth: the crust (↑) and the upper part of the mantle (↑) to a depth of about 100 km. The lithosphere is stiffer than the asthenosphere (↓). **lithospheric** (*adj*).

asthenosphere (*n*) the part of the mantle (↑) from a depth of about 100 km to 250–300 km. It is not as strong and stiff as the lithosphere (↑).

mesosphere (*n*) the part of the mantle (↑) below the asthenosphere (↑), i.e. from a depth of 250–300 km to the core (↑).

atmosphere (*n*) the gases surrounding the Earth. **atmospheric** (*adj*).

地質學(名) 研究地球的科學，即研究地球的形成、地球的構成、地球的歷史以及地球上和地球內部所發生的變化的科學。地質學包括地球物理學(第 11 頁)、礦物學(第 44 頁)、岩石學(第 62 頁)、地層學(第 112 頁)、古生物學(第 98 頁)和構造地質學(第 122 頁)等分科。(形容詞為 geological, geologic)

地球科學 研究地球的一組學科，包括地質學、地球物理學(第 11 頁)、地球化學(第 15 頁)、海洋學(第 34 頁)、氣象學(研究氣候)和天文學(研究各種天體)等與地球有關的學科。

地殼(名) 地球上位於莫霍界面(第 10 頁)以上的部分。其密度小於地幔(↓)。陸地面積巨大的大陸地殼比海洋地殼厚，密度小，年齡則較老。(形容詞為 crustal)

地幔(名) 地球的一部分，在地殼(↑)與地核(↓)之間，即居於莫霍界面(第 10 頁)與古登堡間斷面(第 10 頁)之間。其主要組成可能是 MgO 和 SIO_2，並含少量鈉、鈣和鋁。

地核(名) 地球的中心部分，在古登堡間斷面(第 10 頁)之下，距地表約 2900 公里。人們認為地核幾乎全由鐵組成。地核可再分為外核與內核，外核可能為液態，內核可能為固態，深度約 5100 公里。地核的密度(第 154 頁)大於地幔(↑)一倍以上。

岩石圈(名) 地球外層的固體部分：包括地殼(↑)和地幔(↑)上部，深度約 100 公里。岩石圈比軟流圈(↓)剛硬。(形容詞為 lithospheric)

軟流圈(名) 地幔(↑)的一部分，從約 100 公里深度至 250-300 公里之處。它不像岩石圈(↑)那麼剛硬。

中圈(名) 地幔(↑)的一部分，位於軟流圈(↑)之下，從 250-300 公里深度至地核(↑)之處。

大氣圈(名) 環繞地球的氣體。(形容詞為 atmospheric)

sial (*n*) a term for the parts of the Earth's crust (p.9) made up of rocks containing silica (p.16) and alumina (p.16). **sialic** (*adj*). *See also* **sima** (↓).

矽鋁層；硅鋁層（名） 地殼（第 9 頁）某些部分的術語，由含矽石（第 16 頁）和氧化鋁（第 16 頁）的岩石組成。（形容詞為 sialic）。參見**矽鎂層**（↓）。

sima (*n*) a term for the parts of the Earth's crust (p.9) made up of rocks containing silica (p.16) and magnesium. **simatic** (*adj*). *See also* **sial** (↑).

矽鎂層；硅鎂層（名） 地球地殼（第 9 頁）某些部份的術語，由含矽石（第 16 頁）和鎂的岩石組成。（形容詞為 simatic）。參見**矽鋁層**（↑）。

discontinuity (*n*) a layer or boundary within the Earth that separates parts of the Earth having different properties, e.g. seismic properties. See also Mohorovičić discontinuity (↓) and Gutenberg discontinuity (↓).

間斷面（名） 將地球內部分開為具有不同性質（如地震性質）的部分的一層或一個邊界面。參見莫霍界面（↓）和古登堡間斷面（↓）。

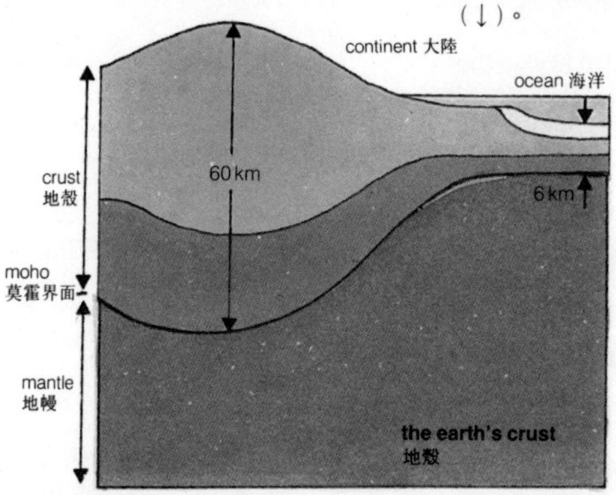

Mohorovičić discontinuity, Moho, M discontinuity a boundary that separates the crust (p.9) above from the mantle (p.9) below. The Moho is at a depth of about 10–70 km below the surface of the continents and about 5–10 km below the ocean floor. There is a difference between the velocities of earthquake waves (p. 12) above and below the Moho.

莫霍洛維奇間斷面，莫霍界面，M 間斷面 上與地殼（第 9 頁）下與地幔（第 9 頁）相分隔的邊界面。莫霍界面距大陸地表約 10-70 公里，距洋底約 5-10 公里。地震波（第 12 頁）在莫霍界面上、下的傳導速度有差異。

Gutenberg discontinuity a boundary that separates the mantle (p.9) from the core (p.9) at a depth of about 2900 km below the Earth's surface. The velocities of earthquake (p.12) waves are different above and below the Gutenberg discontinuity.

古登堡間斷面 分開地幔（第 9 頁）與地核（第 9 頁）的邊界面，距地表約 2900 公里。地震波（第 12 頁）在其上、下的傳導速度有差異。

Weichert–Gutenberg discontinuity = Gutenberg discontinuity (↑).

魏徹特-古登堡間斷面 同古登堡間斷面（↑）。

geophysics (*n*) the study of the physics (p.158) of the Earth, including the hydrosphere (p.34) and the atmosphere (p.9). **geophysical** (*adj*).

地球物理學（名） 研究地球，包括水圈（第 34 頁）與大氣圈（第 9 頁）的物理學（第 158 頁）。（形容詞為 geophysical）

gravity (*n*) the force that pulls a body towards the centre of the Earth. It becomes less with increasing distance from the centre of the Earth and varies according to the mass of the rocks below the surface. For geological purposes, the equivalent values of gravity (*g*) at sea level are calculated from the actual measurements made at a particular place. **gravitational** (*adj*).

重力（名） 將物體吸引向地心的力。重力隨着與地球中心距離的增大而減少，並隨地表下面岩石的質量變化而不同。將在某個特定地點實測的一系列重力數值，換算成在海平面的等效重力數值（*g*），以供地質上使用。（形容詞為 gravitational）

gravitational acceleration the acceleration (p.151) due to gravity (↑); about 9.8 m s^{-2}.

重力加速度 由重力（↑）產生的加速度（第 151 頁），約為 9.8 m s^{-2}。

gal (*n*) a unit for the measurement of gravity (↑); 1 gal is an acceleration (p.151) of 1 cm s^{-2}.

伽（名） 重力（↑）的量度單位：1 伽等於加速度（第 151 頁）1 cm s^{-2}。

milligal (*n*) one-thousandth of a gal (↑); an acceleration (p.151) of 0.01 mm s^{-2}. The milligal is the unit generally used for measuring values of gravity in geophysical work.

毫伽（名） 千分之一伽（↑）： 0.01 mm s^{-2} 的加速度。是地球物理工作上量度重力值的常用單位。

gravity meter, gravimeter an instrument for measuring gravity (↑).

重力儀，重力計 測量重力（↑）的儀器。

gravity anomaly the difference between the value of gravity (↑) measured at a particular place and the value for an imaginary Earth with no variations in density. Gravity anomalies can provide knowledge of variations in the density (p.154) of the rocks below the Earth's surface and are used in studying subsurface structures.

重力異常 在特定地區實測的重力（↑）值與假設地球密度為均一時的理想重力值之間的差異。重力異常可提供地表以下岩石密度變化（第 154 頁）的資料，並藉以研究地下構造。

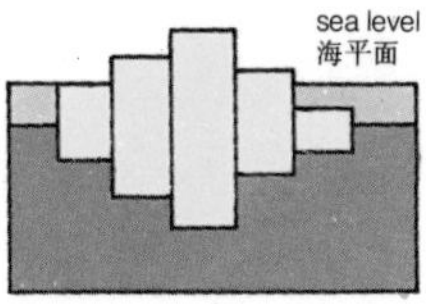

isostasy (*n*) the theory that the Earth's crust (p.9) is near to a state of equilibrium (p.155) without any tendency to move up or down, and that large blocks of the crust behave like blocks floating in a liquid. **isostatic** (*adj*).

地殼均衡說（名） 這一學說認為地殼（第 9 頁）處於接近平衡的狀態（第 155 頁），沒有任何向上或向下移動的傾向，而每一塊大地殼則像漂浮在液體中的地塊。（形容詞為 isostatic）

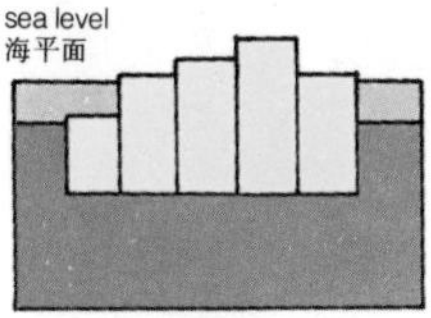

isostacy 地殼均衡

isostatic compensation the means by which differences in the heights of parts of the Earth's crust (p.9) are balanced, either by 'roots' below them or by variations in the density (p.154) of the crust.

均衡補償 地殼（第 9 頁）上各地塊間的高度差異，為其下面"根部"的深度差異所調節，或為其密度差異（第 154 頁）所調節，而處於平衡狀態的方式。

isostatic adjustment vertical movement in the Earth's crust (p.9) resulting from lack of isostatic equilibrium (↑), e.g. a rise in the level of the land surface after the weight of an ice sheet (p.28) has been taken away. The term 'isostatic adjustment' is also used to mean isostatic compensation (↑).

均衡調整 地殼（第 9 頁）由於均衡（↑）失調，產生垂直運動而達致新的均衡。例如，由於冰蓋（第 28 頁）溶化失去重量而導致地面上升。"均衡調整"這個術語也有均衡補償（↑）的意思。

earthquake (*n*) a sudden movement of part of the Earth's crust (p.9); a shock produced in the Earth's crust or mantle (p.9). An earthquake may be caused by movement along a fault (p.128) or by volcanic activity (p.68).
seismology (*n*) the study of earthquakes (↑). **seismological** (*adj*).
seismic (*adj*) relating to earthquakes (↑).
seismograph (*n*) an instrument for studying distant earthquakes (↑).
focus (*n*) the point from which an earthquake (↑) shock comes.
epicentre (*n*) the point on the Earth's surface directly above the focus (↑) of an earthquake (↑).
magnitude (*n*) a measure of the amount of energy (p.155) set free in an earthquake (↑). The magnitude of an earthquake is usually measured on the Richter scale.
intensity (*n*) a measure of the effects of an earthquake (↑) as estimated from the damage done. The scales of intensity generally used are the Mercalli, modified Mercalli, and Rossi–Forel.
P-waves earthquake (↑) waves in which the movements are in the same direction as that in which the waves travel. P-waves are of high frequency (p.156) and short wavelength (p.161).
S-waves earthquake (↑) waves in which the movements are at 90° to the direction in which the waves travel. S-waves are of high frequency (p.156) and short wavelength (p.161).
Rayleigh waves surface waves produced by an earthquake (↑) that give a rolling movement to the ground.
L-waves surface waves produced by an earthquake (↑) that cause horizontal movement at 90° to the direction in which the waves travel. L-waves are of low frequency (p.156) and long wavelength (p.161).
isoseismal, isoseismal line a line joining points at which the intensity (↑) of an earthquake (↑) is the same.
shadow zone the zone in which P-waves (↑) and S-waves (↑) are not received from a distant earthquake (↑). It lies between 103° and 143° from the epicentre (↑) of the earthquake.

地震(名)　局部地殼(第 9 頁)的一種突然運動；地殼或地幔(第 9 頁)發生的震動。地殼沿斷層(第 128 頁)移動，或火山活動(第 68 頁)都可產生地震。

地震學(名)　研究地震(↑)的科學。(形容詞為 seismological)

地震的(形)　與地震(↑)有關的。

地震儀(名)　分析研究遠處地震(↑)用的儀器。

震源(名)　發生地震(↑)衝擊的源點。

震中(名)　地震(↑)震源(↑)正上方的地面。

震級(名)　一次地震(↑)所釋放能量(第 155 頁)的量度。地震震級通常用里氏震級表量度。

烈度(名)　據地震(↑)所造成的災害估量一次地震所產生的破壞或影響的量度。地震烈度通常採用墨卡利烈度表、修訂的墨卡利烈度表和羅西-弗瑞爾烈度表量度。

P 波　震動方向與地震(↑)波傳導方向相同的地震波。P 波具有高頻率(第 156 頁)、短波長(第 161 頁)的特性。

S 波　震動方向與地震(↑)波傳導方向垂直的地震波。S 波具有高頻率(第 156 頁)、短波長(第 161 頁)的特性。

瑞利波　由地震(↑)產生的表面波，它使地表發生滾動運動。

L 波　由地震(↑)產生的一種表面波，它可導致與地震波傳導方向成 90° 的水平運動。L 波具有低頻率(第 156 頁)、長波長(第 161 頁)的特性。

等震的，等震線　聯接地震(↑)相等烈度(↑)各點的線段。

陰影區；震影帶　接收不到遠方地震 P 波(↑)和 S 波(↑)的地帶，位於離震中(↑) 103°–143° 的範圍內。

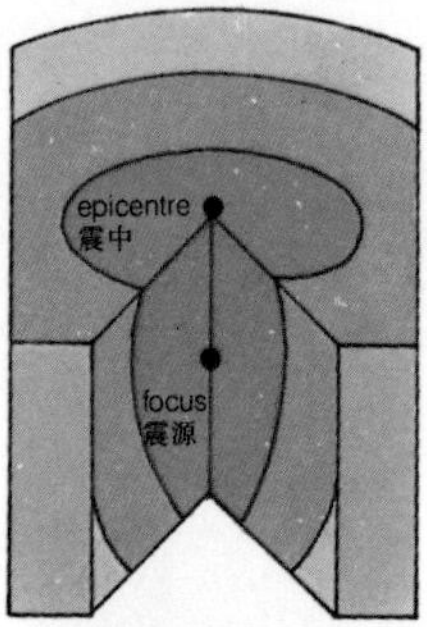

earthquake focus and epicentre
震源和震中

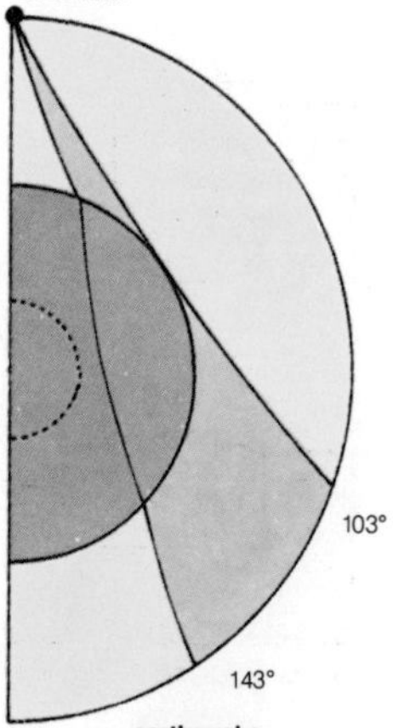

earthquake shadow zone
地震陰影區

microseism (*n*) a very small movement of the Earth's crust (p.9) recorded by a seismograph (↑). Microseisms are caused by the wind, waves, etc.

微震(名) 地震儀(↑)所記錄的一種極輕微的地殼(第 9 頁)運動。 風、浪等因素可產生微震。

tremor (*n*) a small earthquake (↑).

小震(名) 輕微的地震(↑)。

principal shock the main shock in a large earthquake (↑).

主震 一次大地震(↑)中的主要震動。

fore-shock a shock occurring before the principal shock (↑) in a large earthquake (↑).

前震 一次大地震(↑)中發生在主震(↑)前的震動。

after-shock a shock occurring after the principal shock (↑) in a large earthquake (↑).

餘震 一次大地震(↑)中發生在主震(↑)後的震動。

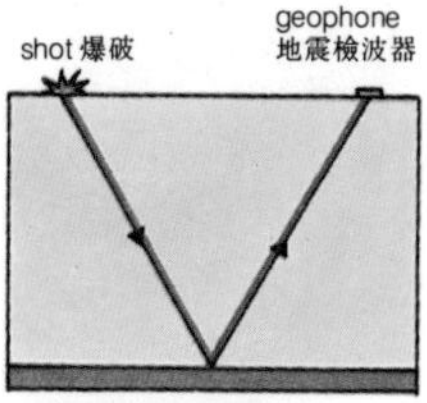

reflection seismology
反射地震學

reflection seismology a method of studying the rocks below the Earth's surface. Explosives are let off at the surface (a *shot*) to produce waves that pass down into the ground. *See diagram*. (A machine may be used instead of explosives.) The waves are *reflected*, i.e. thrown back, by discontinuities (p.10) below the surface and the reflected waves are detected at the surface by a device called a *geophone*. The time taken for the wave to travel from the surface to the discontinuity and back to the geophone is measured and can be used to calculate the depth of the reflecting layer. *See also* **refraction seismology** (↓).

反射地震學 研究地表之下的岩石的一種方法。其法是：在地面引爆炸藥(爆破)以產生震波並將之傳送到地下。見示意圖。(可用機械代替炸藥)。震波被地面下的間斷面(第 10 頁)反射成為反射波，再由安置在地面的地震檢波器接收。測出震波從地面傳播至間斷面再反射回地面檢波器所需的時間，用以計算反射層的深度。參見**折射地震學**(↓)。

refraction seismology a method of studying the rocks below the Earth's surface. As in reflection seismology (↑) explosives are let off at the surface to produce waves that pass down into the Earth. At a certain angle (the *critical angle*) the waves produced will travel along a discontinuity (p.10). Energy (p.155) will then be *refracted*, i.e. its path will be bent, and it will travel upwards from the discontinuity. The depth of the discontinuity can be calculated by measuring the times at which the wave arrives at a number of points at the surface.

折射地震學 研究地表之下的岩石的一種方法，方法和反射地震學(↑)一樣，在地面引爆炸藥產生震波並將之傳送到地下。當震波的入射角等於臨界角時，所產生的震波沿着間斷面(第 10 頁)傳播。然後能量(第 155 頁)發生折射，即其傳播路徑發生彎曲，並從間斷面向上傳播。測出折射波到達地面多個測站的時間就可以計算出間斷面的深度。

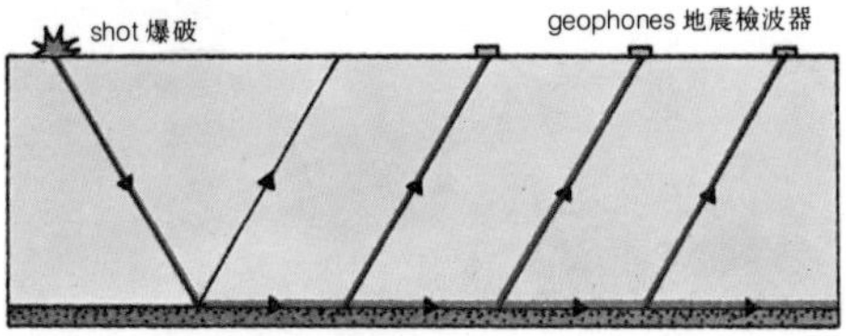

refraction seismology 折射地震學

terrestrial magnetism the magnetic field (p.157) of the Earth. The Earth's magnetic field is *dipolar*, i.e. it has two poles (north and south) like a bar magnet (p.157).

大地磁場 地球的磁場(第 157 頁)。地球的磁場是雙極的，即像磁棒(第 157 頁)一樣有兩個磁極(南極和北極)。

geomagnetism (*n*) (1) the Earth's magnetic field (p.157); (2) the study of the Earth's magnetic properties. **geomagnetic** (*adj*).

地磁(學)(名) (1)地球的磁場(第 157 頁)；(2)研究地球磁場性質的學科。(形容詞為 geomagnetic)

magnetic poles the two points on the Earth's surface to which a compass needle points. The Earth's north and south magnetic poles are not in the same places as the geographical poles and they move with time.

地磁極 在地球的表面上羅盤磁針所指向的兩個點。地球的南、北磁極與地理的南、北極不在同一個位置，它隨時間而移動。

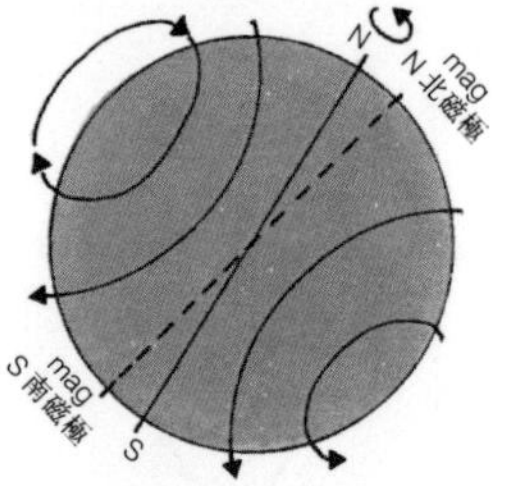

magnetic poles
地磁極

magnetometer (*n*) an instrument for measuring the strength of the Earth's magnetic field.

磁力儀(名) 測量地球磁場強度的儀器。

magnetic anomaly (*n*) a variation or irregularity in the Earth's magnetic field (p.157) as measured at a particular place. Variations caused by differences in the magnetic properties of minerals and rocks are of value in studying the structure (p.122) of the rocks below the surface.

磁異常(名) 在一特定地區測得的地球磁場(第 157 頁)的變化或不規則性。岩石和礦物的磁性不同所引起的磁性變化對研究地下岩石構造(第 122 頁)很有價值。

palaeomagnetism (*n*) 'fossil magnetism'; the magnetism of rocks that is thought to have been there since they were formed. Studies of the magnetization of rocks can provide information about the apparent movements of the Earth's magnetic poles (↑) that have taken place in the geological past and about the movements of continents. **palaeomagnetic** (*adj*).

古地磁(學)(名) "化石磁場"：此種磁性自岩石形成時就已有了。研究岩石磁化作用可獲得過去地質時代地磁極(↑)產生視移動的信息以及有關大陸漂移的信息。(形容詞為 palaeomagnetic)

remanent magnetization the lasting magnetization produced in a material (e.g. a rock) by a magnetic field (e.g. the Earth's magnetic field).

剩餘磁化 磁場(如地球磁場)對某物質(如岩石)所產生的持久磁化作用。

Curie point the temperature at which a material (e.g. a rock) loses any magnetization it has gained.

居里點 指一種物質(如岩石)失去其所獲的任何磁化作用的溫度。

reversed polarity when the Earth's north and south magnetic poles (↑) change places the Earth's magnetic polarity is said to be *reversed*. This has happened many times in the geological past.

極性倒轉 地球的南、北磁極(↑)倒轉換位，稱為極性倒轉。在過去的地質時代曾多次發生這種現象。

palaeopole location the geographical position of one of the Earth's magnetic poles (↑) at some time in the geological past, as shown by magnetic measurements made on rocks. *See also* **polar wander** (p.141).

古地磁極位 由岩石磁性測定所顯示的過去某一地質時期的地球磁極(↑)的地理位置。參見**極移**(第 141 頁)。

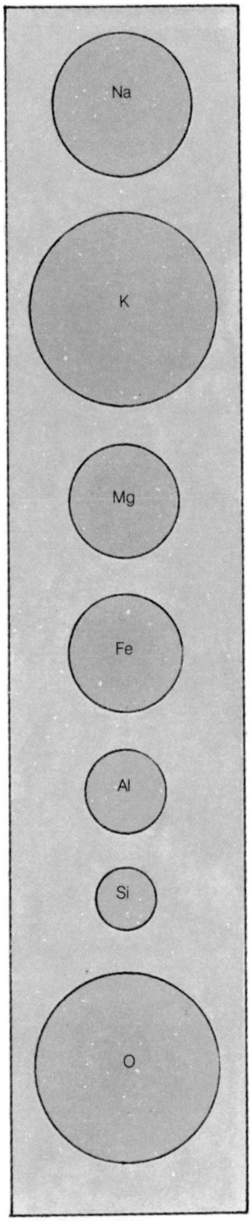

radii of some ions
某些離子的半徑

geochemistry (*n*) the chemistry (p.153) of the Earth, and especially the chemistry of the distribution of the elements (↓) in the various parts of the Earth. **geochemical** (*adj*).

element (*n*) a substance that cannot be broken down by chemical means into simpler substances.

native element an element (↑) occurring in nature in its free state as a mineral (p.44).

trace element an element (↑) occurring in very small quantities.

rare earth element one of a series of metallic elements (↑) which have very similar chemical properties. They occur in group III of the periodic table (p.18) and they all form a 'basic' oxide (an 'earth').

REE = rare earth element(s) (↑).

ion (*n*) an atom (p.152) that has gained or lost one or more electrons (p.155) and has become electrically charged (p.153).

cation (*n*) an ion (↑) with a positive electrical charge (p.153); e.g. a sodium ion is Na^+

anion (*n*) an ion (↑) with a negative electrical charge; e.g. Cl^- (chlorine).

ionic radius (*radii*) a measure of the size of an ion (↑). The ionic radius, usually expressed in ångstrom units (10^{-8} cm), is important in controlling the way in which ions will pack together.

molecule (*n*) the smallest particle of an element (↑) or chemical compound (↓) that can occur in the free state and has the properties of that element or compound. A molecule of an element consists of one or more atoms (p.152); a molecule of a chemical compound (↓) consists of one or more atoms of each of the elements that make up the compound. **molecular** (*adj*).

molecular structure the arrangement in space of the atoms (p.152) in a molecule (↑).

chemical compound a substance made up of two or more elements (↑) in a fixed proportion by weight.

chemical composition the nature of the elements (↑) that make up a substance and the proportions in which they are present in it.

地球化學（名） 地球的化學（第 153 頁），特別是地球不同部位中元素（↓）的分佈化學。（形容詞為 geochemical）

元素（名） 不能用化學方法再細分成更簡單物質的一種物質。

自然元素 在自然界中以游離態作為礦物（第 44 頁）存在的一種元素（↑）。

微量元素 存在量極少的元素（↑）。

稀土元素 化學性質極相似的一系列金屬元素（↑）之一，列於週期表（第 18 頁）第 III 族，可以形成"鹼性"氧化物（一種"鹼土"）。

稀土元素（↑）之英文縮寫為 REE。

離子（名） 獲得或失去一個或多個電子（第 155 頁）而帶電荷之原子（第 152 頁）。

陽離子（名） 帶正電荷（第 153 頁）的離子（↑）。例如 Na^+ 為鈉離子。

陰離子（名） 帶負電荷的離子（↑），例如 Cl^- 為氯離子。

離子半徑 離子（↑）大小的尺度。離子半徑通常以埃（10^{-8} cm）為單位，是支配離子相互堆集的重要因素。

分子（名） 能以游離狀態存在並具元素（↑）或化合物（↓）的特性的最小微粒。元素的分子由一個或多個原子（第 152 頁）組成；化合物（↓）的分子則由構成該化合物的各元素的一個或多個原子組成。（形容詞為 molecular）

分子結構 一個分子（↑）中的各原子（第 152 頁）的空間排列。

化合物 由兩個或多個元素（↑）按固定重量比例構成的一種物質。

化學成分 構成一種物質的各元素（↑）的屬性和其比例。

oxide (*n*) a chemical compound (p.15) of oxygen and another element (p.15). Oxides may be solids, liquids, or gases. Solid oxides are present in the Earth's crust. Many ores (p.145) are oxides; e.g. haematite, Fe_2O_3 (p.48).

alumina (*n*) the oxide (↑) of aluminium, Al_2O_3.

alkali (*n*) (1) a chemical compound (p.15) that will dissolve (p.155) in water to produce a solution (p.159) that will neutralize (↓) an acid. (2) In geology, and especially in petrology (p.62), the word 'alkali' is used for rocks and minerals (p.44) that contain a large proportion of the alkali metals (↓). **alkaline, alkalic** (*adj*).

halide (*n*) one of the group of highly reactive (↓) elements (p.15) in Group VII of the periodic table (p.18): fluorine (F), chlorine (Cl), bromine (Br), and iodine (I). Halides occur naturally only as chemical compounds (p.15); e.g. sodium as the chloride (NaCl): halite (p.52).

sulphide (*n*) a chemical compound (p.15) of the non-metallic element sulphur (S) with another element, e.g. lead sulphide (PbS), the mineral galena (p.50).

carbonate (*n*) a chemical compound (p.15) containing the elements carbon (C) and oxygen (O) in the form of the CO_3 group containing one atom of carbon and three atoms of oxygen; e.g. calcium carbonate ($CaCO_3$), calcite (p.51).

sulphate (*n*) a chemical compound (p.15) containing the elements sulphur (S) and oxygen (O) in the form of the SO_4 group containing one atom of sulphur and four atoms of oxygen; e.g. barium sulphate, $BaSO_4$: barytes (p.52).

phosphate (*n*) a chemical compound (p.15) of the non-metallic element phosphorus (P) with oxygen (O) and another element; e.g. the mineral apatite (p.49).

silica (*n*) the oxide (↑) of silicon (Si), SiO_2. Quartz (p.55) is the most common natural form of silica.

silicate (*n*) a chemical compound (p.15) of silicon (Si), oxygen (O), and a metal or metals. The silicates are the most important group of rock-forming minerals (p.53).

hydroxyl (*n*) the group consisting of an oxygen atom joined to a hydrogen atom –OH.

氧化物(名)　氧與其他元素(第 15 頁)組成的化合物(第 15 頁)。氧化物可以是固態的，液態的或氣態的。地殼中存在的氧化物是固態的。許多礦石(第 145 頁)是氧化物，如赤鐵礦，Fe_2O_3(第 48 頁)。

礬土；鋁氧；氧化鋁(名)　鋁的氧化物(↑)，Al_2O_3。

(強)鹼(名)　(1)可溶解(第 155 頁)於水成為溶液(第 159 頁)並可中和(↓)酸的一種化合物(第 15 頁)；(2)地質學上，特別是岩石學(第 62 頁)上，"alkali"這個詞指含鹼金屬(↓)比例大的岩石和礦物(第 44 頁)。(形容詞為 alkaline, alkalic)

鹵素(名)　一類極活潑(↓)的元素，排列於元素週期表(第 18 頁)第 VII 族，包括氟(F)、氯(Cl)、溴(Br)、碘(I)。自然界的鹵族元素僅以化合物(第 15 頁)形式存在，如氯化鈉(NaCl)即石鹽(第 52 頁)。

硫化物(名)　非金屬元素硫(S)與其他金屬元素組成的化合物(第 15 頁)。例如硫化鉛(PbS)：其礦物名為方鉛礦(第 50 頁)。

碳酸鹽(名)　一種含有由碳(C)元素和氧(O)元素構成 CO_3 根形式，即含一個碳原子三個氧原子的化合物(第 15 頁)。例如碳酸鈣($CaCO_3$)，其礦石為方解石(第 51 頁)。

硫酸鹽(名)　一種含有由硫(S)元素和氧(O)元素構成 SO_4 根形式，即含一個硫原子和四個氧原子的化合物(第 15 頁)。例如硫酸鋇 $BaSO_4$，其礦石為重晶石(第5 2 頁)。

磷酸鹽(名)　一種含有由非金屬元素磷(P)、氧(O)和其他元素組成的化合物(第 15 頁)。例如礦物磷灰石(第 49 頁)。

矽石；矽氧；二氧化矽(名)　矽(Si)的氧化物(↑)，SiO_2。石英(第 55 頁)是矽石最常見的自然形式。

矽酸鹽；硅酸鹽(名)　由矽(Si)、氧(O)和另外一種或幾種金屬元素組成的化合物(第 15 頁)。矽酸鹽是最重要的造岩礦物(第 53 頁)。

羥基(名)　由一個氧原子連接一個氫原子組成的基團(-OH)。

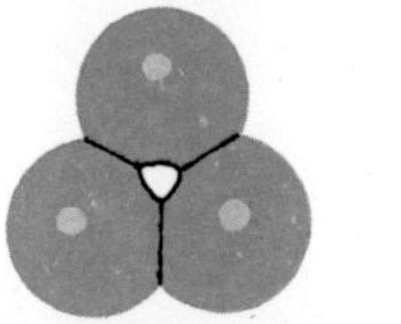

carbonate ion CO_3^{2-}
碳酸根離子 CO_3^{2-}

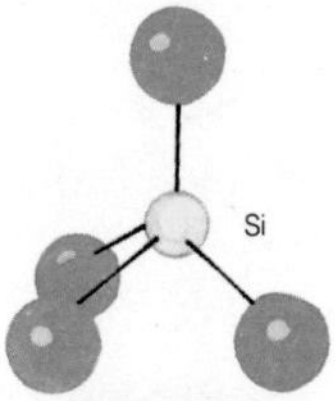

SiO_4 tetrahedron
SiO_4 四面體

the four oxygen atoms are at the corners of the tetrahedron:
四個氧原子在四面體的四個角上

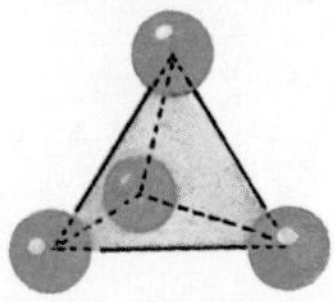

chemical reaction a chemical change that takes place when two or more substances are brought together.

化學反應 兩種或多種物質放在一起所發生的一種化學變化。

reactive (*adj*) readily entering into chemical reactions (↑); chemically active. **react** (*v*); **unreactive** (*adj*), not reactive.

活性的(形) 易參與化學反應(↑)；化學活潑的。(動詞為 react，形容詞 unreactive，意為不活潑的；惰性的)

chemical equilibrium a state in which no further chemical reaction (↑) can take place in a chemical system.

化學平衡 化學系統中不再發生進一步化學反應(↑)的狀態。

neutralization (*n*) the chemical reaction (↑) between an acid and a base (p.152) in which a salt (↓) is formed. **neutralize** (*v*), **neutral** (*adj*).

中和作用(名) 酸和鹼(第 152 頁)之間發生化學反應(↑)生成鹽(↓)的作用。(動詞為 neutralize，形容詞為 neutral)

salt (*n*) (1) a chemical compound (p.15) formed when the hydrogen of an acid is taken away and a metal is put in its place. *See also* **neutralization** (↑). (2) common salt, sodium chloride (NaCl).

鹽(名) (1)酸中含的氫被一種金屬元素取代所形成的合物(第 15 頁)。參見 **中和作用**(↑)；(2)指食鹽，即氧化鈉(NaCl)。

sodium chloride NaCl structure
氯化鈉 NaCl 的結構

inorganic (*adj*) obtained from minerals (p.44), i.e. from the Earth; not organic (↓).

無機的(形) 從礦物(第 44 頁)中獲得的，亦即從地球獲得的；不是有機的(↓)。

organic (*adj*) describes chemical compounds (p.15) of carbon with hydrogen, some also containing oxygen, nitrogen, and other elements, found in living things. (Carbonates are not, however, regarded as organic compounds.)

有機的(形) 描述見於生物體中的碳與氫的化合物(第 15 頁)，有些也含氧、氮和其他元素(惟碳酸鹽不當作有機化合物)。

metal (*n*) an element (p.15) whose atoms form positive ions (p.15) and which is generally a good conductor (p.153) of heat and electricity. **metallic** (*adj*).

金屬(名) 指其原子可形成陽離子(第 15 頁)的元素(第 15 頁)。金屬通常是電與熱的良導體(第 153 頁)。(形容詞為 metallic)

alkali metals the metals (↑) lithium (Li), sodium (Na), potassium (K), rubidium (Rb), and caesium (Cs).

鹼金屬 指鋰(Li)、鈉(Na)、鉀(K)、銣(Rb)和銫(Cs)等金屬(↑)。

non-metal (*n*) an element (p.15) that does not have the general properties of a metal (↑). **non-metallic** (*adj*).

非金屬(名) 指不具金屬(↑)通性的元素(第 15 頁)。(形容詞為 nonmetallic)

analysis (*n*) the use of chemical or physical (p.158) methods to find out the chemical composition (p.15) of a substance. **analyse** (*v*).

分析(名) 用化學或物理(第 158 頁)方法測定一種物質的化學成分(第 15 頁)。(動詞為 analyse)

alloy (*n*) a material made up of two or more metals (↑) or a metal and a non-metal (↑). The composition of an alloy can vary slightly. Brass and steel are examples of alloys.

合金(名) 由兩種或多種金屬(↑)，或由金屬(↑)與非金屬(↑)所構成的一種材料。合金的成分可稍作改變。黃銅和鋼為合金的例子。

NiFe (*n*) an alloy (↑) of the metals (↑) nickel (Ni) and iron (Fe) that was thought to be the material of which the Earth's core (p.9) is made.

鎳鐵(體)(名) 金屬(↑)鎳(Ni)和鐵(Fe)的合金(↑)，人們認為是構成地核(第 9 頁)的物質。

geochemical cycle the path followed by an element (p.15) during a series of geological changes, e.g. from magma (p.62) to rock and back to magma.

地球化學旋迴 在一系列地質變化中一種元素(第 15 頁)所經歷的途徑。例如從岩漿(第 62 頁)到岩石再回到岩漿的過程。

lithophile (*adj*) a lithophile element (p.15) is one that tends to collect in stony (silicate (p.16)) matter. Lithophile elements have a strong tendency to combine chemically with oxygen.

親石的；親氧的(形) 描述一種趨向於聚集在石質(矽酸鹽(第 16 頁))物質中的元素(第 15 頁)。親石元素與氧的化學結合傾向很強烈。

siderophile (*adj*) a siderophile element (p.15) is one that tends to collect in metallic iron. Siderophile elements have only a weak tendency to combine chemically with oxygen and sulphur.

親鐵的(形) 描述一種趨向於聚集在金屬鐵中的元素(第 15 頁)。親鐵元素與硫、氧的化學結合傾向很弱。

chalcophile (*adj*) a chalcophile element (p.15) is one that has a strong tendency to combine with sulphur.

親銅的；親硫的(形) 描述一種與硫結合傾向很強的元素(第 15 頁)。

abundances of elements the relative amounts of various elements in the Earth, the Sun, and other stars can be calculated and expressed as abundances. *Terrestrial abundances* relate to the Earth; *cosmic abundances* to the stars.

元素豐度 各種元素在地球、太陽或其他星球上的相對含量可以測算，並以元素豐度表示。陸地豐度與地球有關，宇宙豐度則與其他恆星有關。

periodic table the elements (p.15), when arranged in rows in a table in order of increasing atomic number (p.152), form groups in the columns of the table. The elements in any group have similar chemical properties. (See p.158).

週期表 將各元素(第 15 頁)按其原子序數(第 152 頁)遞增順序橫列排成表，表中各縱列稱為族。同一族的元素都具有相似的化學性質(第 158 頁)。

hydrolysis (*n*) a chemical reaction (p.17) between water and another chemical compound (p.15). In geology, the word *hydrolysis* is used especially for processes in which minerals react with water, either as liquid water or as vapour (steam). For example, orthoclase (p.56) reacts with water to form kaolinite, a clay mineral (p.61). **hydrolyse** (*v*).

水解作用(名) 水與其他化合物(第 15 頁)之間的化學反應(第 17 頁)，在地質學上，"水解作用"這個詞特別用於指礦物與水(包括液態水和氣態水(蒸氣))之間的反應。例如，正長石(第 56 頁)與水反應生成一種黏土礦物(第 61 頁)高嶺石。(動詞為 hydrolyse)

volatile (*n*) an element (p.15) or chemical compound (p.15) that is dissolved (p.155) in a magma at high temperature (p.160) or pressure. When the magma cools or the pressure is lowered, the volatiles come out of solution as gases. Water and carbon dioxide (CO_2) are examples. **volatile** (*adj*).

揮發物(名) 在高溫(第 160 頁)或高壓下溶解(第 155 頁)於岩漿中的一種元素(第 15 頁)或化合物(第 15 頁)。岩漿冷却或壓力下降時，揮發物呈氣態從溶液中逸出。水與二氧化碳(CO_2)就是其例。(形容詞為 volatile)

precipitation, chemical the process by which a solid material, a **precipitate**, is produced in a liquid (usually water) by a chemical reaction (p.17). **precipitate** (*v*).

化學沉澱作用 因化學反應(第 17 頁)而在液體(通常為水)中生成的固體物質，**稱沉澱物**，此過程稱為化學沉澱作用。(動詞為 precipitate)

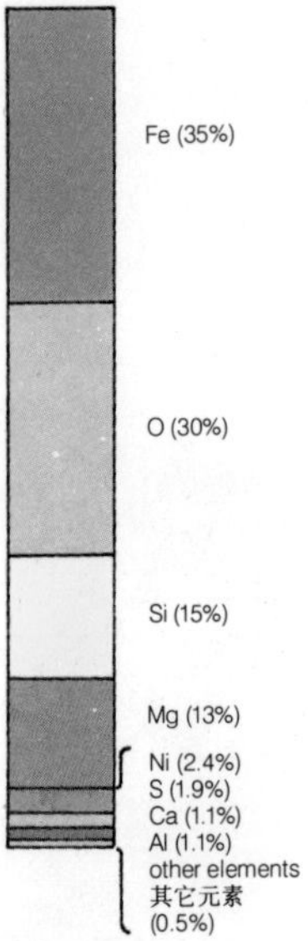

chemical composition of the Earth
地球的化學成分

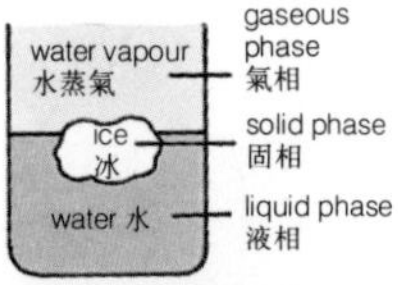

three-phase system: one component
三相系統：單組元

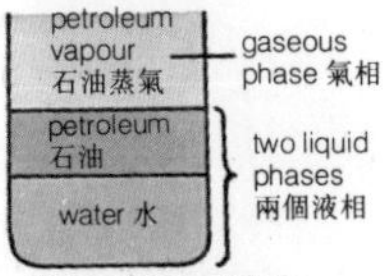

three-phase system: two components
三相系統：雙組元

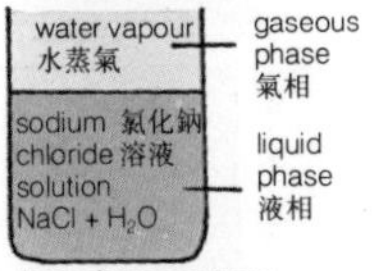

two-phase system: two components
二相系統：雙組元

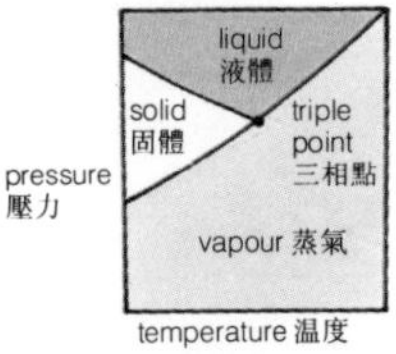

triple point 三相點
relationships between solid, liquid and vapour for water, three phases, solid, liquid and vapour, can exist together at the triple point
水的固體、液體和蒸氣之間的關係，固體、液體和蒸氣三個相在三相點處共存。

phase (*n*) a part of a chemical system that is physically different from the other parts and can be separated from them by purely physical means.

相(名) 某一化學系統的一個部分，它在物理上有別於其他部分並能以純物理方法分離。

component (*n*) any single chemical element (p.15) or compound (p.15) in a chemical system of two or more components.

組元(名) 雙組元或多組元化學系統中任何單一的化學元素(第 15 頁)或化合物(第 15 頁)。

open system a chemical system that can gain or lose material or energy. *See also* **closed system** (↓).

開放系統 可以獲得或失去物質或能量的化學系統。參見**封閉系統**(↓)。

closed system a chemical system that cannot gain or lose material or energy. *See also* **open system** (↑).

封閉系統 不能獲得或失去物質或能量的化學系統。參見**開放系統**(↑)。

isochemical (*adj*) an *isochemical process* is one in which no material is brought in from outside.

等化學的(形) 指沒有從外部帶入物質的化學過程。

supercooling (*n*) the cooling of a liquid below the temperature at which it would normally freeze. If a magma (p.62) is poured out at the Earth's surface and cools rapidly, the minerals formed may be different from those that would have appeared if the magma had crystallized slowly at depth. **supercooled** (*adj*).

過冷卻(名) 液體在正常凝固溫度以下的冷却。岩漿(第 62 頁)流出地表並迅速冷却時，所形成的礦物不同於岩漿在地殼深處緩慢結晶而成的礦物。(形容詞為 supercooled)

nucleation of crystals if small crystal grains (p.72) are put into a supercooled (↑) liquid, crystallization (p.40) may begin. The grains act as *nuclei* (*sing. nucleus*) for crystallization.

晶體的晶核作用 細小的晶粒(第 72 頁)在過冷却(↑)的液體中所產生的結晶作用(第 40 頁)。小晶粒作為結晶作用的晶核。

isotope (*n*) certain elements have atoms (p.152) of more than one kind, called isotopes. All have the same atomic number (p.152) and almost the same chemical properties, but the atomic weights (p.152) of isotopes are different. **isotopic** (*adj*).

同位素(名) 具有不止一種形式的原子(第 152 頁)的某些元素稱為同位素。同位素的原子序數(第 152 頁)相同，化學性質相近，原子量(第 152 頁)則不同。(形容詞為 isotopic)

radioactivity (*n*) the property shown by certain elements (p.15) of changing into other elements by emitting (giving out) charged particles. **radioactive** (*adj*).

放射性(名) 某些元素(第 15 頁)可藉發射出帶電荷的粒子而本身蛻變為另一種元素的性質。(形容詞為 radioactive)

half-life the time taken for half the atoms in a piece of radioactive (↑) material to disintegrate (break up) into atoms of another element or isotope (↑).

半衰期 放射性(↑)物質中的半數原子蛻變為另一種元素或同位素(↑)原子所需的時間。

parent element an element (p.15) that breaks down by radioactive (↑) decay to yield a daughter element (↓).

母元素 可藉放射性(↑)衰變而產生子元素(↓)的一種元素(第 15 頁)。

daughter element an element (p.15) produced by the radioactive (↑) decay (breakdown) of a parent element (↑).

子元素 由母元素(↑)經放射性(↑)衰變而產生的一種元素(第 15 頁)。

weathering (*n*) the process by which rocks at or near the surface of the Earth are broken up by the action of wind, rain, and changes in temperature. The effects of plants and animals are usually also included. Weathering is part of the process of erosion (↓). It includes mechanical weathering (↓) and chemical weathering (↓). **weather** (*v*), **weathered** (*adj*).

風化(名) 地表或近地表的岩石，經風吹雨刷和溫度變化作用而破壞的過程。通常也包括植物與動物的影響。風化是侵蝕(↓)過程的一部分。可分為機械風化(↓)和化學風化(↓)。(動詞為 weather，形容詞為 weathered)

unweathered (*adj*) not weathered (↑).

未風化的(形) 沒有風化的(↑)。

erosion (*n*) (1) the wearing away of rocks: the effect of weathering (↑) and corrasion (p.24); (2) the processes by which soil and rock are removed from any part of the Earth's surface: part of the process of denudation (p.32), including weathering (↑), solution (p.159), corrasion (p.24), and transport (p.21). **erode** (*v*), **eroded** (*adj*).

侵蝕(名) (1)岩石磨損是風化(↑)與刻蝕(第 24 頁)作用的結果；(2)從地表任何部分移去土壤和岩石的過程，為剝蝕作用(第 32 頁)的一部分。剝蝕作用包括風化(↑)、溶解(第 159 頁)、刻蝕(第 24 頁)和搬運(第 21 頁)的作用。(動詞為 erode，形容詞為 eroded)

mechanical weathering weathering (↑) produced by forces that break up the rock physically. These forces usually result from changes in temperature, e.g. insolation (↓); water freezing in cracks in a rock and forcing it apart; the growth of roots in cracks in the rock.

機械風化 物理作用力使岩石破壞的風化(↑)作用。這種力通常源自：岩石溫度變化，例如曝曬(↓)的效應；水在岩石裂隙中凍結迫使岩石裂開；岩石裂隙中植物根的生長作用。

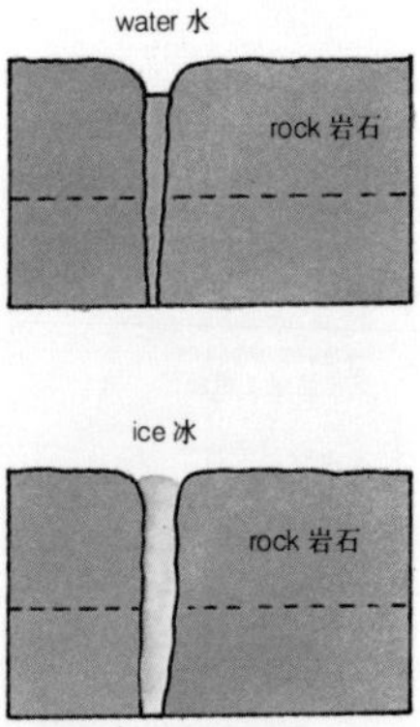

mechanical weathering
effect of water freezing in cracks
機械風化
水在裂縫內結冰的效應

insolation (*n*) the effect of the sun's heat on rocks at the Earth's surface, especially the effect of changes in temperature on the mechanical weathering (↑) of rocks. Heating by the sun during the day causes rocks to expand (p.156). When they cool at night they contract (p.154). This causes the rock to break up. **insolate** (*v*).

曝曬(名) 太陽熱力對地表岩石的影響，特別是溫度變化對岩石機械風化(↑)的影響。日間太陽的熱力使岩石膨脹(第 156 頁)，夜間時冷却又使岩石收縮(第 154 頁)，從而造成岩石破裂。(動詞為 insolate)

exfoliation (*n*) the formation and breaking off of shells or sheets from the bare surfaces of rocks, especially granite and other igneous rocks (p.62). **exfoliate** (*v*).

頁狀剝落(名) 岩石的裸露表面形成殼狀或片狀層並脱落，常見於花崗岩或其他火成岩(第 62 頁)。(動詞為 exfoliate)

abrasion (*n*) the wearing away of a rock by rubbing, e.g. by small particles of rock. **abrade** (*v*), **abraded** (*adj*).

磨蝕(名) 岩石因磨擦而磨損，例如受細小的岩石顆粒的磨擦而磨損。(動詞為 abrade，形容詞為 abraded)

chemical weathering weathering (↑) caused by chemical action, usually when water is present. For example, rain water containing carbon dioxide (CO_2) in solution (p.159) will dissolve (p.155) limestone (p.86).

化學風化 因化學作用而致的風化(↑)，通常有水參與。例如，含二氧化碳(CO_2)溶液(第 159 頁)的雨水，可溶解(第 155 頁)石灰岩(第 86 頁)。

corrosion (*n*) the eating away of rocks by chemical action. See also corrasion (p.24). **corrode** (*v*), **corroded** (*adj*).

熔蝕；溶蝕；腐蝕(名) 岩石因化學作用而被侵蝕。參見"刻蝕"(第 24 頁)。(動詞為 corrode，形容詞為 corroded)

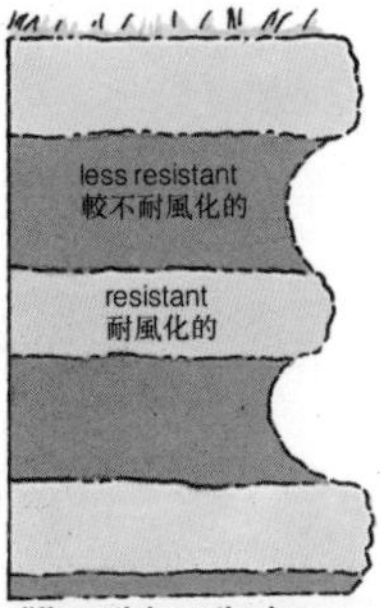

differential weathering
差異風化

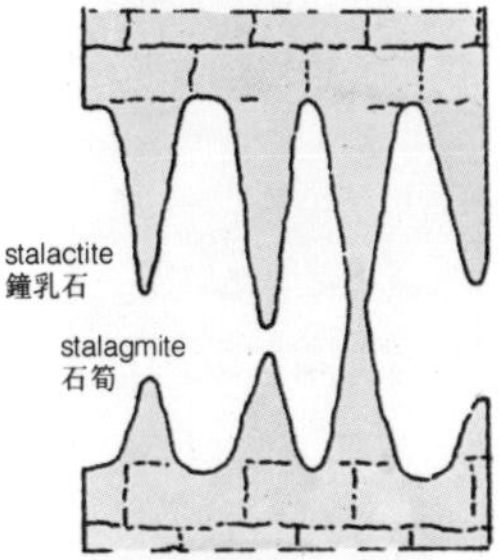

stalactites, stalagmites
鐘乳石，石筍

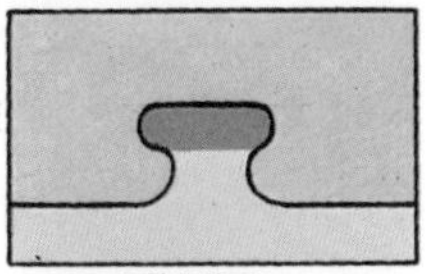

zeugen 風蝕桌狀石

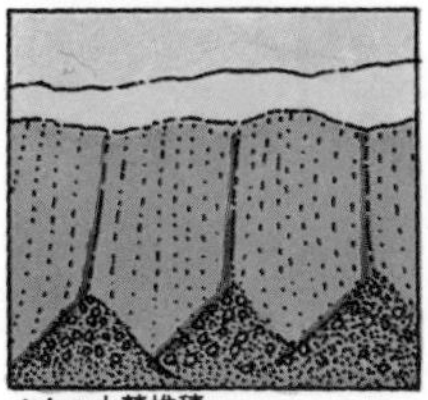

talus 山麓堆積

differential weathering, differential erosion the process by which an uneven surface is developed where some rocks are worn away less rapidly than others and remain standing out in an exposure (p.122).

joint (*n*) a break or fracture (p.122) in a rock along which no movement has taken place.

joint set a series of joints (↑) that are more or less parallel to each other.

joint system two or more sets of joints (↑) that cut across each other.

fissure (*n*) a large crack or break in a body of rock. **fissured** (*adj*).

tension gash a joint (↑) that has opened up during deformation (p.122). It may contain minerals, e.g. quartz (p.55).

sink-hole, swallow-hole a hole in limestone (p.86) country into which water flows. Sink-holes are formed when the roof of a cave falls in or by the solution (p.159) of limestone at the surface.

stalactite (*n*) a deposit (p.80) of calcium carbonate, $CaCO_3$, hanging from the roof of a cave. *See also* **stalagmite** (↓).

stalagmite (*n*) a deposit of calcium carbonate, $CaCO_3$, standing up from the floor of a cave. *See also* **stalactite** (↑).

calcareous tufa, calc tufa a deposit (p.80) of calcium carbonate, $CaCO_3$, precipitated (p.18) from solutions (p.159). It is found in limestone (p.86) regions and round springs.

zeugen (*n*) a mass of harder rock resting on a pillar of softer rock.

talus (*n*) a loose heap of weathered (↑) pieces of rock at the foot of a steep slope. The pieces of rock may be of any size or shape.

scree (*n*) = talus (↑).

angle of repose the steepest angle or slope at which loose material such as pieces of rock or sand will remain without sliding down.

rock glacier a mass of pieces of rock flowing slowly down a slope like a glacier (p.28).

transport, transportation (*n*) the carrying-away of sediment (p.80) and other rock material on the surface of the Earth by gravity (p.11), moving water, ice, or air. **transport** (*v*).

差異風化，差異侵蝕 某些岩石的風化侵蝕比另一些岩石慢而突出於露頭(第 122 頁)從而產生凹凸不平的表面的過程。

節理(名) 岩石中沒有發生錯動而產生的破裂或裂隙(第 122 頁)。

節理組 一系列相互間近於平行的節理(↑)。

節理系 兩組或兩組以上相互橫截的節理(↑)。

裂隙(名) 岩體中較大的破裂或斷裂。(形容詞為 fissured)

張裂隙 形變(第 122 頁)過程中張開的節理(↑)。其中可能含礦物，例如含石英(第 55 頁)。

落水洞，溶岩硐 在石灰岩(第 86 頁)地區，水自此流入地下的孔洞。落水洞是由洞頂塌陷或地面的石灰岩溶解(第 159 頁)而形成的。

鐘乳石(名) 從溶洞頂垂下的一種碳酸鈣($CaCO_3$)沉積物(第 80 頁)。參見**石筍**(↓)。

石筍(名) 自溶洞底板向上增長的一種碳酸鈣($CaCO_3$)沉積物。參見**鐘乳石**(↑)。

石灰華，鈣華 從溶液(第 159 頁)中沉澱(第 18 頁)出的一種碳酸鈣($CaCO_3$)沉積物(第 80 頁)，在石灰岩(第 86 頁)地區的泉眼附近形成。

風蝕桌狀石；風蝕柱(名) 覆於較軟岩石柱上的硬岩塊。

山麓堆積(名) 陡坡脚下鬆散堆積的風化(↑)岩石碎塊。碎塊的大小或形狀不一。

山麓碎石；岩屑堆(名) 同山麓堆積(↑)。

靜止角；安息角 斜坡的最陡角度，在此斜坡上鬆散物質如岩石碎塊或砂粒可保持靜止而不下滑。

石冰川 像冰川(第 28 頁)那樣沿斜坡向下緩慢流動的岩石碎塊。

搬運，搬運作用(名) 地球表面的沉積物(第 80 頁)和其他岩石物質由重力(第 11 頁)、流水、冰或空氣運走的作用。(動詞為 transport)

aeolian weathering weathering (p.20) caused by the action of the wind. A strong wind blowing sand or other hard, sharp particles against a rock can wear it away.

風力風化 由風吹作用產生的風化(第 20 頁)。強風挾帶砂粒及其他堅銳顆粒吹刮岩石而使岩石磨蝕。

eolian weathering (US) = aeolian weathering (↑).

風力風化(↑)的美語為 eolian weathering。

ventifact (*n*) a stone shaped by the action of the wind.

風稜石(名) 由風磨蝕所形成特有形狀的石塊。

desert varnish a thin, shiny coating that forms on stones in deserts. It is bluish-black in colour and consists mainly of iron and manganese oxides (p.16).

沙漠岩漆(名) 覆蓋於沙漠岩石上，由風磨平，表面發亮的藍黑色薄膜，主要含鐵和錳的氧化物(第 16 頁)。

dune (*n*) a heap or bank of sand piled up by the wind into a regular shape.

沙丘(名) 由風吹堆積而成形狀有規律的沙堆或沙堤。

barchan dune a dune (↑) shaped in plan like a crescent. The ends of the dune point in the direction in which the wind generally blows. Barchans may be up to 400 m wide and 30 m high.

新月形沙丘；彎月沙丘 平面上呈新月形的沙丘(↑)。沙丘兩端指向風通常吹的方向。新月形沙丘可寬達 400 米，高達 30 米。

longitudinal dune a long, narrow dune (↑), up to 80 km long and 200 m or more high, with its length parallel to the general direction in which the wind blows.

縱向沙丘 一種狹長的沙丘(↑)，可長達 80 公里，高達 200 米或以上，砂丘的延伸方向是風吹的總方向。

seif (*n*) a longitudinal dune (↑) with a long sharp edge at its top. One side of a seif is rounded; the other is a steep slip face. Seifs occur in chains.

劍沙丘；直線沙丘 一種狹長的沙丘(↑)，頂部具刀鋒狀的脊，一側圓滑，另一側是陡峭的滑落面。劍沙丘呈鏈狀連串出現。

whaleback dune a very large longitudinal dune (↑) with a flat top. There may be barchans (↑) or seifs (↑) on top of a whaleback dune.

鯨背沙丘 平頂的巨大縱向沙丘(↑)，頂部可能出現新月形沙丘(↑)或劍沙丘(↑)。

transverse dune a dune (↑) with its length at 90° to the direction in which the wind generally blows.

橫向沙丘 長軸方向與風吹總方向成 90° 角的一種沙丘(↑)。

parabolic dune a dune (↑) shaped in plan like the path of a ball thrown in the air (i.e. a *parabola*). The points of the dune face the direction from which the wind blows. Parabolic dunes are formed where there is thick grass or other plants covering the sand. The sand is blown away from an area without plants but the sand on either side is held back by the plants there. A parabolic dune is thus formed.

拋物線沙丘 平面形狀如在空中拋球徑跡(即拋物線)的一種沙丘(↑)。沙丘的兩個端點面向主要的風向。拋物線沙丘形成於被草叢或其他植物覆被的沙地。無植被地段的沙被風吹走，其兩邊的沙則為植被固定，從而形成拋物線沙丘。

oghurd (*n*) a large mountainous dune (↑).

星狀沙丘(名) 一種大的山狀沙丘(↑)。

loess (*n*) an unconsolidated (p.84) deposit (p.80) of silt (p.88), usually unstratified (p.80), carried by the wind.

黃土(名) 一種由風帶來的未固結的(第 84 頁)粉砂(第 88 頁)沉積物(第 80 貝)，通常不成層(第 80 頁)。

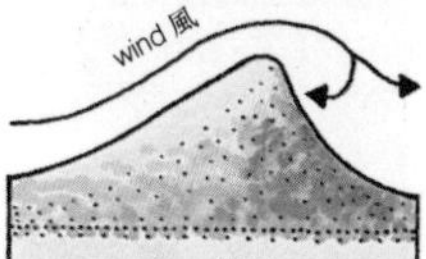

dune 沙丘

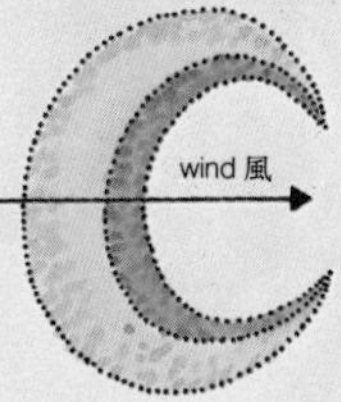

barchan dune (plan)
新月形沙丘(平面)

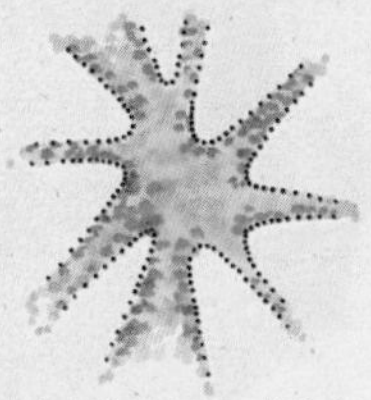

oghurd (star) dune 星狀沙丘

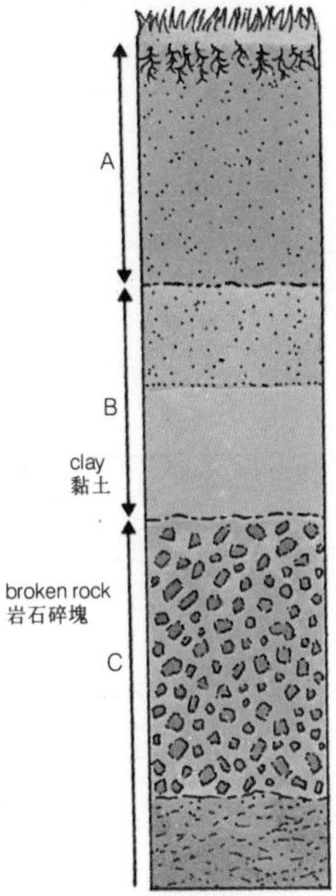

soil profile
(residual soil)
土壤剖面
（殘積土）

soil (*n*) the material produced by the effects of weathering and the action of plants and animals on the rocks at the Earth's surface.
subsoil (*n*) the layer of broken rock between the soil (↑) and the unweathered solid rock, or *bedrock*, below.
pedology (*n*) the study of soils; soil science.
regolith (*n*) a layer of pieces of loose rock resting on solid rock (bedrock).
mantle rock = regolith (↑).
soil profile the three soil horizons (↓) A, B, and C.
soil horizon one of the layers into which the soil is divided.
A horizon the top layer of the soil. It is dark in colour and contains organic (p.17) material formed by the decay of vegetable matter.
B horizon the subsoil; the layer below the soil surface. It usually contains more clay and iron oxides (p.16) than the A and C horizons.(↑ ↓).
C horizon the lowest layer of the soil, resting on unweathered (p.20) solid rock. It consists of loose, slightly weathered pieces of rock.
Cca horizon a white layer of calcium carbonate, $CaCO_3$, beneath the B horizon (↑).
K horizon a thick hard layer containing 50% or more of calcium carbonate, $CaCO_3$, that forms beneath the B horizon (↑) in very dry regions.
caliche = K horizon (↑).
A2 horizon a whitish layer between the A and B horizons (↑). Most of the iron oxides (p.16) have been removed from the A2 horizon by water moving down through the soil.
duricrust (*n*) a hardened layer formed in the soils of very dry (semi-arid) regions by the precipitation (p.18) of salts (p.17) from water in the soil.
bauxite (*n*) a residual (p.33) deposit (p.80) formed under very hot, wet conditions. It contains hydrated (p.157) aluminium oxides (p.16). Bauxite is an important ore (p.145) of aluminium.
laterite (*n*) a residual (p.33) deposit (p.80) formed under very hot, wet conditions, especially from igneous rocks (p.62). It contains hydrated (p.157) iron oxides (p.16). **laterization** (*n*).

土壤(名)　地球表面的岩石經風化作用及動、植物的作用所形成的一種物質。

底土；亞層土(名)　界於土壤(↑)與土壤下方未風化的固體岩石(或稱基岩)之間的一層破碎的岩石層。

土壤學(名)　研究土壤的學科；土壤科學。
風化層；浮土(名)　覆蓋於固體岩石(基岩)上的一層鬆散岩石碎塊。
覆蓋層　同風化層(↑)。
土壤剖面　分A、B、C三個土壤層(↓)。
土壤層　被劃分出來的各個土壤單元之一。

A層　土壤分層的頂層。色深，所含有機(第17頁)質為植物質腐爛分解而成。

B層　亞層土：A層下的土壤層，常比A層(↑)、C層(↓)含較多黏土和氧化鐵(第16頁)。

C層　土壤分層的最下一層，在未風化(第20頁)固體岩石之上，由鬆散的輕微風化岩石碎塊組成。

C鈣層　白色的碳酸鈣($CaCO_3$)層，在B層(↑)之下。

K層　含碳酸鈣($CaCO_3$)至少50%的厚硬土層，在極乾燥地區的B層(↑)之下形成。

鈣質層　同K層(↑)。

A2層　A層和B層(↑)之間略帶白色的土層。此層中含的大部分氧化鐵(第16頁)已被向下流過土壤的水除去。

鈣質殼；硬殼(名)　在極乾旱(半乾旱)地區土壤中形成的一種硬化土層；是由土壤中的水所含的鹽(第17頁)沉澱(第18頁)而成。

鋁土礦(名)　在極濕熱環境下形成的一種殘留(第33頁)沉積物(第80頁)。其中含水合的(第157頁)鋁氧化物(第16頁)。鋁土礦是一種重要的鋁礦石(第145頁)。

紅土(名)　在極濕熱環境下形成，尤其是由火成岩(第62頁)形成的一種殘留(第33頁)沉積物(第80頁)。其中含水合的(第157頁)鐵氧化物(第16頁)。(名詞 laterization 意為紅土化作用)

load (*n*) the material moved by a stream. It may float in the water, be pushed along the stream bed, or be dissolved (p.155) in the water.

河流泥沙；流水移運物(名)　河流攜運的物質。這些物質在水中漂移；或沿河床前移；或溶解(第 155 頁)於流水中。

suspension load the part of the load (↑) of a stream that is carried along in the water above the stream bed. *See also* **traction load** (↓).

懸移質　為河床上方的水流所沖走的河流泥沙(↑)部分。參見**底移質**(↓)。

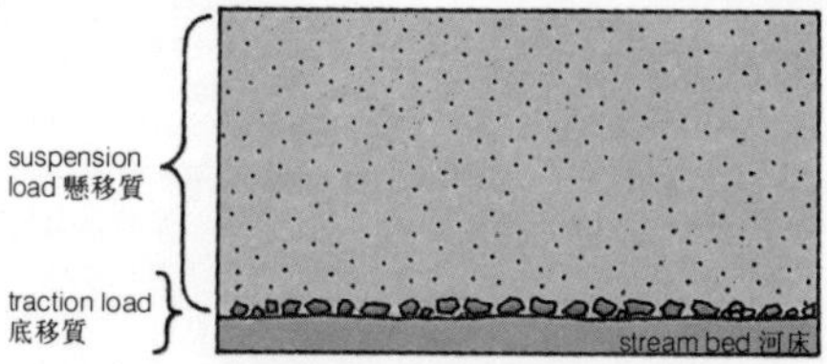

suspension and traction loads of a stream
河流的懸移質和底移質

traction load the part of the load (↑) of a stream that is carried along on the stream bed. *See also* **suspension load** (↑).

底移質　為水流沿着河床所攜運前移的河流泥沙(↑)部分。參見**懸移質**(↑)。

corrasion (*n*) the wearing away of rocks by the rubbing action of particles carried by a stream.

刻蝕；流蝕(名)　河流所攜運的顆粒使岩石受磨擦而損蝕。

alluvium (*n*) mud, sand, gravel (p.87), and other materials moved by streams and deposited (p.80) by them. **alluvial** (*adj*).

沖積層；沖積物(名)　被河流攜運並沉積(第 80 頁)的泥沙、礫石(第 87 頁)和其他物質。(形容詞為 alluvial)

alluvial fan a cone-shaped pile of alluvium (↑) deposited (p.80) where the gradient (↓) of a stream becomes less (e.g. at the base of a steep slope). Alluvial fans are common in dry regions.

沖積扇　在河流坡降(↓)減小(例如陡坡的坡脚)處沉積(第 80 頁)的錐形沖積層(↑)。沖積扇常見於乾燥區域。

stream gradient the slope of a stream bed as measured down the valley.

河流坡降　沿河谷方向測量的河床坡度。

longitudinal profile, long profile a curved line representing the way in which the height of the valley floor changes along the course of a stream.

縱向剖面，河流縱斷面　表示谷底高度沿河流水道變化的一條曲線。

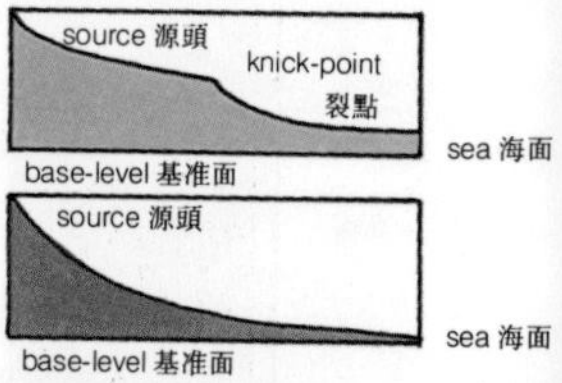

stream profiles 河流剖面
(exaggerated) （誇大示意圖）

Thalweg (*n*) = longitudinal profile (↑).

最深河谷底線(名)　同縱向剖面(↑)。

stream profile = longitudinal profile (↑).

河流剖面　同縱向剖面(↑)。

rejuvenation (*n*) if a region is uplifted (p.125) or the sea-level falls, the streams are *rejuvenated* and cut down into the land again. **rejuvenate** (*v*).

回春作用(名)　一個區域隆起(第 125 頁)或海平面下降，使河流回春並再次向下切割基岩。(動詞為 rejuvenate)

knick-point (*n*) the point at which the old longitudinal profile (↑) of a stream meets a new one. A knick-point is the result of rejuvenation (↑).

裂點；轉折點(名)　河流新、舊縱向剖面(↑)的交會點。裂點是河流回春作用(↑)的結果。

base level the imaginary level surface to which the longitudinal profile (↑) of a stream is related.

基准面　和河流縱向剖面(↑)相關的假想水準面。

mature (*adj*) a mature stream is one that has reached its full growth. **maturity** (*n*).

壯年的(形)　描述一條生長成熟的河。(名詞為 maturity)

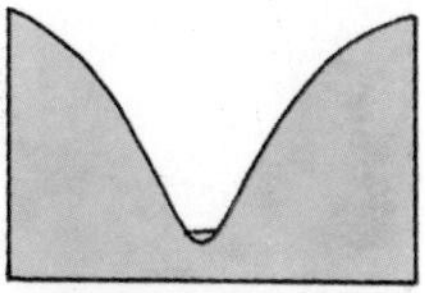

V-shaped valley V 形河谷

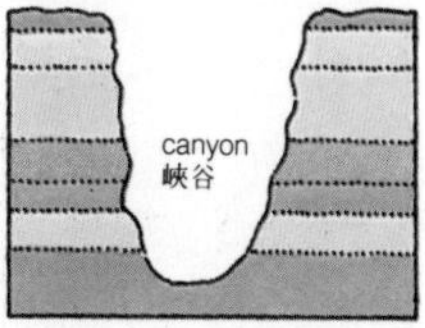

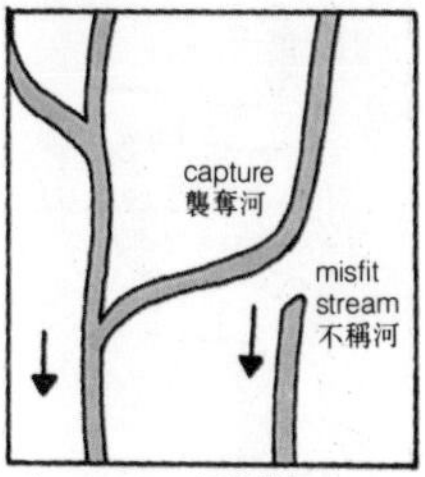

river capture 河流襲奪

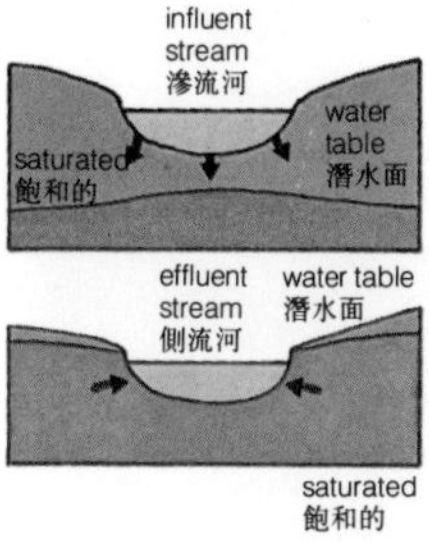

V-shaped valley a valley with steep sides and in cross-section like a letter V. V-shaped valleys are characteristic of young streams.

canyon (*n*) a deep valley with sides that are vertical, or nearly so, which has been cut by a river. Canyons are usually formed by rejuvenation (↑).

waterfall (*n*) a point in the course of a stream where the water descends more or less vertically without support.

river terrace a flat area that borders a river valley. It slopes towards the sea at about the same angle as the river. A terrace marks the level of the floor of an earlier valley.

capture (*n*) a stream that is actively eroding (p.20) may cut back and reach the upper part (the headwaters) of another stream, thus capturing these waters and turning them aside into its own course. **capture** (*v*).

beheaded (*adj*) a beheaded stream is one that has had its upper part (its headwaters) captured (↑) by another stream. **behead** (*v*).

misfit stream a stream that is too small to have eroded (p.20) the valley in which it flows. River capture (↑) is a possible cause.

watershed (*n*) (1) the line that divides two areas from which water flows into two separate streams; a narrow area of ground between two such areas; (2) the area from which water flows into a particular stream system.

drainage system a stream or river together with the streams that flow into it.

influent (*adj*) an influent stream is one that flows above the water table (p.146) and thus adds to the supply of water below ground. Influent streams are common in dry regions. *See also* **effluent** (↓).

effluent (*adj*) an effluent stream is one that flows at the level of the water table (p.146) and receives water from it. *See also* **influent** (↑).

dry valley (*n*) a valley without a stream. Stream capture is one cause of dry valleys.

wadi (*n*) a valley in which a stream flows from time to time. Wadis are common in deserts.

arroyo (*n*) = wadi (↑).

V 形河谷 谷坡陡峻、橫截面呈 V 形的河谷。V 形河谷是幼年河谷的特徵。

峽谷(名) 被河流切割成兩側谷坡垂直或近於垂直的深谷。峽谷通常是由回春作用(↑)形成。

瀑布(名) 河流水道上河水近於垂直下瀉的一個地點。

河成階地；河岸階地 與河谷毗連，坡度與河流坡度大致相同並向海傾斜的一塊平地。階地標誌河流早期的谷底。

襲奪河；掠奪河(名) 具強烈侵蝕作用(第 20 頁)，不斷地向河源擴展至達到另一條河的上源(上游)，並掠奪其河水而使之改向流入本身水道的一條河流。(動詞為 capture)

斷頭的(形) 指河流的上源(上游)被另一條河襲奪(↑)。(動詞為 behead)

不稱河 流量太小難侵蝕(第 20 頁)其河谷的小河。河流襲奪(↑)可能是其成因。

分水界；流域(名) (1)劃分兩個相鄰流域的界線，此線兩側的水流入兩個獨立的河流；這兩個相鄰流域間的狹窄地帶；(2)水流匯入特定水系的區域。

水系 江河及其支流合稱水系。

滲流的(形) 描述在潛水面(第 146 頁)以上流動的河流，河水可補給地下水。滲流河常見於乾旱地區。參見**側流的**(↓)。

側流的；潛水補給的(形) 描述河水流經潛水面(第 146 頁)，並由之獲得地下潛水補給的河流，參見**滲流河**(↑)。

乾谷(名) 沒有河流的谷地。河流襲奪是乾谷的一個成因。

間歇乾谷；旱谷(名) 不時有河流流經的乾溝。常見於沙漠地帶。

西班牙語稱旱谷(↑)為 arroyo。

flood-plain the flat area on either side of a stream over which it spreads when too much water is flowing for the stream channel to be able to carry all of it.

meander (*n*) the curved path of a river, especially in a flood-plain (↑). **meandering** (*adj*), **meander** (*v*).

incised meander a meander (↑) that has been cut down (incised) in the flood-plain (↑). The river then flows in a twisting channel with steep sides.

ox-bow lake a lake shaped like a new moon that has been formed when a meander (↑) has been cut off from the main stream by continuing erosion.

氾濫平原 因大量水湧入河道，超出河道的承載能力而溢出河床，在兩岸堆積而成的平坦地帶。

曲流；河曲(名) 河流的曲徑，常見於氾濫平原(↑)。(形容詞為 meandering，動詞為 meander)

深切河曲；刻蝕曲流 在氾濫平原(↑)向下刻蝕(深切)的河曲(↑)。河流在陡峭而迂迴曲折的河道流淌。

牛軛湖 因河流不斷侵蝕，河曲(↑)被水流切斷，與主流分開而形成的新月狀湖。

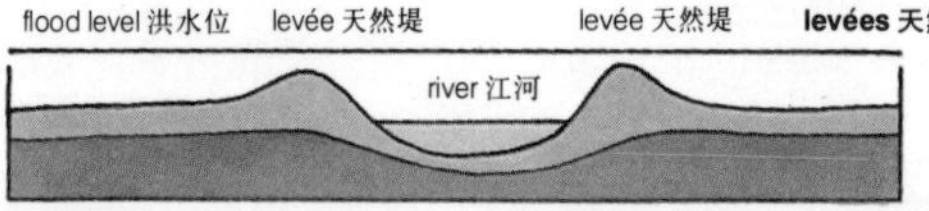

levées 天然堤

levée (*n*) a natural wall or embankment that has been formed of sediment (p.80) deposited at the sides of a river when it has flowed over its banks. Levées are usually found in the flood-plain (↑) and they contain the river while the flow of water is not too large.

天然堤(名) 洪水期河水溢出河岸，沉積物(第80頁)沉積在河兩岸所形成的天然牆或堤防。天然堤常見於氾濫平原(↑)，若水量不太大則河流挾持於其中。

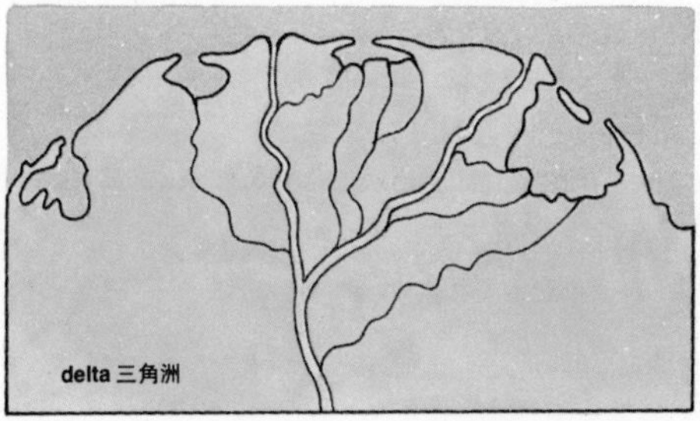
delta 三角洲

delta (*n*) sediment (p.80) laid down at the mouth of a river where it enters the sea or a lake. The word comes from the Greek letter delta (△). **deltaic** (*adj*).

distributary (*n*) one of the branches into which a river divides in a delta (↑) or elsewhere.

三角洲(名) 在河流入海或入湖的河口處，以希臘字母△狀分布的泥沙沉積(第80頁)。(形容詞為 deltaic)

汊流(名) 一條河流在三角洲(↑)或別處分出的許多支河道。

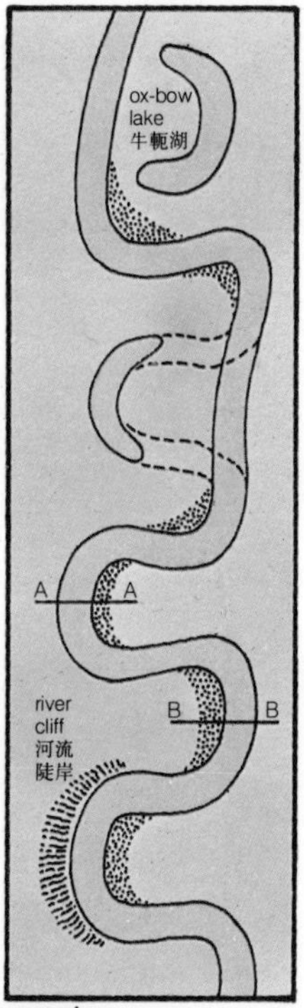

meanders
曲流

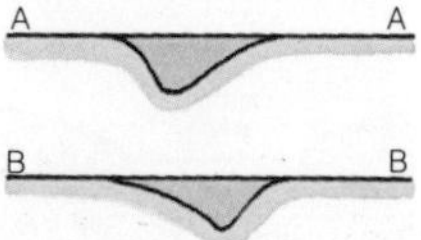

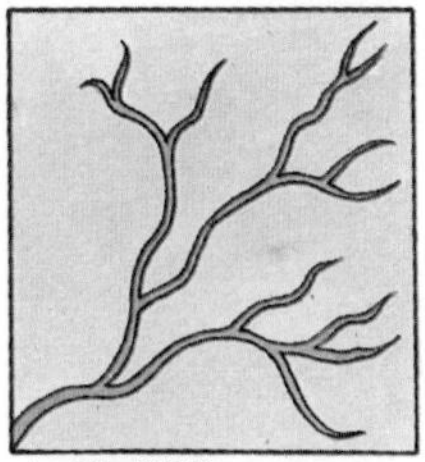

dendritic 樹枝狀

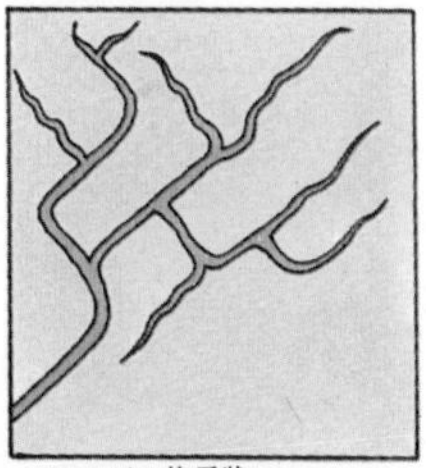

rectangular 格子狀

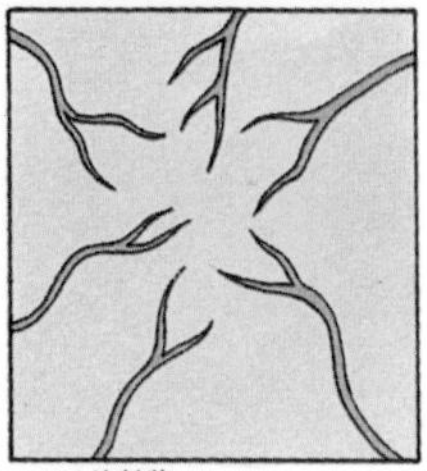

radial 放射狀

drainage patterns
水系類型

drainage pattern the arrangement of a system of streams as seen in plan. Common types of drainage patterns are: *dendritic*, in which the streams are arranged like the branches of a tree; *trellis* or *rectangular*, in which the streams flow in two directions at 90° to each other, parallel to the strike and dip (p.123) of the rocks; and *radial*, in which the streams flow out in all directions from a central point.

水系類型　平面所見的河流系統排列狀態。常見的水系類型有：樹枝狀水系，諸河流的排列似樹枝狀；格狀或長方水系，諸河流以互成 90° 相交的兩個方向流動，與岩層的走向和傾向(第 123 頁)平行；放射狀水系，諸河流從一個中心點向各個方向流出。

consequent, subsequent and obsequent streams
順向河、後成河和逆向河

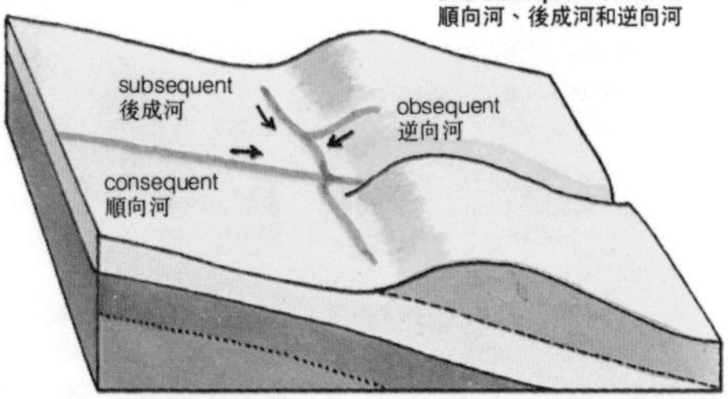

consequent (*adj*) a consequent stream is one that flows in the same direction as the downward slope on which it was originally formed.

順向的；順斜的(形)　描述河流的流向與該河流原先形成的下斜坡向一致。

subsequent (*adj*) a subsequent stream is one that flows into a consequent (↑) stream.

後成的；順層的(形)　描述流入順向河(↑)的河流。

obsequent (*adj*) an obsequent stream is one that flows in the opposite direction to the original consequent stream (↑).

逆向的；反斜的(形)　描述河流的流向與原先的順向河(↑)流向相反。

inconsequent (*adj*) an inconsequent drainage system is one that does not fit the geological structure (p.122) of the region.

非順向的；非順斜的(形)　水系與所流經區域的地質構造(第 122 頁)不相適應。

superimposed (*adj*) superimposed drainage is a form of inconsequent drainage (↑). It results when a drainage system is formed on a younger series (p.113) of rocks that rest with an angular unconformity (p.118) on older rocks. The drainage is fitted to the younger series of rocks and is then superimposed when the streams cut down through them to the older rocks, whose structure is not related to the drainage system.

疊置的；上層遺留的(形)　描述非順向水系(↑)的一種形式河流。一個水系在年青岩系(第 113 頁)中形成，而年青岩系角度不整合(第 118 頁)於老岩系之上，此時水系與年青岩系地質構造相適應，當水系下切達到老岩系時，原來的水系遺留於老岩系之上，而水系類型與老岩系的地質構造沒有關連。

antecedent (*adj*) an antecedent stream or drainage system cuts across a geological structure (p.122) that has formed across its course.

先成的(形)　指河流或水系穿過橫越該河流或水系所形成的地質構造(第 122 頁)。

ice sheet a mass of ice in the form of a sheet covering a large area of the Earth's surface. Ice sheets can stretch across continents and can cover mountains.
glacier (*n*) a large mass of ice formed from snow that has packed together and moves slowly down a slope under its own weight.
mountain glacier a glacier (↑) that flows in a mountain valley with solid rocks standing above the highest levels of ice and snow.
valley glacier = mountain glacier (↑).
alpine glacier = mountain glacier (↑).
piedmont glacier a glacier that is formed where a mountain glacier (↑) leaves the valley and spreads out across a plain.
crevasse (*n*) a deep crack in a glacier. Crevasses are caused by stresses (p.122) in the ice that result from movement.
Bergschrund (*n*) a crevasse (↑) at the head of a glacier (↑) in a cirque (p.31).
glaciation (*n*) the action of glaciers (↑) and ice-sheets (↑), including erosion (p.20) and deposition (p.80).
glaciology (*n*) the study of glaciers (↑) and ice-sheets (↑). **glaciological** (*adj*).
glacier lake a lake, usually in a valley, in which the water is held back by a glacier (↑).
ice-dammed lake = glacier lake (↑).
glacial lake = glacier lake (↑).
glacial period a period of time in the Earth's history when ice in the form of glaciers (↑) spread into areas that were free from it at other times.
ice age = glacial period (↑).
interglacial period a period between two glacial periods (↑) when the temperature was higher and the ice moved back towards the poles.

冰蓋；冰野 形狀像床單，覆蓋大片地球表面面積的一大片冰。冰蓋可以橫跨幾個大陸，並可以覆蓋山脈。
冰川；冰河(名) 由積雪所形成，並因其自身重量而順坡緩慢下移的大冰體。
高山冰川 在山谷中流動並有基岩高出最高冰雪面之上的冰川(↑)。
山谷冰川 同高山冰川(↑)。
阿爾卑斯型冰川 同高山冰川(↑)。
山麓冰川 高山冰川(↑)離開山谷向外伸展橫越平原形成的一種冰川。
冰裂隙(名) 因冰川運動在冰層內產生應力(第122頁)而在冰川中產生的深裂隙。
冰後隙；壁前大縫(名) 冰川(↑)源頭處冰斗(第31頁)中的冰裂隙(↑)。
冰川作用(名) 冰川(↑)和冰蓋(↑)的作用，包括侵蝕(第20頁)和沉積(第80頁)。
冰川學(名) 研究冰川(↑)和冰蓋(↑)的學科。(形容詞為 glaciological)
冰川湖 通常位於山谷中的一類湖泊，湖水為冰川(↑)所阻。
冰塞湖 同冰川湖(↑)。
冰河湖 同冰川湖(↑)。
冰期 地球歷史上，冰體以冰川形式推進到當時並沒有冰體的其他地區的一段時間。
冰河時代 同冰期(↑)。
間冰期 兩個冰期(↑)之間的一段時間，其時溫度較高，故冰向兩極後退。

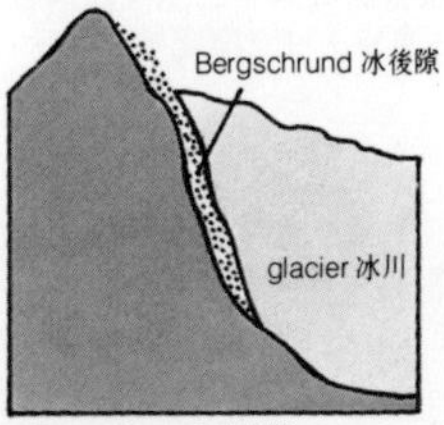

Bergschrund 冰後隙

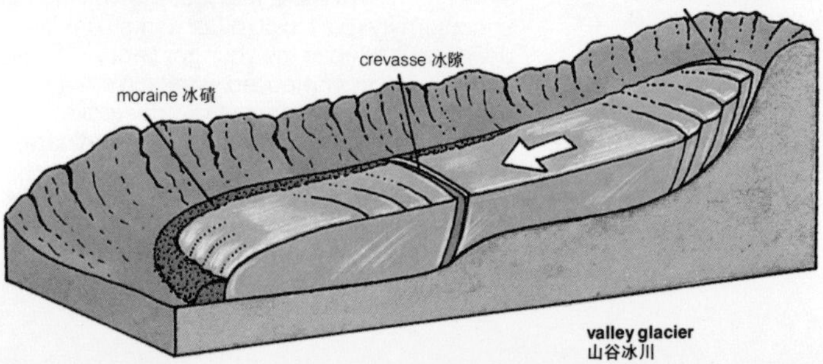

valley glacier
山谷冰川

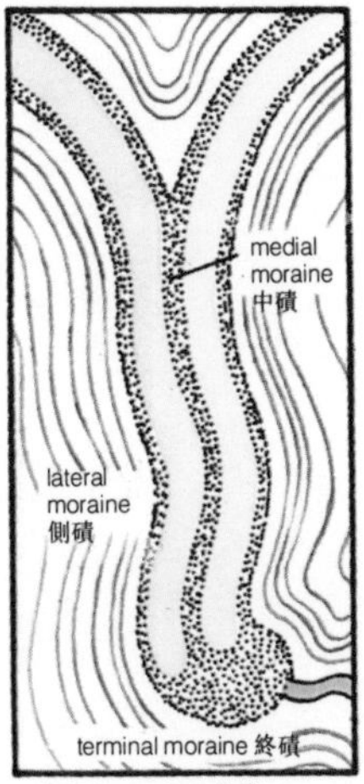

glacial moraines 冰磧

moraine (*n*) accumulated rock material that is being transported, or has been transported, by ice. **morainic** (*adj*).
lateral moraine a moraine (↑) formed by rock material falling on the sides of a glacier (↑) from the sides of the valley.
medial moraine a moraine (↑) formed by the joining together of lateral moraines (↑) when two glaciers (↑) meet in the same valley.
englacial moraine a moraine inside the ice of a glacier (↑).
terminal moraine a moraine (↑) left where the ice melts at the lower end, or *snout*, of a glacier (↑).
end-moraine = terminal moraine (↑).
boulder clay a deposit (p.80) left behind after the melting of an ice-sheet (↑): a fine clay containing pebbles (p.87) and boulders (p.87) of subangular (p.83) shape.
till (*n*) = boulder clay (↑).
tillite (*n*) an indurated (p.84) till (↑).
glacial striae long marks made on rocks that have been under a glacier (↑). Striae are made by pieces of rock that are carried along with the ice.
glacial striations = glacial striae (↑).

冰磧（名） 正被或已被冰體搬運的積聚的岩石物質。（形容詞為 morainic）

側磧；側冰磧 岩石物質從山谷邊部落在冰川（↑）兩邊而形成之冰磧（↑）。

中磧；中冰磧 兩冰川（↑）在同一山谷相會，兩道側磧（↑）滙合一起所形成之冰磧（↑）。

內磧；內冰磧 冰川（↑）冰層內部之冰磧。

終磧；端冰磧 冰在冰川（↑）較低的一端（即冰川鼻）溶化後遺留下之冰磧（↑）。

尾磧 同終磧（↑）。

泥礫；漂礫土 冰蓋（↑）溶化後留下的沉積物（第 80 頁），為含次稜角（第 83 頁）狀礫石（第 87 頁）和漂礫（第 87 頁）的細黏土。

冰磧物（名） 同泥礫（↑）。

冰磧岩（名） 一種固結的（第 84 頁）冰磧物（↑）。

冰川擦痕 冰川（↑）下的岩石上，由冰川攜帶的岩塊刻成的長刻痕。

冰川條痕 同冰川擦痕（↑）。

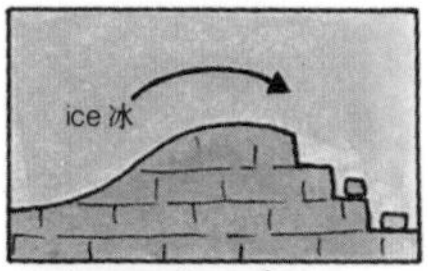

roches moutaonnées
羊背石

roches moutonnées (*n.pl., French*) rock forms produced by glacial erosion (p.20). They have a rounded shape, less steep on the side from which the ice travelled, and are usually oval in plan. Roches moutonnées are generally found in groups with a parallel arrangement. **roche moutonnée** (*sing.*).

羊背石；羊背岩（名、複） 冰川侵蝕（第 20 頁）所造成的岩石形態。在冰川流動方向相反的一側，羊背石呈圓形且較平緩，而在平面上則通常呈橢圓形。羊背石一般成群出現，平行排列。（單數為 roche moutonnée）

permafrost (*n*) ground that is frozen all the time.
frost heave the process by which soil and rocks are lifted up by the freezing of water below the surface.
frost wedging the action of frost in forcing rocks apart.
solifluction (*n*) the flow of wet material at the surface that takes place when the ground in a permafrost (↑) area is partly unfrozen.
pingo (*n*) a small hill formed by ice action in a permafrost (↑) region.
esker (*n*) a long, narrow hill of gravel and sand in a region that was once covered by ice. Eskers usually follow a twisting course, with many bends. They are probably formed by water from streams flowing through the ice. *See* **kame** (↓).

永久凍土（名） 一直凍結的土地。
凍脹 土壤和岩石因地面下的水凍結而被升高的作用。
冰楔作用 冰凍膨脹迫使岩石裂開的作用。
解凍泥流（名） 永久凍土（↑）區的土地，因局部解凍而在地表出現的潮濕泥石物質流。
冰核丘；平鍋（名） 在永久凍土（↑）區內因冰的作用而形成的小丘。
蛇形丘；蛇狀丘（名） 位於曾為冰覆蓋區域，由礫石和砂所組成的狹長山丘。蛇形丘通常依循多彎曲的河道伸展。它們很可能是由穿越冰體的河水形成。參見**冰礫阜**（↓）。

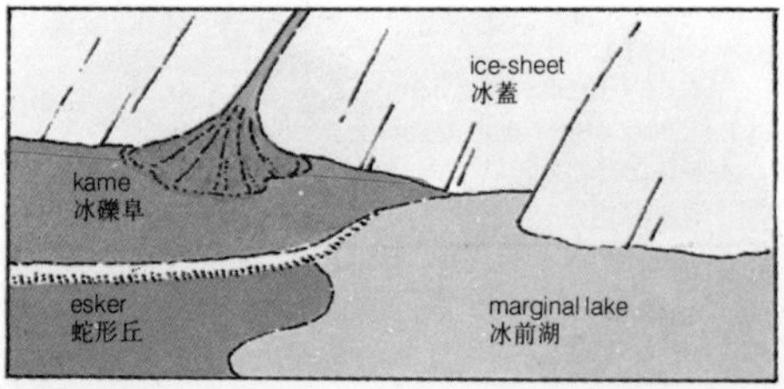

kame (*n*) a hill, usually conical in shape, made of glacial deposits (p.80) and formed at the edge of an ice-sheet (p.28) by water flowing from the ice. The word 'kame' is also used for mounds or ridges formed of glacial deposits.
drumlin (*n*) an oval mound of boulder clay (p.29). Drumlins usually occur in groups.
marginal lake a lake formed at the edge of an ice-sheet (p.28).
glaciofluvial deposits deposits (p.80) formed by water that comes from a glacier or ice-sheet (p.28).
fluvio-glacial deposits = glaciofluvial deposits (↑).
outwash fan glaciofluvial deposits (↑) that have formed at the front of an ice-sheet or large glacier (p.28).

冰礫阜；冠丘（名） 在冰蓋（第 28 頁）邊緣處，由冰體流出的水所形成並由冰川沉積物（第 80 頁）組成的、通常為錐形的山丘。"kame" 這個詞也用於由冰川沉積物所形成的土丘或山脊。
鼓丘；冰磧丘（名） 橢圓形的泥礫（第 29 頁）土丘，常成群出現。
冰前湖；冰川邊緣湖 在冰蓋（第 28 頁）邊緣處形成的湖。
冰水沉積 來自冰川或冰蓋（第 28 頁）的水所形成的沉積物（第 80 頁）。
冰水沉積（↑）英文亦拼作 fluvio-glacial deposits。
冰水扇形地；外洗扇 冰蓋或大冰川（第 28 頁）前緣處形成的冰水沉積（↑）。

outwash plain an area of more or less level ground at the edge of a glacier (p.28) where glaciofluvial deposits (↑) are formed.

冰水沉積平原；外洗平原 在冰水沉積物(↑)形成的冰川(第 28 頁)邊緣處大致平坦的地面區域。

glacial erratic a large piece of rock, or boulder, that has been carried by ice for some distance and has then come to rest where the ice has melted.

冰川漂礫；冰川漂石 被冰體搬運一定距離而在冰體融化之處停留下來的巨大石塊或漂礫。

drift (*n*) glacial (p.28) or glaciofluvial (↑) deposits.

冰磧(名) 冰川(第 28 頁)或冰水(↑)沉積。

periglacial (*adj*) on the borders of an ice-sheet (p.28). The word 'periglacial' is used both for the geographical area and for the physical conditions in it.

冰緣的(形) 描述在冰蓋(第 28 頁)的邊緣。這個詞既指地理區域也指該區域的自然條件。

U-shaped valley a valley that in cross-section is shaped like a letter U. The shape is produced by glacial (p.28) erosion (p.20).

U 形谷 橫斷面形狀似 U 字形的山谷。這種形狀由冰川(第 28 頁)侵蝕(第 20 頁)所造成。

hanging valley a smaller valley that joins a larger valley high above the floor of the larger valley. Hanging valleys are usually the result of glacial (p.28) erosion (p.20).

懸谷 與一條較大山谷相會而其谷底遠高於較大山谷的谷底的一條小山谷。懸谷通常由冰川(第 28 頁)侵蝕(第 20 頁)所造成。

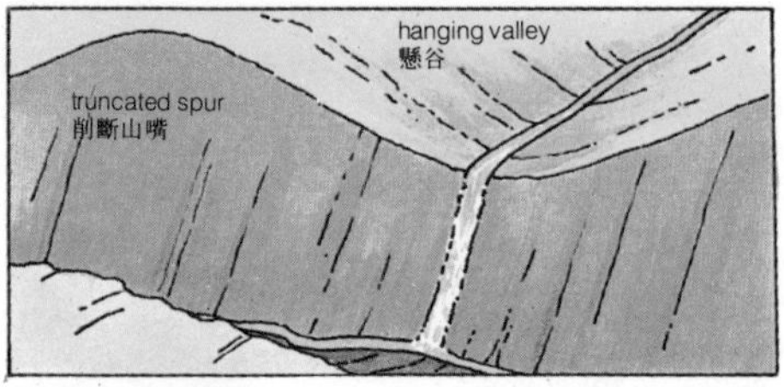

hanging valley 懸谷

truncated spur a hill at the side of a valley whose end has been cut off by the action of a glacier (p.28).

削斷山嘴 在山谷一側，其一端已為冰川(第 28 頁)作用所截斷的山丘。

fjord, fiord (*n*) a glaciated (p.28) valley of which part is below the sea.

峽灣(名) 部分位於海面以下的冰蝕(第 28 頁)谷。

cirque (*n*) a hollow of rounded shape at the head of a mountain valley. Cirques are formed by ice action.

冰斗(名) 在山谷谷源上的一個圓形凹地。冰斗是由冰蝕作用形成的。

corrie (*n*) = cirque (↑).

冰斗(↑)亦稱 corrie。

cwm (*n*) = cirque (↑).

冰斗(↑)亦稱 cwm。

arête (*n*) a sharp-edged ridge formed by glacial (p.28) erosion (p.20), commonly between two cirques (↑).

刃脊；冰蝕脊(名) 冰川(第 28 頁)侵蝕(第 20 頁)所形成的刀刃形山脊，通常位於兩個冰斗(↑)之間。

geomorphology (*n*) the study of land forms; the study of the surface forms of the Earth and their development. **geomorphological** (*adj*).

地貌學；地形學(名) 研究地形的學科；研究地球表面形態及其發育的學科。(形容詞為 geomorphological)

erosional cycle the idea that the erosion (p.20) of a region can be understood as a process that repeats itself again and again – a cycle. Each cycle begins with uplift (p.125) of the land; hills and valleys (relief) (↓) are then formed. At the end of the cycle the hills are worn down and a level plain, a *peneplain* (↓), is left. Rivers are the means by which the erosional cycle is brought about in parts of the world where the weather is neither very hot nor very cold. In regions where the rainfall is very small the cycle ends with a pediplain (↓), which may be covered by a thin layer of alluvium (p.24).

侵蝕旋迴 將一個地區的侵蝕(第 20 頁)理解為一個不斷重複(即旋迴)的過程的一種概念。每一旋迴以陸地的隆起(第 125 頁)為始，接着形成山丘和山谷(地勢)(↓)。旋迴之末，山丘被夷為平地，稱為準平原(↓)。在地球上那些不太冷也不太熱的地區，侵蝕旋迴以河流侵蝕的方式體現。在降雨量很小的區域，旋迴以山麓侵蝕面平原(↓)結束，平原上可以覆蓋一層薄的沖積層(第 24 頁)。

cycle of erosion = erosional cycle (↑).

侵蝕循環 同侵蝕旋迴(↑)。

denudation (*n*) the lowering of the land surface by all the processes of erosion (p.20); the laying bare of the rocks by the carrying away of the material covering them. **denude** (*v*), **denuded** (*adj*).

剝蝕作用(名) 由一切侵蝕(第 20 頁)過程而致的陸地表面降低；覆蓋於岩石上的物質被帶走使其裸露。(動詞為 denude，形容詞為 denuded)

degradation (*n*) the wearing down of the rocks to a lower level. **degrade** (*v*), **degraded** (*adj*).

陵夷作用(名) 岩石剝蝕到一個較低的平面。(動詞為 degrade，形容詞為 degraded)

aggradation (*n*) the opposite of degradation (↑): the building up of a surface by deposition (p.80). **aggrade** (*v*).

加積作用(名) 陵夷作用(↑)的反義詞；藉沉積作用(第 80 頁)而造成一個平面。(動詞為 aggrade)

gradation (*n*) the combined effects of aggradation (↑) and degradation (↑) on the Earth's crust. Gradation can be divided into three processes: the erosion (p.20) of the surface; the transport (p.21) of the eroded material; the deposition (p.80) of the eroded material.

均夷作用(名) 陵夷作用(↑)和加積作用(↑)對地殼的綜合效應。均夷作用分為三個過程：表面的侵蝕(第 20 頁)；受侵蝕物質的搬運(第 21 頁)；受侵蝕物質的沉積(第 80 頁)。

peneplain, peneplane (*n*) 'almost a plain'. A land surface that has been eroded to a nearly level plain.

準平原(名) "幾乎是平原"。被侵蝕成為幾乎是水平的一塊地面。

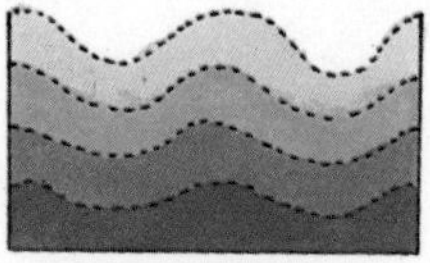

development of a peneplain
準平原的發育過程

pediplain (*n*) a kind of peneplain (↑). An erosion surface of large area formed by the joining up of two or more pediments (↓).

山麓侵蝕面平原(名) 一種準平原(↑)。兩片或多片山前侵蝕平原(↓)連成的大面積侵蝕面。

relief (*n*) the shape of part of the Earth's surface as it is shown in differences of height and steepness of slopes.

地勢(名) 地表部分的形狀，正如高度及邊坡陡度的種種差別所表現。

plateau (*n*) (*plateaux*) a flat or nearly flat area of high ground with steep sides that stands above the country round about it.

高原(名) 一塊平的或接近平的高地，其邊緣陡峭，高出於四周的地區。(複數為 plateaux)

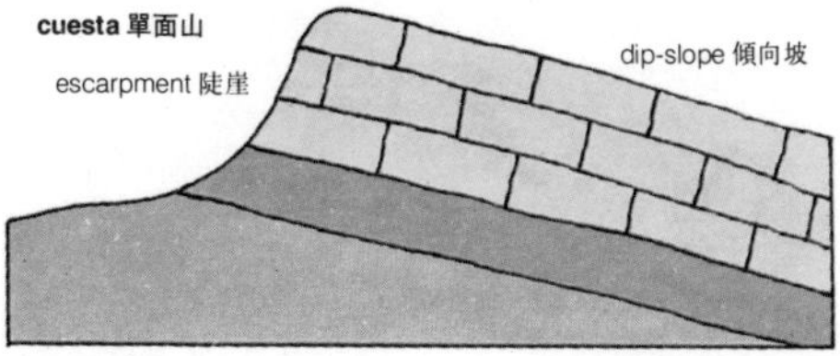

cuesta (*n*) a long, narrow hill with a steep slope (*escarpment* (↓)) on one side and a gentle slope (*dip-slope* (↓)) on the other.

單面山(名) 一側為陡坡(陡崖(↓))，另一側為緩坡(傾向坡(↓))的狹長山丘。

escarpment (*n*) a steep slope or cliff on one side of a cuesta (↑).

陡崖(名) 單面山(↑)一側的陡坡或懸崖。

dip-slope (*n*) the gentle slope on one side of a cuesta (↑), corresponding to the dip (p.123) of the rocks.

傾向坡(名) 單面山(↑)一側的緩坡，與岩層的傾斜(第 123 頁)相對應。

pediment (*n*) a gently sloping surface produced by the erosion of cliffs or steep slopes. A pediment is usually cut in solid rock with a thin layer of sediment (p.80) resting on it.

山前侵蝕平原；麓原(名) 懸崖或陡坡受侵蝕後所造成的緩斜地面。麓原通常切入基岩中，其上覆有薄層沉積物(第 80 頁)。

residual (*adj*) remaining above the general level of an area of land that has been worn down to a plain; as, e.g. a monadnock (↓) or inselberg (↓).

殘留的(形) 餘留在已剝蝕成平原陸地的一般水平面之下的。例如，殘丘(↓)或島山(↓)。

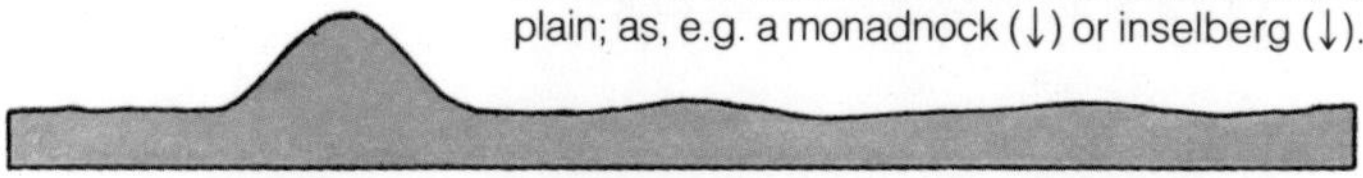
monadnock 殘丘

monadnock (*n*) a residual (↑) hill or mountain standing by itself above a peneplain (↑).

殘丘(名) 單獨突出於準平原(↑)之上的殘留(↑)山丘。

inselberg (*n*) an 'island mountain'. A residual (↑) hill or mountain with steep sides and a round top that stands by itself in a plain. A type of monadnock (↑).

島山(名) "島形的山"。單獨突出於平原，帶有圓頂和陡峭邊緣的殘留(↑)山丘。屬殘丘(↑)的一種。

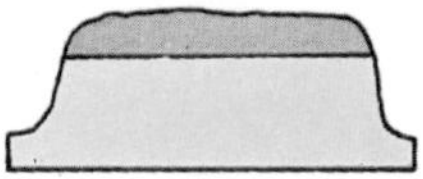
mesa 方山

mesa (*n*) an area of high, flat land (tableland) with steep sides. It has horizontal beds of hard rock at the top.

方山；平頂山(名) 高而邊緣陡峭的平地(臺地)區域，其頂部有平坦的硬岩層。

butte (*n*) a small flat-topped hill formed by the erosion (p.20) of a mesa (↑).

孤山(名) 方山(↑)受侵蝕後(第 20 頁)所形成的小平頂山丘。

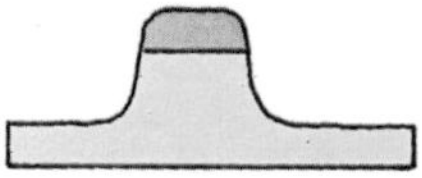
butte 孤山

karst topography an uneven type of countryside found in limestone (p.86) areas. Groundwater (p.146) makes its way through the rocks and dissolves (p.155) them, and streams flow below the surface.

岩溶地形；喀斯特地形 在石灰岩(第 86 頁)地區形成的一類不平坦的地形。地下水(第 146 頁)貫穿並溶解(第 155 頁)岩石，因而有地下流流動。

térra rossa 'red earth'. A residual (↑) deposit (p.80) formed in karst topography (↑) by the solution (p.159) of limestone (p.86).

鈣質紅土 亦稱"紅壤"。在岩溶地形(↑)區內由於石灰岩(第 86 頁)溶解(第 159 頁)而形成的殘留(↑)沉積(第 80 頁)。

hydrosphere (*n*) all the water in, on, or above the Earth's surface: in the oceans, rivers, or lakes, under the ground or in the air.

水圈(名) 所有在地表下、地面上或大氣中的水：包括海洋、江河、湖泊、地下及大氣中的水。

oceanography (*n*) the study of the oceans. **oceanographic, oceanographical** (*adj*).

海洋學(名) 研究海洋的學科。(形容詞為 oceanographic，oceanographical)

marine (*adj*) of the sea, in the sea, or formed by the sea.

海洋的；海成的(形) 指屬海洋的，在海中的，或在海形成的。

continental shelf the nearly level part of the sea floor, next to a land mass, over which the sea is not more than 180–200 m deep.

大陸架 緊鄰陸塊且其海底近於平坦的部分，大陸架的海水深度不超過 180–200 米。

shelf (*n*) short for *continental shelf* (↑).

陸架(名) 大陸架(↑)的簡稱。

neritic (*adj*) relating to shallow seas, especially to the waters over the continental shelf (↑).

淺海的(形) 與淺海有關的，特別是與大陸架(↑)上方的海域有關的。

neritic zone the part of the sea floor from low tide mark (p.37) to the outer edge of the continental shelf (↑).

淺海帶 由低潮標(第 37 頁)至大陸架(↑)外緣的海底部分。

epicontinental sea a shallow sea over part of a continental shelf (↑).

陸緣海 大陸架(↑)上方部分的淺海。

continental slope the sloping part of the sea floor from the outer edge of the continental shelf (↑). It slopes at 3–6°.

大陸坡 海底從大陸架(↑)的外緣向外傾斜的部分，其坡度為 3°–6°。

bathyal (*adj*) relating to the sea and the sea floor of the region of the continental slope (↑) from about 200 m to 2000 m depth.

半深海的(形) 與大陸坡(↑)上深度約 200 米至 2000 米的海區範圍的海及海底有關的。

marginal plateau (*plateaux*) a level part of the sea floor at a greater depth than the continental shelf – usually at depths between 240 m and 2400 m.

陸緣高原 深於大陸架的海底平坦部分，通常深度為 240 米至 2400 米。

abyssal (*adj*) of the deep ocean and its floor; at more than 2000 m depth.

深海的(形) 指深洋及深洋底的；深度超過 2000 米。

abyssal plain a level area of the floor of the deep ocean.

深海平原 深洋底的平坦區域。

abyssal hills relatively small hills that rise up from the floor of the deep ocean to heights of about 1000 m.

深海丘陵 深洋底上隆起的較小的丘陵，高度 1000 米。

continental rise the gently sloping surface between the continental slope (↑) and the abyssal plain (↑).

大陸隆 大陸坡(↑)和深海平原(↑)之間的緩傾斜面。

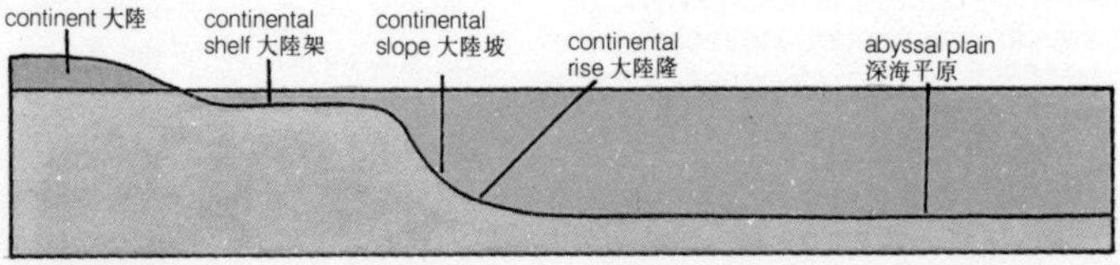

the oceans 海洋

submarine (*adj*) below the sea.

海底的(形)　海的下面。

submarine canyon a valley with steep sides cut in the continental shelf (↑) or continental slope (↑). Some submarine canyons also cut into the continental rise (↑).

海底峽谷　切入大陸架(↑)或大陸坡(↑)並具有陡壁的溝谷。有些海底峽谷也切入大陸隆(↑)中。

ocean trench a long, narrow area, much deeper than the rest of the ocean floor, that runs parallel to an island arc (p.135) or a continental margin (p.157).

洋溝；海溝　比周圍洋底深得多，與島弧(第 135 頁)或陸緣(第 157 頁)平行的狹長地帶。

oceanic ridge a mountain range under the ocean that is long and relatively narrow.

洋脊　長而窄的洋底山脈。

mid-oceanic ridge an oceanic ridge (↑) in the middle of an ocean, e.g. the mid-Atlantic Ridge.

洋中脊　大洋中央的洋脊(↑)。例如大西洋中脊。

mid-oceanic ridge in section
洋中脊割面

submarine plateau a generally flat area that is higher than the rest of the ocean floor.

海底高原　高出周圍洋底並大致平坦的區域。

seamount (*n*) a mountain under the sea rising from the floor of the ocean but not reaching sea-level; most are of volcanic (p.68) origin.

海山(名)　自洋底崛起而未達海面的海底山岳；海山大多數起源於火山(第 68 頁)。

guyot (*n*) a seamount (↑) with a flat top. Most guyots are thought to be volcanoes (p.68) that have been eroded by wave action.

海底平頂山(名)　平頂的海山(↑)。一般認為海底平頂山大多數是火山(第 68 頁)受波浪侵蝕作用形成的。

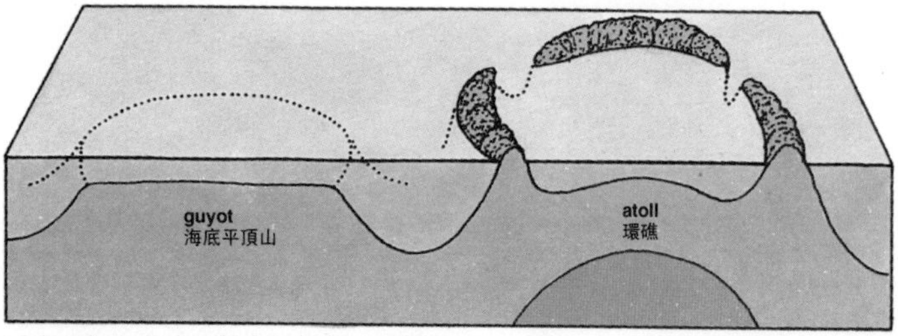

atoll (*n*) an island or chain of islands formed from coral reefs (p.38) shaped like a ring round an area of water (a lagoon).

環礁(名)　由珊瑚礁(第 38 頁)形成的島或列島鏈，呈環狀將一水域(瀉湖)環繞起來。

isobath (*n*) a line joining points on the sea-bed (or other surface) that are all at the same depth below sea-level or some other horizontal surface: a submarine contour.

等深線(名)　海床或其他面上深度相等的各點連成的曲線，即海底等高線。

wave base the depth in the sea below which there is no effective erosion (p.20) or transport (p.21) of material by waves: about 100m below sea level.

turbidity current a mass of water carrying sediment (p.80) that travels with violent movement down a slope under water. The sediment it carries makes the turbidity current denser than the seawater around it, and its higher density causes it to move down the slope. The sediment deposited by a turbidity current is called a **turbidite**.

波蝕底面 海中的某一深度，在此深度以下不存在波浪對物質的有效侵蝕(第 20 頁)和搬運(第 21 頁)。此深度約在海面下 100 米。

濁流 在水底攜帶沉積物(第 80 頁)並以激烈運動方式順坡向下流動的水體。由於濁流攜有沉積物，其密度比周圍海水高，因而順坡向下移動。濁流所沉積的沉積物稱為**濁積岩**。

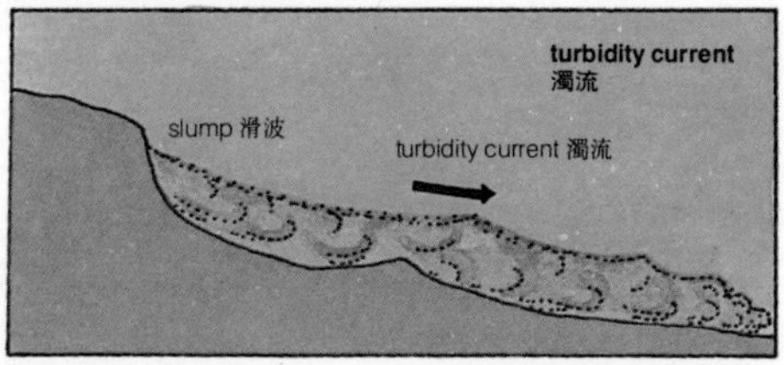

density current = turbidity current (↑).

tsunami (*n*) a sea wave caused by an earthquake under the sea.

eustatic (*adj*) *eustatic* movements are world-wide changes in sea level.

photic zone the part of the sea that the light of the sun reaches: down to about 200 m.

ooze (*n*) a deposit (p.80) in the abyssal zone (p.34) of the oceans. Oozes contain the hard parts (skeletons) of tiny animals living in the sea. There are two types of ooze: *calcareous* (p.86) *oozes*, formed at depths from 2000 m to 3900 m, and *siliceous* (p.87) *oozes*, formed at depths greater than 3900 m.

Globigerina ooze a calcareous (p.86) deposit (p.80) formed on the floor of the deep ocean. It is made up largely of the sheels of Foraminifera (p.104), especially *Globigerina*.

Radiolarian ooze a siliceous (p.87) deposit (p.80) formed on the floor of the deep ocean. It is made up largely of the skeletons of Radiolaria (p.104).

red clay a deposit (p.80) formed in the deepest parts of the oceans, at depths of more than 5000 m. It is made up of fine material carried by the wind, including volcanic (p.68) dust.

密度流 同濁流(↑)。

海嘯(名) 海底地震引起的海浪。

海面升降的(形) 海面升降運動是指全球性海平面的變化。

透光層 指陽光能達到的那部分海水，陽光可達水深約 200 米。

軟泥；海泥(名) 海洋深海區域(第 34 頁)中的沉積(第 80 頁)。其中含有海生細小生物的堅硬部分(骨骼)。軟泥分兩類：形成於 2000 米至 3900 米深處的鈣質(第 86 頁)軟泥和形成於 3900 米以下深度的矽質(第 87 頁)軟泥。

抱球蟲軟泥 一種在深洋底形成的鈣質(第 86 頁)沉積(第 80 頁)，主要是由有孔蟲目(第 104 頁)的殼，特別是抱球蟲的殼所組成。

放射蟲軟泥 一種在深洋底形成的矽質(第 87 頁)沉積(第 80 頁)，主要是由放射蟲目(第 104 頁)的骨骼組成。

紅黏土 形成於海洋最深處(深度超過 5000 米)的沉積(第 80 頁)，是由風所攜帶的細粒物質所組成，其中包括火山(第 68 頁)塵。

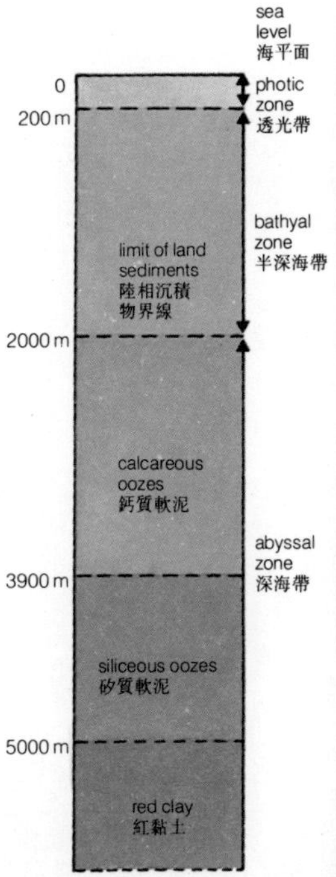

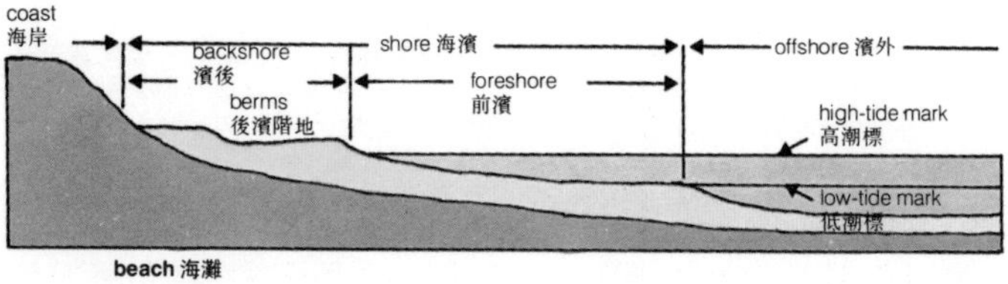

coast (*n*) the part of the land that is next to the sea and on which waves have a direct effect.

coastline (*n*) (1) the line that divides the coast (↑) from the shore (↓); (2) the line that divides the land from the sea. *See also* **shoreline** (↓).

tide (*n*) the rise and fall of the water in the oceans, caused by the gravitational (p.11) pull of the sun and moon. **tidal** (*adj*).

tide mark the point on a shore (↓) or cliff reached by the water of the sea at high or low tide (↑).

shore (*n*) the region from low tide mark (↑) to the highest point reached by waves or tides (↑).

shoreline (*n*) the line at which the land meets the sea. *See also* **coastline** (↑).

off-shore (*adj*) from low tide mark (↑) to the edge of the continental shelf (p.34).

berm (*n*) the backshore: the part of the shore that is level or slopes towards the land.

backshore (*n*) = berm.

foreshore (*n*) the zone between low tide mark (↑) and the berm or backshore (↑).

littoral (*adj*) of the shore (↑); between high and low tide marks.

beach (*n*) the shore (↑), especially where it is formed of sand or pebbles (p.87).

shingle (*n*) small rounded stones and loose pebbles (p.87); material characteristic of beaches (↑).

cliff (*n*) a high and very steep rock face, nearly vertical, especially one on the sea shore (↑).

sea cave a cave that has its entrance in a sea cliff (↑). A sea cave is likely to be formed where a cliff is made of stratified sediments (pp.80, 81).

wave-cut platform a level or nearly level area of rock formed by wave action below a sea-cliff. It may be bare or covered by a beach (↑).

notch, wave-cut notch a V-shaped cut at the base of a sea-cliff formed by wave action.

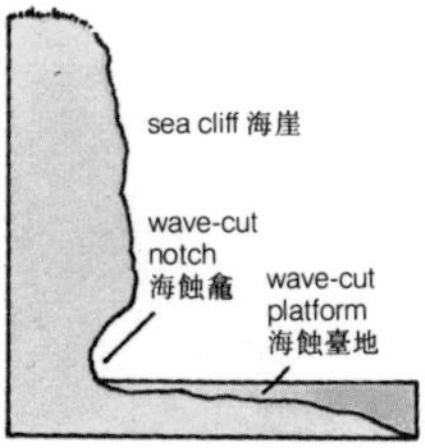

海岸(名)　與海相鄰且直接受海浪影響的陸地部分。

海岸線(名)　(1)海岸(↑)與海濱(↓)的分界線；(2)陸地和海洋的分界線。參見**海濱線**(↓)。

潮；潮汐(名)　太陽和月亮的萬有引力(第 11 頁)所造成的海水漲落。(形容詞為 tidal)

潮標　在高潮或低潮(↑)時海水到達海濱(↓)或懸崖的標誌點。

海濱(名)　從低潮標(↑)到波浪或潮汐(↑)所達最高點之間的區域。

海濱線(名)　陸地和海洋交會的界線。參見**海岸線**(↑)。

濱外的；離岸的(形)　指低潮標(↑)至大陸架(第 34 頁)邊緣之間的區域。

後濱階地(名)　濱後，即平坦或朝陸地傾斜的海濱部分。

濱後；後濱(名)　同後濱階地。

前濱(名)　低潮標(↑)和後濱階地(濱後)(↑)之間的區域。

海岸區的；潮汐區的(形)　指海濱的(↑)：高、低潮標之間的。

海灘(名)　海濱(↑)，尤指由砂或礫石構成的海濱。

灘礫(名)　小圓石和鬆散礫石(第 87 頁)：海灘(↑)所特有的物質。

海崖；懸崖(名)　高而極陡的岩石壁，近於直立，尤指海濱(↑)之岩石壁。

海蝕洞　入口在海崖(↑)的洞穴。海崖由層狀沉積物(第 80、81 頁)構成之處很可能形成海蝕洞。

浪蝕臺地；海蝕臺　在海崖下由海浪作用形成的平坦或近於平坦的岩石區域。岩石區域可以是裸露的，也可以被海灘(↑)覆蓋。

海蝕龕　在海崖底部由海浪作用形成的 V 形溝。

stack (*n*) a mass of rock with steep sides standing off shore (p.37) by itself like a small island. Stacks are formed when a sea cliff (p.37) is eroded (p.20).

海蝕柱（名） 如小島般獨自矗立於濱外（第 37 頁）並具陡峭邊緣的岩塊。海崖（第 37 頁）受侵蝕（第 20 頁）後形成海蝕柱。

stack 海蝕柱

arch (*n*) a mass of rock in the shape of an arch. Arches are commonly formed by coastal erosion (p.20) of stratified rocks (p.80) where harder beds (p.80) rest on weaker beds.

海蝕穹（名） 形態像拱橋的岩塊。海蝕穹通常由軟岩層上覆有硬岩層（第 80 頁）的層狀岩石（第 80 頁）受海岸侵蝕（第 20 頁）所形成。

spit (*n*) a long, narrow deposit (p.80) of sediment (p.80) that stands out from a coast.

沙嘴；岬（名） 從海岸向外伸展的狹長沉積物（第 80 頁）。

bar (*n*) a long, narrow deposit (p.80) of sand or gravel (p.87) that stands at or not far below sea level, either at the mouth of a river or parallel to a beach (p.37).

沙壩；沙洲（名） 在河口或平行於海灘（第 37 頁）處，露出海面或在海面下不太深處的狹長砂礫（第 87 頁）沉積（第 80 頁）。

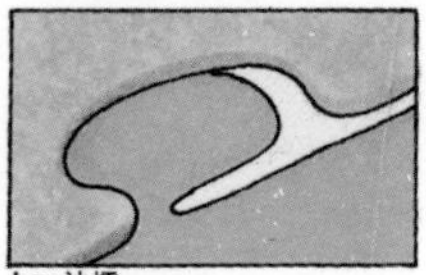
bar 沙壩

barrier island a long island or beach (p.37) parallel to a shoreline. There is usually a lagoon (↓) between the barrier island and the coast. The island is usually only one or two metres above sea level. A barrier island may be formed by the growth of a spit (↑) or by the submergence (↓) of a coast after dunes (p.22) have been formed.

堡礁島 和海濱線平行的長條形島或海灘（第 37 頁）。海岸和堡礁島之間通常有瀉湖（↓）。堡礁島通常僅高出海面一、二米。堡礁島可由沙嘴（↑）生長而成，也可由海岸沙丘（第 22 頁）形成之後下沉（↓）而造成。

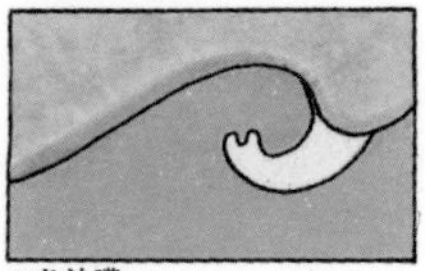
spit 沙嘴

reef (*n*) a narrow ridge of rock at or near the surface of the water.

礁（名） 在水面或近水面處的狹窄岩脊。

coral reef a reef (↑) made up of corals (p.105) or other organisms; a type of bioherm (p.101).

珊瑚礁 由珊瑚（第 105 頁）或其他有機體構成的礁（↑）；生物岩礁（第 101 頁）的一種。

barrier reef a long, narrow coral reef (↑), parallel to the shore and separated from it by a lagoon (↓).

堡礁；堤礁 和海濱平行，且其間被瀉湖（↓）隔開的狹長珊瑚礁（↑）。

lagoon (*n*) (1) a body of shallow water between a barrier island (↑) and the shore; (2) a body of water inside the coral (p.105) reefs (↑) of an atoll (p.35). **lagoonal** (*adj*).

礁湖；瀉湖（名） (1)堡礁島（↑）和海濱之間的淺水區的水體；(2)環礁（第 35 頁）的珊瑚（第 105 頁）礁（↑）內的水體。（形容詞為 lagoonal）

sabhka (*n*) a broad, flat coastal area with a thin covering of salt (NaCl). Sabhkas lie above the high tide mark (p.37) and are not often covered by the sea.

鹼灘；鹽沼（名） 有薄鹽（NaCl）層覆蓋的廣濶而平坦的海岸區域。鹼灘位於高潮標（第 37 頁）之上，且不常被海水淹沒。

marine swamp a wet area on a coast. The water moves slowly if at all. Plants grow in large numbers and plant remains make up a large part of a marine swamp. In the geological past marine swamps were common in the Carboniferous period (p.114). They can be seen today on the eastern and southern coasts of the United States.

沿海沼澤 海岸上的潮濕地區，其中的水即使有流動也是很緩慢。區域內植物生長茂盛，植物遺骸佔據沿海沼澤的大部分。在地質史上，沿海沼澤在石炭紀（第 114 頁）普遍存在。目前在美國東海岸和南海岸仍可見到沿海沼澤。

primary coast a coast (p.37) formed by the sea coming to rest against a land form that has been shaped by terrestrial (p.81) activities: erosion, deposition, volcanism, fault movements. *See also* **secondary coast** (↓).

原生海岸 海水淹過由陸地(第 81 頁)活動所形成的某種地形而形成的一類海岸(第 37 頁)，陸地活動包括侵蝕、沉積、火山作用、斷層活動等等。參見**次生海洋**(↓)。

secondary coast a coast (p.37) that has been shaped by the sea or by marine (p.34) organisms (p.98). *See also* **primary coast** (↑).

次生海岸 由海洋活動或海生(第 34 頁)生物體(第 98 頁)的活動所形成的海岸(第 37 頁)。參見**原生海洋**(↑)。

submerged coast a coast (p.37) that is the result of a rise in sea level, or of downward movement of the land. The sea usually reaches inland in embayments (↓), which may go far in from the coast.

下沉海岸 海平面上升或陸地下沉所形成的海岸(第 37 頁)。海水通常浸入內陸很遠而形成海灣(↓)。

drowned coast = submerged coast (↑).

淹沒海岸 同下沉海岸(↑)。

embayment (*n*) an inward curve in a shoreline (p.37) forming an open *bay*.

海灣(名) 海濱線(第 37 頁)向內彎曲而形成的開濶港灣。

emergent coast a coast (p.37) that is the result of a fall in sea level or of the rising of the land. Emergent coasts commonly have marine terraces (↓) with beaches (p.37), wave-cut platforms (p.37), and old sea cliffs (p.37) standing above sea level.

上升海岸 海平面下降或陸地上升所形成的海岸(第 37 頁)。上升海岸一般有海蝕階地(↓)伴有海灘(第 37 頁)、浪蝕臺地(第 37 頁)和矗立於海面上的古海崖(第 37 頁)。

raised beach 上升海灘

raised beach a beach that is above sea level and is separated from the present beach. Raised beaches are characteristic of emergent coasts (↑).

上升海灘 高出海面並與現海灘分開的海灘。上升海灘為上升海岸(↑)所特有。

marine terrace an old marine (p.34) beach (p.37) or wave-cut platform (p.37) now standing some distance above sea level.

海蝕階地 現出海面以上一定高度的舊海(第 34 頁)灘(第 37 頁)或浪蝕臺地(第 37 頁)。

ria (*n*) an embayment (↑) or arm of the sea that has been shaped by stream erosion (p.20) before being filled by the sea.

里亞式海灣(名) 受河流侵蝕(第 20 頁)後海水淹沒而形成的海灣(↑)。

ria coast a submerged coast (↑) in which valleys formed by stream erosion (p.20) have been filled by the sea.

里亞式海岸 河流侵蝕(第 20 頁)所形成的谷地被海水注入而成的一種下沉海岸(↑)。

ria coast 里亞式海岸
S.W. Ireland愛爾蘭西南部

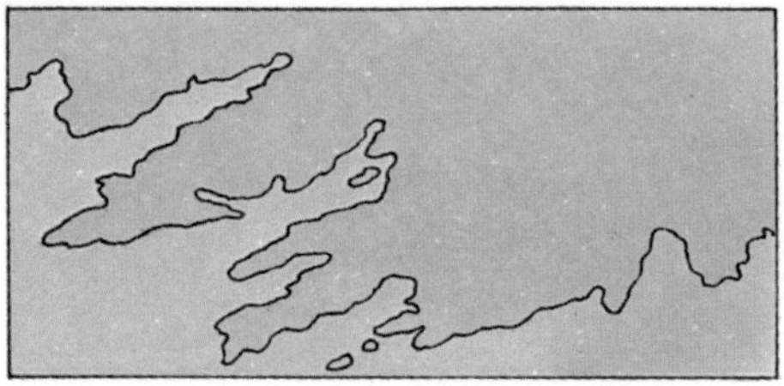

crystal (*n*) a body with surfaces that are smooth, flat, and regularly arranged. The regular shape of a crystal results from the regular arrangement of the atoms (p.152) of which it is made. **crystalline** (*adj*); *see also p.44.*

crystallize (*v*) to form crystals (↑). **crystallization** (*n*).

crystallography (*n*) the study of crystals (↑). **crystallographic** (*adj*).

crystal lattice the regular arrangement of atoms (p.152) in three dimensions in a crystalline solid.

unit cell the smallest complete piece of a crystal lattice (↑) that shows the arrangement of the atoms (p.152) in a crystal. The unit cell contains a number of atoms arranged in a regular way. It is repeated in three dimensions to form the crystal lattice.

face (*n*) a single, flat surface on a crystal (↑). In crystallography it is not the sizes of faces that are important but the angles between them – the interfacial angles (↓).

interfacial angle the angle between two faces (↑) of a crystal (↑). It is measured between lines at 90° to the crystal faces. These lines are called *normals*.

normal (*n*) *see* interfacial angle (↑).

goniometer (*n*) an instrument for measuring the interfacial angles (↑) of crystals (↑). There are two types: the *contact goniometer*, in which two straight arms are placed on the crystal faces to be measured; and the *reflecting goniometer*, in which a beam of light is used to make the measurement.

form (*n*) in crystallography, a form is a group of crystal faces that are related to a single face by the symmetry elements (p.42) of a particular crystal class. For example, eight faces make up a pyramid form in the cubic or tetragonal system (p.43); six faces make up the form of a hexagonal prism. A closed form (e.g. a pyramid) can enclose space by itself; an open form (e.g. a prism) cannot.

zone (*n*) a set of crystal faces (↑) meeting at edges parallel to each other is called a *zone*. **zonal** (*adj*).

晶體（名） 有多個平滑、呈規則排列表面的一種物體。晶體呈現規則形狀是由於組成晶體的原子（第 152 頁）成規則排列的結果。（形容詞為 crystalline）；參見第 44 頁。

使結晶（動） 使形成晶體（↑）。（名詞為 crystallization）

結晶學（名） 研究晶體（↑）的學科。（形容詞為 crystallographic）

晶體點陣；晶格 結晶固體中的原子（第 152 頁）在三維空間的規則排列。

晶胞 顯示晶體中諸原子（第 152 頁）排列最小而完整的晶格（↑）單位。晶胞包含有許多規則排列的原子。晶胞在三維空間重複堆疊而形成晶格。

晶面（名） 晶體（↑）上的單個平面。在結晶學上，晶面的大小無關緊要，晶面之間的角度——晶面角（↓）才是重要的。

晶面角；交互面夾角 晶體（↑）的兩個晶面（↑）之間的角度。它是兩個晶面的垂線之間的夾角。這些垂線稱法線。

法線（名） 見晶面角（↑）。

測角儀（名） 測量晶體（↑）之晶面角（↑）的儀器。有兩種：接觸測角儀，將兩直臂置於待測晶面上即可測得晶面角；反射測角儀，用一道光束來測量。

晶形；格式（名） 在結晶學中，晶形由一組晶面構成，這一組晶面均按其所屬晶體類別的對稱要素（第 42 頁）而與一個單面相聯係。例如，在等軸立方晶系和四方晶系（第 43 頁）中由八個晶面組成一個錐形，而六方柱晶形則由六個晶面構成。閉合晶形（如錐體）本身可將空間封閉，而開放形（如柱體）則不行。

晶帶（名） 一組晶面（↑），它們相交所成的晶稜互相平行，則這組晶面名之晶帶。（形容詞為 zonal）

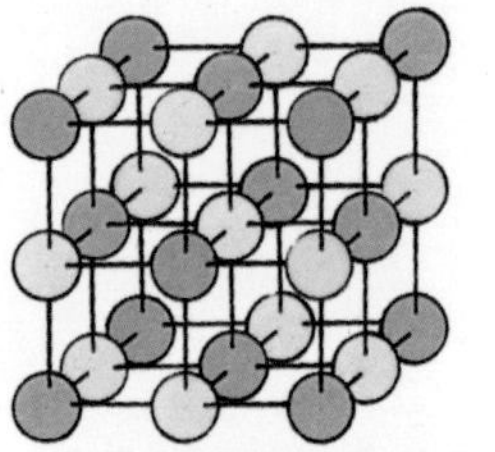

unit cell of sodium chloride (NaCl)
氯化鈉（NaCl）的晶胞
each Na^+ ion is surrounded by six Cl^- ions and each Cl^- ion by six Na^+ ions
每個 Na^+ 離子為 6 個 Cl^- 離子圍着，而每個 Cl^- 離子為 6 個 Na^+ 離子圍着

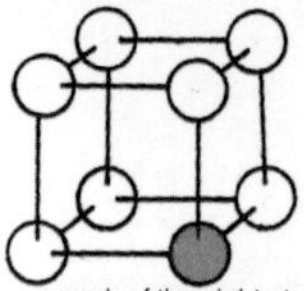

unit cell containing one atom
含一個原子的晶胞
each of the eight atoms shown is common to eight unit cells 圖示 8 個原子的每一個都為相鄰八個晶包共有

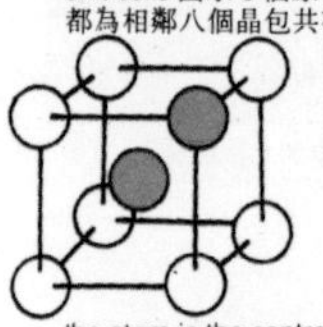

Unit cell containing two atoms
含有 2 個原子的晶胞
the atom in the centre belongs only to the unit cell shown
中心的原子僅屬於圖示的晶胞

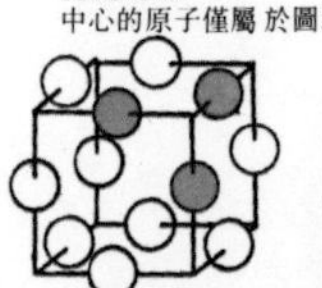

unit cell containing three atoms
含 3 個原子的晶胞
each of the twelve atoms shown belongs to four unit cells
圖示 12 個原子中每一個均屬相鄰 4 個晶胞

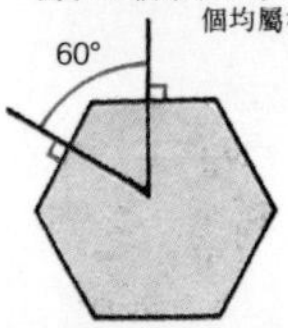

interfacial angle
晶面角

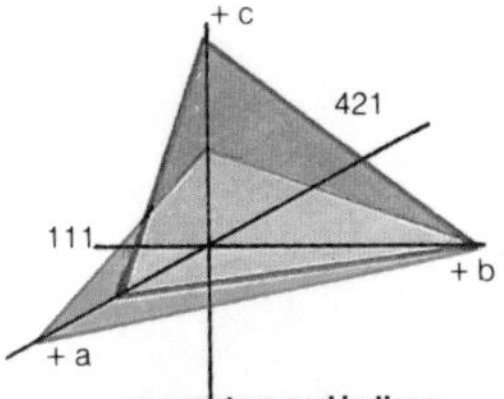

parameters and indices
軸單位比和晶面指數
blue: unit form, 111
red: plane with parameters ½, 1, 2: indices 421
藍：單位晶形，111
紅：軸單位比分別為 ½，1，2：即晶面指數為 421 的面

crystallographic axis (*axes*) one of the set of three or four imaginary lines used for measuring and describing the forms of crystals (↑). **axial** (*adj*).

晶軸　指用於描述和測量晶形(↑)所設想的三條一組或四條一組的直線之一，(複數為 axes，形容詞為 axial)

intercept (*n*) the distance between the point where a plane, such as a crystal face (↑), cuts a crystallographic axis (↑) and the point where the axes meet each other (the origin).

截距(名)　指一個面如晶面(↑)與晶軸(↑)之交點至各晶軸交會點(原點)之間的距離。

parameter (*n*) the ratios of the intercepts (↑) that the unit form (↓) makes with the crystallographic axes (↑) are called the *parameters*.

參數；軸單位比(名)　各晶軸(↑)與單位晶形(↓)相交所構成的幾個截距(↑)的連比值。

axial ratio = parameter (↑).

軸率　同軸單位比(↑)。

unit form a crystal form (↑) is chosen as the unit form to provide the parameters (↑) for measuring other forms present in the same crystal.

單位晶形　一種晶形(↑)被選定作為單位晶形，目的在於求出軸單位比(軸率)(↑)以作為測量同一晶體中存在的其他晶形的依據。

indices (*n.pl.*) the crystallographic indices are obtained by calculating in turn the value of the *reciprocal* of each parameter (↑), i.e. the value of 1 divided by the parameter, and then turning the figures obtained into whole numbers. The indices are used for describing crystals.

晶面指數(名、複)　晶面指數是依次計算每一軸率(↑)的倒數值而得出的，亦即將軸率除 1，再將其數值轉換為整數。晶面指數用以描述晶體。

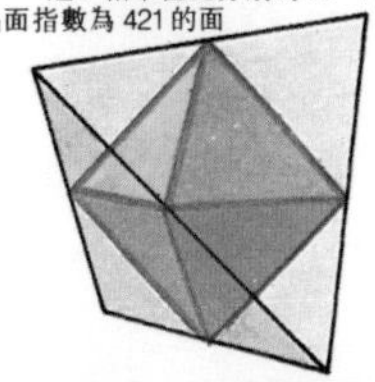

tetrahedron and octahedron
(grey) (red)
四面體（灰色）和**八面體**（紅色）

cube (*n*) a solid with six faces (↑), each of which is a square, at 90° to each other.

立方體(名)　由六個互成 90° 的正方形晶面(↑)構成的立體。

tetrahedron (*n*) (*tetrahedra*) a solid with four faces (↑), each of which is an equilateral triangle (a triangle with all sides equal in length). **tetrahedral** (*adj*).

四面體(名)　由四個等邊三角形(三角形的各邊等長)晶面(↑)構成的立體。(複數為 tetrahedra，形容詞為 tetrahedral)

octahedron (*n*) (*octahedra*) a solid with eight faces (↑), each of which is an equilateral triangle (a triangle with all sides equal in length). **octahedral** (*adj*).

八面體(名)　由八個等邊三角形(三角形的各邊等長)晶面(↑)構成的立體。複數為 octahedra，形容詞為 octahedral)

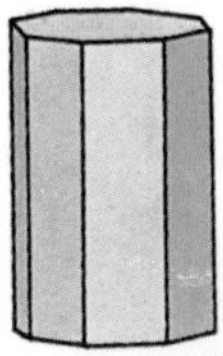

prism faces **稜柱面**
(blue)（藍色）

prism (*n*) a crystal face (↑) that cuts the horizontal crystallographic axes (↑) and is parallel to the vertical (*c*) axis. **prism, prismatic** (*adj*).

(稜)柱(名)　切割水平晶軸(↑)並與垂直晶軸(C)平行的一種晶面。(形容詞為 prism，prismatic)

pyramid (*n*) an open crystal form (↑) consisting of faces that meet at a point.

(稜)錐(名)　由交會於一點的幾個晶面組成的一種開放晶形(↑)。

pinacoid (*n*) an open crystal form (↑) consisting of two faces parallel to each other.

軸面(名)　由兩個互相平行面組成的一種開放晶形(↑)。

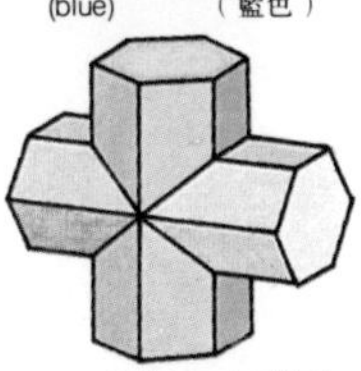

twin crystal 孿晶

twin crystal two crystals (↑) of the same substance joined together in such a way that a crystallographic (↑) direction or crystallographic plane is shared by the two parts. Twins may be *simple, penetrating* (in which one crystal appears to pass through the other), *repeated*, or *compound* (*complex*). **twinning** (*n*), **twinned** (*adj*).

孿晶；雙晶　由同一種物質形成兩個晶體(↑)以共用結晶(↑)方向或晶面的方式連合在一起。孿晶(雙晶)可以是簡單的、貫穿的(一個晶體穿過另一個)、重複的或複合的(複雜的)。(名詞為 twinning，形容詞為 twinned)

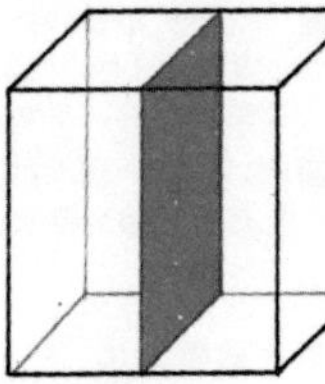

plane of symmetry
對稱面

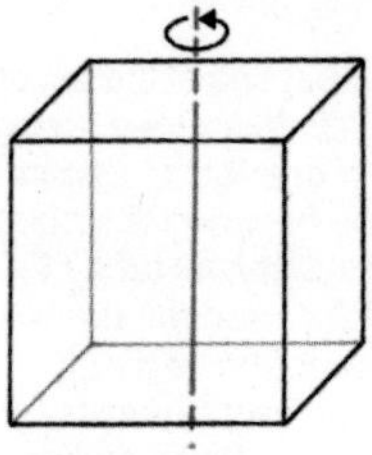

axis of symmetry
對稱軸

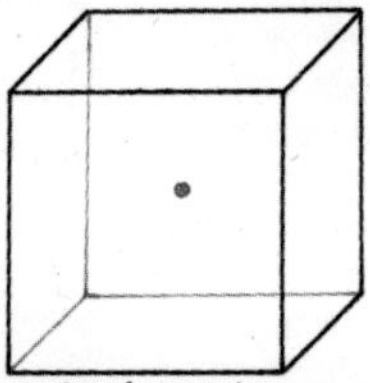

centre of symmetry
對稱中心

symmetry (*n*) in crystallography (p.40), the exact agreement of faces (p.40) on opposite sides of a crystal (p.40). Crystallographic symmetry depends on the angles between the faces (p.40) of a crystal.

對稱(名) 結晶學(第 40 頁)上指晶體(第 40 頁)中相對側的兩晶面完全吻合。結晶學的對稱依賴於晶體各晶面(第 40 頁)間的角度。

symmetry element a plane (↓), axis (↓), or centre (↓) of symmetry.

對稱要素 指對稱面(↓)、對稱軸(↓)、或對稱中心(↓)。

plane of symmetry an imaginary flat surface that divides a body, such as a crystal (p.40), into two halves, each of which is like the other as reflected in a mirror.

對稱面 一個假想的平面，它將一個物體，例如晶體(第 40 頁)等分成兩半部，而此兩半如鏡像般完全一樣。

axis of symmetry an imaginary line on which a crystal may be turned so that it comes into positions that are crystallographically the same two or more times in one complete turn of 360°. An axis of symmetry can be twofold, threefold, fourfold, or sixfold (but not fivefold).

對稱軸 晶體可繞之旋轉的一條假想直線。每轉 360° 的一整圈，結晶學上相同的狀態可重複出現兩次或多次。對稱軸可以是二次、三次、四次、六次的(但無五次的)對稱軸。

centre of symmetry a centre of symmetry is present when for each point on the surface of a crystal there is a similar point on the opposite side of the crystal, the two points being on a straight line passing through the centre of the crystal and being at equal distances from the centre.

對稱中心 晶體表面上的每一個點都在相對側的晶面上相應有一個相似點，此兩點的連線如經過晶體中心，且中心至兩點距離相等，則此中心稱為對稱中心。

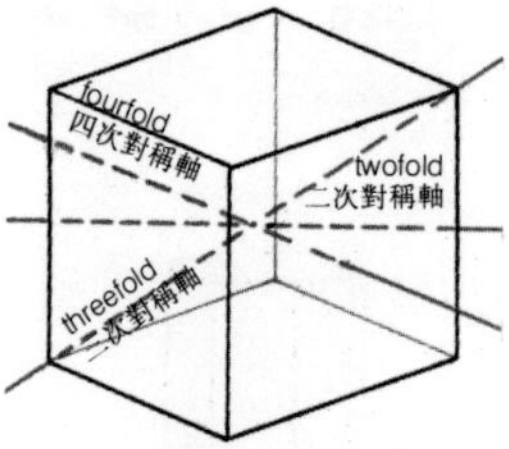

symmetry axes of the cube
立方體的對稱軸

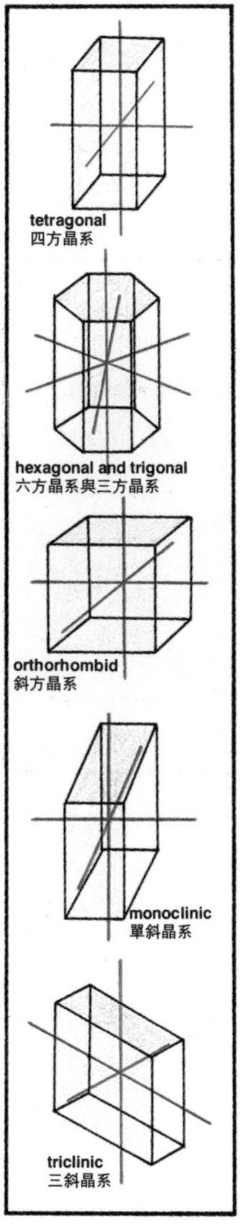

crystal system any crystal (p.40) can be classified as belonging to one of seven crystal systems, each of which has its own symmetry elements (↑). The same set of crystallographic axes (p.41) is used for describing all the crystals belonging to any one system.

cubic system a crystal system (↑) with four threefold axes of symmetry (↑). There are three crystallographic axes (p.41) at 90° to each other; the parameters (p.41) on all three axes are equal.

tetragonal system a crystal system (↑) with one fourfold axis of symmetry (↑). There are three crystallographic axes (p.41) at 90° to each other. The parameters (p.41) on the two horizontal axes are equal but are not equal to the parameter on the vertical axis.

hexagonal system a crystal system (↑) with one sixfold axis of symmetry (↑). There are four crystallographic axes (p.41), three at 120° in the horizontal plane, and one vertical and perpendicular to them. The parameters (p.41) for the three horizontal axes are all equal but are not equal to the parameter for the vertical axis.

trigonal system a crystal system (↑) with one threefold axis of symmetry (↑). The crystallographic axes (p.41) are usually as for the hexagonal system (↑). In the past the trigonal system was included with the hexagonal system.

rhombohedral system = trigonal system (↑).

orthorhombic system a crystal system (↑) with three twofold axes of symmetry (↑). There are three crystallographic axes at 90° to each other. The parameters (p.41) are unequal.

monoclinic system a crystal system (↑) with one twofold axis of symmetry. There are three crystallographic axes (p.41), of which two (one of them the vertical axis) are at 90° to each other. The three parameters (p.41) are unequal.

triclinic system a crystal system with no axes of symmetry. There is only a centre of symmetry. There are three crystallographic axes (p.41), none of them at 90° to each other. The parameters (p.41) are all unequal.

晶系　任何晶體(第 40 頁)均可歸入七個晶系中的一個，而每一晶系都自有其對稱要素(↑)。同屬任一個晶系的所有晶體，均可用同一組晶軸(第 41 頁)來描述。

等軸晶系；立方晶系　一種具有四條三次對稱軸(↑)的晶系(↑)，有三條互成 90° 的晶軸(第 41 頁)，所有三條軸的參數(第 41 頁)都相等。

四方晶系；正方晶系　一種具有一條四次對稱軸(↑)的晶系(↑)。有三條互成 90° 的晶軸(第 41 頁)。兩條水平軸的參數(第 41 頁)相等，但與垂直軸的參數不相等。

六方晶系　具有一條六次對稱軸(↑)的晶系(↑)。有四條晶軸(第 41 頁)，其中三條晶軸在一個水平面上互成 120° 角相交，另一條晶軸直立並與其他三條晶軸垂直。三條水平軸的參數(第 41 頁)相等，而與垂直軸的參數不相等。

三方晶系　具有一條三次對稱軸(↑)的晶系(↑)。其結晶軸(第 41 頁)通常和六方晶系(↑)相同。過去將三方晶系列入六方晶系。

菱形晶系　同三方晶系(↑)。

斜方晶系；正交晶系　具有三條二次對稱軸(↑)的晶系(↑)。有三條互成 90° 的晶軸，其參數(第 41 頁)各異。

單斜晶系　具有一條二次對稱軸的晶系(↑)。有三條晶軸(第 41 頁)，其中兩條(一條為垂直軸)互成 90°。三條軸的參數(第 41 頁)不相等。

三斜晶系　無對稱軸的晶系。只一個對稱中心，有三條晶軸(第 41 頁)互不垂直。參數(第 41頁)互不相等。

mineral (*n*) a substance having a definite chemical composition (p.15), or a definite range of composition, that has been formed naturally and occurs in the Earth's crust. Most minerals have a characteristic crystal form (p.40).

礦物(名) 具一定的化學成分(第 15 頁)，或成分有一定的變動範圍，天然形成並存在於地殼內的物質，礦物大都具特有的晶形(第 40 頁)。

mineralogy (*n*) the study of minerals (↑).

礦物學(名) 研究礦物(↑)的學科。

specific gravity the ratio of the mass of a substance to the mass of an equal volume of water.

比重 物質質量與同體積水的質量之比。

hardness (*n*) the hardness of a mineral (↑) is measured by its ability to make a mark on the surface of another mineral. The scale of hardness that is used is due to Mohs:, ranging from 1, talc (p.61) to 10, diamond (p.48). The surfaces of minerals with a hardness of less than $6\frac{1}{2}$ can be marked (scratched) with a knife. Minerals with a hardness of $2\frac{1}{2}$ or less can be scratched with a finger-nail.

硬度(名) 礦物(↑)的硬度是依據它刻劃另一礦物表面的能力來度量的。目前所用的硬度計是莫氏建立的，其硬度級別自硬度為 1 的滑石(第 61 頁)到硬度為 10 的金剛石(第 48 頁)。硬度低於 $6\frac{1}{2}$ 的礦物，其表面可用刀刻劃。硬度為 $2\frac{1}{2}$ 或低於 $2\frac{1}{2}$ 的礦物可用指甲刻劃。

1 talc 滑石
2 gypsum 石膏
3 calcite 方解石
4 fluorite 螢石
5 apatite 磷灰石
6 orthoclase 長石
7 quartz 石英
8 topaz 黃玉
9 corundum 剛玉
10 diamond 金剛石

Mohs' scale *see* **hardness** (↑).

莫氏硬度 見硬度(↑)。

cleavage (*n*) the cleavage of a mineral (↑) is the property of breaking along clearly marked smooth planes which are parallel to possible crystal faces (p.40). **cleave** (*v*).

解理；分裂性(名) 礦物(↑)的解理是指礦物順着清晰光滑面劈開的性質，而該光滑面則平行於可能存在的晶面(第 40 頁)。(動詞為 cleave)

cleavage plane a flat surface along which a mineral (↑) will cleave (↑).

解理面；分裂平面 礦物(↑)沿之劈開(↑)的平面。

fracture (*n*) the broken surface of a mineral. Its character can be useful in naming minerals.

斷口(名) 礦物的斷開面，其特徵可用於礦物的命名。

streak (*n*) the colour of a mineral when it is in the form of a powder. This colour may be different from its colour in the mass. The streak is seen by rubbing a piece of the mineral on a rough plate called a *streak-plate*.

條痕(名) 礦物粉末呈現的顏色。這種顏色可能有別於礦物塊體的顏色。在"條痕板"的粗糙面上磨擦一片礦物即可見其條痕。

crystallinity (*n*) the degree to which a substance shows crystal form (p.40).

結晶度(名) 物質顯現晶形(第 40 頁)的程度。

crystallized (*adj*) showing well-developed crystals (p.40).

結晶的(形) 顯示發育良好的晶體(第 40 頁)。

crystalline (*adj*) (1) of the nature of a crystal (p.40); with regular atomic (p.152) structure; (2) composed of a mass of imperfectly formed crystal grains (p.72).

晶質的(形) (1)指晶體(第 40 頁)的本性；具有規則的原子(第 152 頁)結構；(2)由大量不完整的晶體顆粒(第 72 頁)所組成的。

cryptocrystalline (*adj*) with crystals (p.40) that can be seen only under the microscope (p.147).

隱晶質的(形) 在顯微鏡(第 147 頁)下才能看見其晶體(第 40 頁)的。

amorphous (*adj*) with no crystal (p.40) structure, as in natural glasses (p.77).

非晶質的；無定形的(形) 不具晶體(第 40 頁)構造的，如在天然玻璃(第 77 頁)中所見。

habit (*n*) the characteristic shapes of crystals (p.40) that are due to variations in the number, size, and shape of the crystal faces (p.40).

習性(名) 晶體(第 40 頁)由於晶面(第 40 頁)數目、大小和形狀的差異而具有的特徵形態。

equant (*adj*) a habit (↑) in which the dimensions of a crystal (p.40) are about the same in all directions.

等量綱的(形) 描述晶體(第 40 頁)尺度在所有方向上都大致相同的一種習性(↑)。

tabular (*adj*) a habit (↑) in which two dimensions of a crystal (p.40) are much greater than the third.

板狀(形) 描述晶體(第 40 頁)在兩個方向的尺度遠遠大於第三方向的尺度的一種習性(↑)。

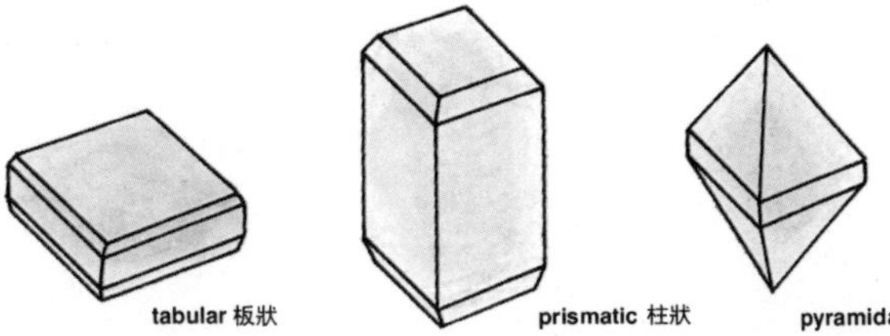

tabular 板狀 prismatic 柱狀 pyramidal 錐狀

prismatic (*adj*) a habit (↑) in which the prism faces (p.41) of a crystal (p.40) are well shown.

柱狀的(形) 描述晶體(第 40 頁)柱面(第 41 頁)十分醒目的一種習性(↑)。

columnar (*adj*) a habit (↑) in which the dimensions of a crystal (p.40) are almost equal in cross-section.

圓柱狀的(形) 描述晶體(第 40 頁)橫斷面上各方向的尺度近乎相等的一種習性(↑)。

botryoidal 葡萄狀

pyramidal (*adj*) a habit (↑) in which the pyramid faces (p.41) of a crystal (p.40) are well developed.

錐狀(形) 描述晶體(第 40 頁)錐面(第 41 頁)發育良好的一種習性(↑)。

lamellar (*adj*) a habit (↑) in which the crystals (p.40) are in the form of thin plates or *lamellae*.

頁片狀(形) 描述晶體(第 40 頁)呈薄板形或頁片狀的一種習性(↑)。

acicular (*adj*) a habit (↑) in which the crystals (p.40) are shaped liked needles.

針狀(形) 描述晶體(第 40 頁)形狀像針的一種習性(↑)。

fibrous (*adj*) a habit (↑) in which the crystals (p.40) are like threads.

纖維狀(形) 描述晶體(第 40 頁)像絲線的一種習性(↑)。

Terms used to describe minerals in the mass:

描述礦物塊體用的術語：

dendritic 樹枝狀

dendritic (*adj*) shaped like a tree.

樹枝狀(形) 狀似一棵樹的。

foliated (*adj*) in the form of thin leaves or *folia*.

葉片狀(形) 形態呈薄葉或薄片的。

botryoidal (*adj*) shaped like a bunch of grapes.

葡萄狀(形) 形狀好像一串葡萄的。

euhedral (*adj*) showing fully developed (p.40) form.

自形的(形) 顯示晶形發育完美的。

subhedral (*adj*) showing some signs of crystal (p.40) form.

半自形的(形) 顯示部分晶體(第 40 頁)形態。

anhedral (*adj*) showing no crystal (p.40) form at all.

他形的(形) 全不顯示晶體(第 40 頁)形態的。

massive (*adj*) not clearly crystalline (p.40).

塊狀的(形) 結晶質(第 40 頁)不清晰的。

solid solution an ion (p.15) can take the place of another ion in a crystal lattice (p.40) if it is of about the same size and has the same charge (p.153). When this happens, the result is a solid solution. For example, the composition of the olivines (p.58) can vary continuously between Fe_2SiO_4 at one end of the series and Mg_2SiO_4 at the other end. A series of this kind is called a *solid solution series*, or *isomorphous series*.

solid solution series *see* **solid solution** (↑).

isomorphous series *see* **solid solution** (↑).

end-member one of the mineral compositions (p.15) at one end of a solid solution series (↑).

zoned crystal a crystal (p.40) of which the chemical composition (p.15) varies from the centre to the outside. There are usually zones or bands in a regular arrangement. Zoning is found where there is a solid solution series (↑).

epitaxis (*n*) the growth of one mineral upon another. **epitaxial** (*adj*).

reaction rim a band or zone on the outside of a crystal (p.40) formed by chemical reaction (p.17) with the material around it.

inclusion (*n*) a piece of one mineral or other substance with another mineral or rock around it on all sides, e.g. a crystal of leucite (p.58) containing augite (p.57) as inclusions.

polymorphism (*n*) a substance that is found in two or more forms having the same chemical composition (p.15) but with different physical properties (e.g. colour, crystal form, hardness) is said to show polymorphism; e.g. aluminium silicate, Al_2SiO_5, as andalusite, sillimanite, and kyanite (p.59).

dimorphism (*n*) a type of polymorphism (↑) in which the substance is found in two different forms; e.g. carbon as diamond and graphite (p.48); calcium carbonate as calcite and aragonite (p.51).

pseudomorph (*n*) a mineral in the form of another mineral. A pseudomorph may be formed by replacement; by the coating of one mineral by another; by the filling up of the space left by another mineral; or by alteration; e.g. pseudomorphs of quartz after fluorite.

固溶體　一個離子(第 15 頁)和另一個離子的大小大致相等、電荷(第 153 頁)相同，則此離子能取代另一離子在晶格(第 40 頁)中的位置，形成固溶體。例如，橄欖石(第 58 頁)的成分可在以 Fe_2SiO_4 為一端和以 Mg_2SiO_4 為另一端這一系列之間連續變化，這種系列稱為固溶體系列或類質同象系列。

固溶體系列　見固溶體(↑)。

類質同象系列；同形系列　見固溶體(↑)。

端員　在固溶體系列(↑)一端的礦物成分(第 15 頁)之一。

環帶狀晶體　化學成分(第 15 頁)自中心往外側有所變化的一種晶體(第 40 頁)，通常形成規則排列的環帶或條帶，環帶狀晶體見於有固溶體系列(↑)存在之處。

取向附生(名)　一個礦物生長在另一礦物之上。(形容詞為 epitaxial)。

反應邊　晶體(第 40 頁)外緣與周圍物質發生化學反應(第 17 頁)所形成的一種條帶或環帶。

包體(名)　包裹一塊礦物或其他物質整體的另一種礦物或岩石。例如，白榴子石(第 58 頁)晶體含有輝石(第 57 頁)作為包體。

(同質)多晶形象；同質多象(名)　一種物質存在兩種或多種形態，其他學成分(第 15 頁)相同而物理性質(如顏色、晶形、硬度)却不同，則稱之為(同質)多晶形現象。例如矽酸鋁(Al_2SiO_5)就有紅柱石、矽線石和藍晶石(第 59 頁)三種形態。

同質二象(名)　一種物質存在兩種不同形態的多晶形現象(↑)。例如，碳有金剛石和石墨(第 48 頁)兩種形態；碳酸鈣有方解石和霰石(第 51 頁)兩種形態。

假象(名)　一種礦物呈別種礦物的形態。假象的形成可以是由於：置換作用；一種礦物為另一種礦物包覆；另一種礦物充填所留下的空間；或由於蝕變作用等原因。例如石英的螢石假象。

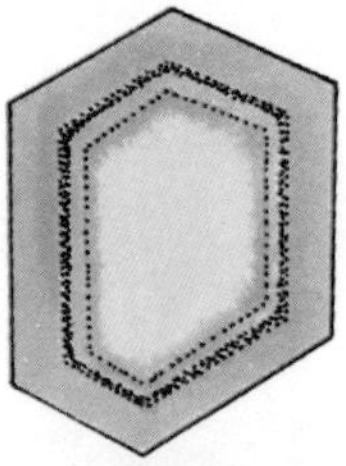

zoned crystal
環帶狀晶體

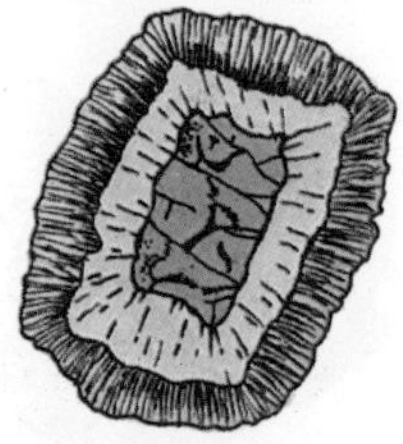

reaction rim
反應邊

optical properties the characters of minerals that depend upon light. Some of these can be studied only with a petrological microscope (p.147).

光學性質 取決於光的礦物特性，其中有些特性只能利用岩石顯微鏡(第 147 頁)來研究。

isotropic (*adj*) having the same physical properties in all directions. A mineral that is optically isotropic has the same refractive index (p.159) in all directions.

各向同性的；無向性的(形) 指各個方向上均具相同之物理性質。光學各向同性的礦物，在各個方向上均具相同的折光率(第 159 頁)。

anisotropic (*adj*) not isotropic (↑); having physical properties that are different in different directions. A mineral that is optically anisotropic thus has a refractive index (p.159) that varies with direction in the crystal. All minerals except those belonging to the cubic system (p.43) are anisotropic.

各向異性的；有方向性的(形) 不是各向同性的(↑)；在不同方向具不同的物理性質。因此，光學各向異性的礦物，其折光率(第 159 頁)因在晶體中的方向而異。除屬等軸(立方)晶系(第 43 頁)外的所有礦物都是各向異性的。

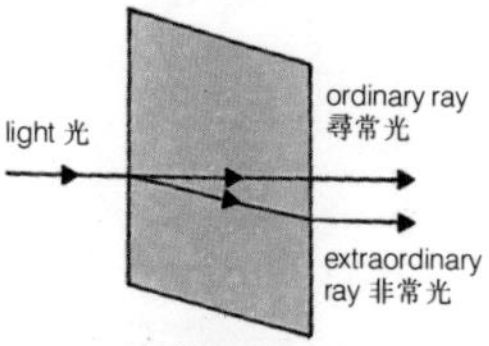

birefringent crystal 雙折射晶體

birefringence **雙折射**

birefringence (*n*) double refraction; the property of having more than one value for the refractive index (p.159); the difference between the largest and smallest values of the refractive index for a given (anisotropic (↑)) mineral. **birefringent** (*adj*).

雙折射(名) 雙重折射。具有一個以上的折光率(第 159 頁)數值的性質；一個給定(各向異性的(↑))礦物的折光率的最大值和最小值之差。(形容詞為 birefrigent)

pleochroism (*n*) a change in the colour of a mineral when a very thin slice is turned round in polarized light (p.158). Pleochroism is caused by the fact that light that is vibrating in certain directions can pass through the mineral but light vibrating in other directions cannot. **pleochroic** (*adj*).

多色性(名) 極薄的一片礦物在偏振光(第 158 頁)中旋轉時，礦物顏色的變化。多色性的成因是由於以某些方向振動的光可以透過礦物而以其他方向振動的光則不能透過。(形容詞為 pleochroic)

dichroism (*n*) = pleochroism (↑).

二色性(名) 同多色性(↑)。

iridescence (*n*) a show of colours caused by optical interference (p.158) in the mineral. Iridescence is seen in calcite (p.51), mica (p.55), and other minerals. **iridescent** (*adj*).

暈色；彩虹色(名) 礦物中因光干涉(第 158 頁)引起的彩色顯示。暈色可見於方解石(第 51 頁)、雲母(第 55 頁)和其他礦物中。(形容詞為 iridescent)

schiller (*n*) a play of light that is seen on the surfaces of certain minerals when they are held at a particular angle to the light. It can resemble metallic lustre (↓). Schiller is shown by certain pyroxenes and feldspars (pp.57, 56) and by other minerals.

閃光(名) 某些礦物面對光保持一特定角度時所見的光的閃晃。閃光可類似金屬光澤(↓)。某些輝石和長石(第 57、56 頁)及其他礦物可顯現閃光。

lustre (*n*) the character of the light reflected by a mineral. It may be *metallic* (like a metal); *vitreous* (like broken glass); *resinous* (like resin, the sticky material that comes out of trees); *pearly* (like pearls, the round white jewels that are found in certain shell-fish); *silky* (like silk); *adamantine* (like diamond, p.48). **lustrous** (*adj*).

光澤(名) 礦物反射光線的特性。光澤包括有金屬的(如金屬)；玻璃質的(如碎玻璃)；樹脂狀的(如樹脂，係樹木流出的黏性物質)；珍珠狀的(如珍珠，係一種產自某些貝殼類的白色球狀珠寶)；絹絲狀的(如絲)；金剛石似的(如金剛石，第 48 頁)。(形容詞為 lustrous)

carbon minerals the element carbon (C) is known in three different forms: graphite (↓), diamond (↓), and amorphous carbon (soot, etc.). The molecular structures (p.15) of the two minerals graphite and diamond are different and their properties differ as a result.

graphite (*n*) carbon with a layer structure. Very soft, with perfect cleavage; grey to black in colour. Graphite occurs in metamorphic (p.90) and igneous (p.62) rocks. *See also* **diamond** (↓).

diamond (*n*) pure carbon with a structure in which the atoms are connected to each other throughout the crystal lattice (p.40). Diamond is very hard and is used for jewellery and for cutting. It occurs in ultrabasic rocks (p.75) and alluvial (p.24) deposits (p.80). *See also* **graphite** (↑).

native gold pure gold (Au) or gold plus silver. Usually found as yellow grains in alluvial (p.24) deposits (p.80).

native sulphur sulphur (S); not always pure. Found at hot springs, round volcanoes, and as bedded deposits (p.80) with gypsum (p.52).

haematite, hematite (*n*) iron oxide, Fe_2O_3. It occurs as fibrous (p.45) crystals (p.40) or in masses of rounded shape, grey to black or reddish-black. An important ore (p.145) of iron.

limonite (*n*) a mixture of iron oxides (p.16) and hydroxides (chemical compounds (p.15) containing the OH group). Amorphous (p.44); yellow or reddish-brown to black in colour. Limonite is formed by the weathering (p.20) of minerals that contain iron. **limonitic** (*adj*).

cassiterite (*n*) tin oxide, SnO_2. Twin crystals (p.41) are common. Cassiterite occurs in acid igneous rocks (p.74), in veins (p.145), and in alluvial (p.24) deposits (p.80). An important ore (p.145) of tin.

cuprite (*n*) copper oxide, Cu_2O. Cuprite is found in the zone of weathering (p.20) of copper ores (p.145). It is itself an ore of copper.

uraninite (*n*) uranium oxide, UO_2, but not usually pure. Occurs as a primary mineral in granites (p.76) and pegmatites (p.79) and in hydrothermal veins (p.145).

pitchblende (*n*) a variety of uraninite (↑).

含碳礦物 已知碳(C)元素有三種不同的形態：石墨(↓)、金剛石(↓)和非晶形碳(煤煙等)。石墨和金剛石這兩種礦物的分子結構(第 15 頁)是不同的，因而其性質各異。

石墨(名) 具層狀結構的碳，質軟，解理完全，灰到黑色。石墨賦存於變質(第 90 頁)岩和火成(第 62 頁)岩中。參見**金剛石**(↓)。

金剛石；金剛鑽(名) 具有某種構造的純碳，此構造中整個晶格(第 40 頁)的各原子相互連結。金剛石極硬，用於製造寶石和切削材料。金剛石產於超基性岩(第 75 頁)和沖積(第 24 頁)礦床(第 80 頁)中。參見**石墨**(↑)。

自然金 純金(Au)或含銀的金。通常呈黃色顆粒產於沖積(第 24 頁)礦床(第 80 頁)中。

自然硫 硫(S)：未必是純的。見於溫泉和火山附近，呈層狀礦床(第 80 頁)與石膏(第 52 頁)共生。

赤鐵礦(名) 成分為鐵的氧化物(Fe_2O_3)，以纖維狀(第 45 頁)晶體(第 40 頁)或以圓形團塊形式存在，呈灰至黑色或紅黑色。是一種重要的鐵礦石(第 145 頁)。

褐鐵礦(名) 成分為鐵的氧化物(第 16 頁)與鐵的氫氧化物(含 OH 基的化合物(第 15 頁))的混合物，非晶形(第 44 頁)：呈黃或紅棕到黑色。褐鐵礦是由含鐵的礦物風化(第 20 頁)形成的。(形容詞為 limonitic)

錫石(名) 成分為錫的氧化物(S_nO_2)。常為孿晶(第 41 頁)。錫石產於酸性火成岩(第 74 頁)、礦脈(第 145 頁)和沖積(第 24 頁)礦床(第 80 頁)中，是一種重要的錫礦石(第 145 頁)。

赤銅礦(名) 成分為銅的氧化物(Cu_2O)。赤銅礦見於各種銅礦石(第 145 頁)的風化(第 20 頁)帶中。它本身也是一種銅礦石。

晶質鈾礦(名) 成分為鈾的氧化物(UO_2)，通常是不純的。以原生礦物形式賦存於花崗岩(第 76 頁)、偉晶岩(第 79 頁)和熱液礦脈(第 145 頁)中。

瀝青鈾礦(名) 晶質鈾礦(↑)的一個變種。

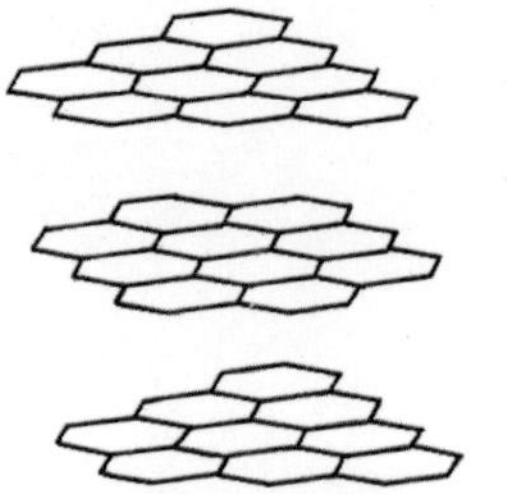

the structure of graphite
石墨的結構

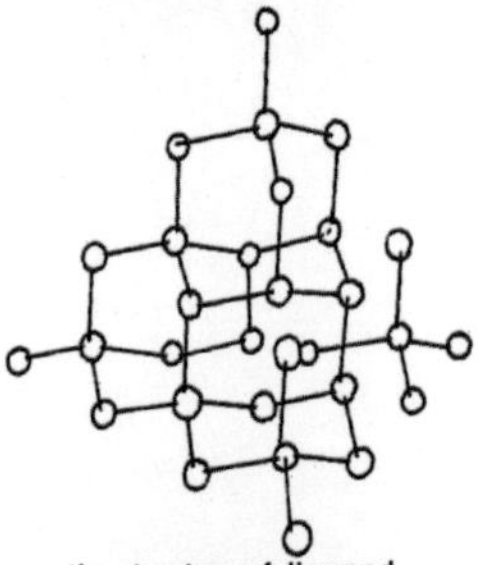

the structure of diamond
金剛石的結構

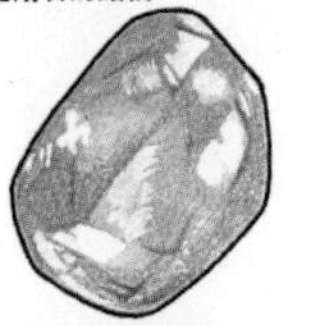

diamond
金剛石

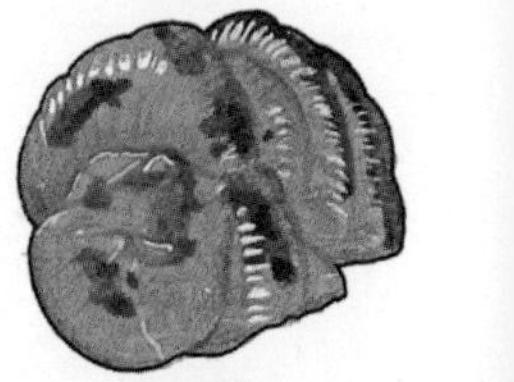

haematite
赤鐵礦

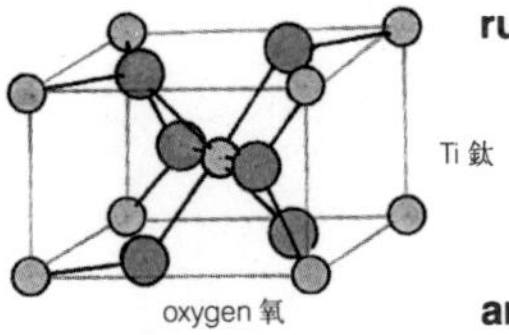

structure of ruitle, TiO_2
金紅石（TiO_2）的結構

corundum
剛玉

rutile (*n*) titanium dioxide, TiO_2. Crystals are commonly acicular (p.45); twin crystals (p.41) are knee-shaped (*geniculate*). An accessory mineral (p.75) in many kinds of igneous (p.62) and metamorphic (p.90) rocks and in sediments (p.80). *See also* **anatase** (↓).

anatase (*n*) titanium dioxide, TiO_2. Occurs in metamorphic rocks (p.90) and in veins (p.145). *See also* **rutile** (↑).

ilmenite (*n*) an oxide of iron and titanium, $FeTiO_3$. Occurs as an accessory mineral (p.75) in basic igneous rocks (p.74), in veins, and in detrital deposits (p.85). It is the chief ore (p.145) of the metal titanium.

corundum (*n*) aluminium oxide, Al_2O_3. Common in barrel-shaped crystals; occurs in igneous rocks (p.62) and as a product of contact metamorphism (p.90). Very hard; used as an abrasive and for jewellery.

spinels (*n.pl.*) metallic oxides (p.16) occurring in basic and ultrabasic igneous rocks (pp.74, 75). Members of the spinel group include magnetite, Fe_3O_4, an important ore (p.145) of iron, which occurs in a variety of igneous and metamorphic rocks (p.90); spinel (in the narrower sense), $MgAl_2O_4$, found in crystalline limestones and schists (p.97); and chromite, $FeCr_2O_4$, which occurs in ultrabasic igneous rocks (p.75) and is an important ore (p.145) of chromium.

magnetite (*n*) *see* **spinels** (↑).

chromite (*n*) *see* **spinels** (↑).

pyrolusite (*n*) manganese dioxide, MnO_2; dark grey in colour. Pyrolusite is found in sediments (p.80) as nodules (p.84) – e.g. on the floor of the deep ocean – and in lakes.

apatite (*n*) a calcium phosphate with small amounts of fluorine, chlorine, and OH. Crystals are common. Apatite occurs in igneous rocks (p.62) as an accessory mineral; in metamorphic rocks (p.90), especially metamorphosed limestones (p.86); as veins (p.145); and as bedded deposits (p.80).

pyromorphite (*n*) a lead compound, $(PbCl)Pb_4(PO_4)_3$; green, yellow, or brown in colour. It occurs with lead ores (p.145).

金紅石（名） 成分為二氧化鈦（TiO_2）。晶體常呈針狀（第 45 頁）；孿晶（第 41 頁）呈膝狀（膝狀彎曲的）。在多種火成岩（第 62 頁）、變質岩（第 90 頁）和沉積岩（第 80 頁）中是一種附生礦物（第 75 頁）。參見**銳鈦礦**（↓）。

銳鈦礦（名） 成分為二氧化鈦（TiO_2）。賦存於變質岩（第 90 頁）和岩脈（第 145 頁）中。參見**金紅石**（↑）。

鈦鐵礦（名） 成分為鐵和鈦的氧化物（$FeTiO_3$）。在基性火成岩（第 74 頁）、岩脈和碎屑沉積（第 85 頁）中以附生礦物（第 75 頁）產出。是金屬鈦的主要礦石（第 145 頁）。

剛玉（名） 成分為鋁的氧化物（Al_2O_3）。常見其桶狀晶體；賦存於火成岩（第 62 頁）中，並成為接觸變質作用（第 90 頁）的一種產物出現。質極硬，可用作磨料和製造寶石。

尖晶石族（名、複） 產於基性及超基性火成岩（第 74、75 頁）中的金屬氧化物（第 16 頁）。尖晶石族礦物包括：磁鐵礦（Fe_3O_4），是一種重要的鐵礦石（第 145 頁），存於多種火成岩和變質岩（第 90 頁）中；尖晶石（狹義的，成分為 $MgAl_2O_4$），見於結晶灰岩和片岩（第 97 頁）中；和鉻鐵礦（$FeCr_2O_4$），賦存於超基性火成岩（第 75 頁）中，是一種重要的鉻礦石（第 145 頁）。

磁鐵礦（名） 見**尖晶石族**（↑）。

鉻鐵礦（名） 見**尖晶石族**（↑）。

軟錳礦（名） 成分為二氧化錳（M_nO_2），深灰色，以結核形式（第 84 頁）（例如在深洋底）存在於沉積物（第 80 頁）以及湖泊中。

磷灰石（名） 含少量氟、氯和 OH 的一種磷酸鈣。其晶體常見。磷灰石作為一種附生礦物存在於火成岩（第 62 頁）中；存在於變質岩（第 90 頁），特別是變質灰岩（第 86 頁）中；作為礦脈（第 145 頁）和作為層狀礦床（第 80 頁）產出。

磷酸氯鉛礦（名） 鉛的化合物 $(PbCl)Pb_4(PO_4)_3$；綠、黃或棕色。與鉛礦石（第 145 頁）共生。

galena (*n*) lead sulphide, PbS. Crystals are cubes and combinations of cubes and octahedra (p.41). Grey in colour with metallic lustre (p.47). Occurs in veins (p.145). An important ore (p.145) of lead.

cinnabar (*n*) mercury sulphide, HgS. Red in colour. Found in volcanic areas (p.68). An important ore (p.145) of mercury.

pyrite (*n*) iron sulphide, FeS_2. Yellow in colour with metallic lustre (p.47). Occurs in igneous rocks (p.62) as an accessory mineral (p.75); in veins (p.145); and in large deposits and in sediments formed under anaerobic conditions (p.81).

chalcopyrite (*n*) copper iron sulphide, $CuFeS_2$. Crystals are commonly twinned (p.41); usually massive (p.45). Yellow in colour with metallic lustre (p.47). Softer than pyrite. Occurs in veins (p.145). An important ore (p.145) of copper.

chalcocite (*n*) copper sulphide, Cu_2S. Usually massive (p.45). Crystals are commonly twinned (p.41). Occurs in veins (p.145) or beds (p.80) together with other copper minerals. An important ore (p.145) of copper.

sphalerite, zinc blende zinc sulphide, ZnS. Usually massive (p.45). Crystals are commonly twinned (p.41). Occurs with galena (↑) in hydrothermal and replacement deposits (p.145), in lodes and in veins (p.145). An important ore of zinc.

zinc blende *see* **sphalerite** (↑).

molybdenite (*n*) molybdenum sulphide, MoS_2. Usually occurs in plates or scales; very soft. Found in hydrothermal veins (p.145) and in acid igneous rocks (p.74). The chief ore of molybdenum.

realgar (*n*) arsenic monosulphide, AsS. Usually massive (p.45) or granular; red or orange. Occurs with orpiment (↓) in veins (p.145), in deposits of hot springs, and round volcanoes.

orpiment (*n*) arsenic trisulphide, As_2S_3. Usually foliaceous or massive (p.45); yellow in colour. Occurs in veins (p.145) and round hot springs.

arsenopyrite (*n*) iron arsenosulphide, FeAsS. Found in hydrothermal veins (p.145) with lead and silver; also round volcanoes.

方鉛礦(名) 成分為硫化鉛(PbS)。其晶體是立方體、立方體與八面體(第 41 頁)的聚形。呈灰色，有金屬光澤(第 47 頁)。賦存於礦脈(第 145 頁)中。是一種重要的鉛礦石(第 145 頁)。

辰砂；銀朱(名) 成分為硫化汞(HgS)。呈紅色。見於火山地區(第 68 頁)。是一種重要的汞礦石(第 145 頁)。

黃鐵礦(名) 成分為硫化鐵(FeS_2)。呈黃色，具金屬光澤(第 47 頁)。以附生礦物(第 75 頁)存在於火成岩(第 62 頁)中；也存在於礦脈(第 145 頁)、大礦床以及在嫌氧條件下(第 81 頁)形成的沉積物中。

黃銅礦(名) 成分為銅和鐵的硫化物($CuFeS_2$)。其晶體常為孿晶(第 41 頁)，通常為塊狀(第 45 頁)。黃色，具金屬光澤(第 47 頁)。硬度比黃鐵礦軟。賦存於礦脈(第 145 頁)中，是一種重要的銅礦石(第 145 頁)。

輝銅礦(名) 成分為硫化銅(Cu_2S)。常為塊狀(第 45 頁)。其晶體常為孿晶(第 41 頁)。與其他銅礦物一起賦存於礦脈(第 145 頁)或礦層(第 80 頁)中。是一種重要的銅礦石(第 145 頁)。

閃鋅礦 成分為硫化鋅(ZnS)。一般為塊狀(第 45 頁)。其晶體常為孿晶(第 41 頁)。與方鉛礦(↑)共存於熱液和交代礦床(第 145 頁)中，也存在於礦脉和岩脈(第 145 頁)中。是一種要的鋅礦石。

閃鋅礦(↑)英文亦稱 zinc blende。

輝鉬礦(名) 成分為二硫化鉬(MoS_2)。通常以板狀體或片狀體存在；質極軟。見於熱液礦脈(第 145 頁)和酸性火成岩(第 74 頁)中。為主要的鉬礦石。

雄黃(名) 成分為硫化砷(AsS)。通常為塊狀(第 45 頁)或粒狀；呈紅或橙色。與雌黃(↓)共存於礦脈(第 145 頁)中，也存在於温泉沉積中及火山附近。

雌黃(名) 成分為三硫化二砷(As_2S_3)。通常為頁片狀或塊狀(第 45 頁)，呈黃色。存在於礦脈(第 145 頁)中和温泉附近。

砷黃鐵礦；毒砂(名) 成分為砷硫化鐵(FeAsS)。與鉛和銀一起見於熱液礦脈(第 145 頁)中和火山周圍。

galena
方鉛礦

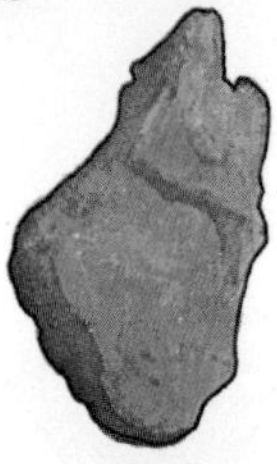

cinnabar
辰砂

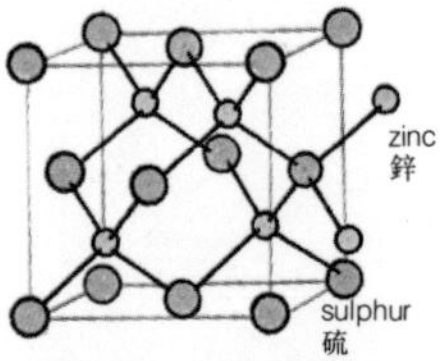

structure of sphalerite, ZnS
閃鋅礦（ZnS）的結晶構造

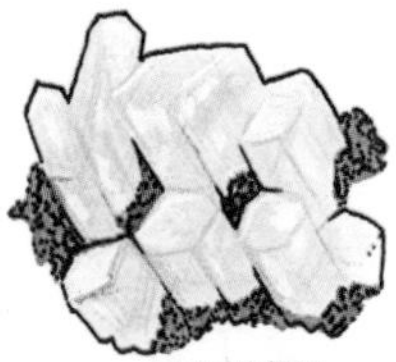

calcite 方解石

malachite 孔雀石

azurite 藍銅礦

calcite (*n*) calcium carbonate, $CaCO_3$. Good crystals are common, with perfect cleavage (p.44). Occurs as limestone and marble; in veins (p.145); and as stalactites and stalagmites (p.21). *See also* **aragonite** (↓).

aragonite (*n*) calcium carbonate, $CaCO_3$. Prismatic crystals (p.41) are common, often twinned (p.41) with perfect cleavage (p.44). Less stable than calcite (↑) and is converted to calcite by heat or pressure. It occurs in sedimentary rocks (p.80).

dolomite (*n*) calcium magnesium carbonate, $CaMg(CO_3)_2$. Crystals show perfect cleavage (p.44). Occurs in beds formed by the alteration of limestone (p.86) and in veins (p.145).

malachite (*n*) hydrated basic copper carbonate, $CuCO_3.Cu(OH)_2$. Bright green in colour. Usually occurs massive (p.45). Found with copper deposits. An ore (p.145) of copper.

azurite (*n*) hydrated basic copper carbonate, $2CuCO_3.Cu(OH)_2$; compare with malachite (↑). Monoclinic (p.43). Occurs as crystals but is usually massive (p.45) or earthy. Deep blue in colour. Azurite occurs with other oxidized copper minerals such as malachite (↑) in the zone of weathering (p.20) of copper deposits. An ore (p.145) of copper.

siderite (*n*) iron carbonate, $FeCO_3$. Crystals commonly have curved faces (p.40); also massive (p.45) and oolitic (p.86). Occurs in sediments as metasomatic deposits (p.90) and as a vein mineral (p.145). Impure forms of siderite are clay-ironstone (p.88) and black-band ironstone (p.88), which are important ores (p.145) of iron.

magnesite (*n*) magnesium carbonate, $MgCO_3$. Crystals are rare; usually in massive (p.45) and fibrous (p.45) forms. Occurs as veins (p.145) in serpentine (p.61) and replacing dolomite (↑) and limestone (p.86).

smithsonite (*n*) zinc carbonate, $ZnCO_3$. Massive, botryoidal, encrusting, stalactitic, granular, or earthy (p.45); in beds and veins (p.145).

witherite (*n*) barium carbonate, $BaCO_3$. Crystals show repeated twinning (p.41); also massive. Occurs with galena and barite in veins (p.145).

方解石(名)　成分為碳酸鈣($CaCO_3$)。常見良好的晶體，解理(第 44 頁)完全。以石灰岩和大理岩形式產出；賦存於脈體(第 145 頁)中；也以石鐘乳和石筍(第 21 頁)形式產出。參見**文石**(↓)。

文石；霰石(名)　成分為碳酸鈣($CaCO_3$)。常見柱狀晶體(第 41 頁)，一般為孿晶(第 41 頁)，解理(第 44 頁)完全。不如方解石(↑)穩定，受熱或受壓可轉變為方解石。賦存於沉積岩(第 80 頁)中。

白雲石(名)　成分為碳酸鈣鎂($CaMg(CO_3)_2$)。其晶體顯示解理完全(第 44 頁)。存在於由石灰岩(第 86 頁)蝕變形成的岩層中，也存在脈體(第 145 頁)中。

孔雀石(名)　成分為水合鹼式碳酸銅($CuCO_3 \cdot Cu(OH)_2$)。鮮綠色。常呈塊狀(第 45 頁)。產於銅礦床中。是一種銅礦石(第 145 頁)。

藍銅礦(名)　成分為水合鹼式碳酸銅($2CuCO_3 \cdot Cu(OH)_2$)；其化學式與孔雀石(↑)略不同。屬單斜晶系(第 43 頁)。以晶體形式產出，通常為塊狀(第 45 頁)或土狀。呈深藍色。藍銅礦與其他氧化的銅礦物如孔雀石(↑)一起存在於銅礦床的風化(第 20 頁)帶中。是一種銅礦石(第 145 頁)。

菱鐵礦(名)　成分為碳酸鐵($FeCO_3$)。其晶體的晶面(第 40 頁)常彎曲；也呈塊狀(第 45 頁)和鮞狀(第 86 頁)。在沉積岩中以交代礦層(第 90 頁)出現和以脈石礦物(第 145 頁)存在於脈體中、泥鐵礦(第 88 頁)和黑泥鐵礦(第 88 頁)為不純的菱鐵礦，二者均為重要的鐵礦石(第 145 頁)。

菱鎂礦(名)　成分為碳酸鎂($MgCO_3$)。常呈塊狀(第 45 頁)和纖維狀(第 45 頁)，其晶體罕見。在蛇紋石(第 61 頁)中以脈狀(第 145 頁)產出，並交代白雲石(↑)和石灰岩(第 86 頁)。

菱鋅礦(名)　成分為碳酸鋅($ZnCO_3$)。呈塊狀、葡萄狀、皮殼狀、鐘乳石狀、粒狀或土狀(第 45 頁)；產於岩層和脈體(第 145 頁)中。

毒重石(名)　成分為碳酸鋇($BaCO_3$)。晶體顯示重複雙晶(第 41 頁)；也呈塊狀，在礦脈(第 145 頁)中與方鉛礦和重晶石共存。

gypsum (*n*) calcium sulphate, $CaSO_4.2H_2O$. Crystals are common, often twinned (p.41). Colourless and very soft. An evaporite (p.85).
selenite (*n*) a crystallized (p.44) variety of gypsum (↑).
baryte, barite (*n*) barium sulphate, $BaSO_4$. Crystals are common. Often occurs as a vein mineral (p. 145).
barites = baryte (↑).
anhydrite (*n*) calcium sulphate, $CaSO_4$. Usually occurs in fibrous or lamellar form (p.45) or as granules. An evaporite; occurs in sedimentary rocks with gypsum (↑) and halite (↓).
celestite, celestine (*n*) strontium sulphate, $SrSO_4$. Usually occurs as tabular (p.45) crystals (p.40) resembling barite (↑); also massive (p.45) and fibrous (p.45). White or pale blue in colour. Celestite occurs in sedimentary rocks (p.80) and in hydrothermal veins (p.145). It is a source of strontium.
celestine = celestite (↑).
wolframite (*n*) tungstate of iron and manganese, $(Fe,Mn)WO_4$. Occurs as tabular crystals (p.45) or prismatic crystals (p.45); also massive (p.45). Wolframite is found in pneumatolytic (p.63) veins and hydrothermal veins (p.145).
scheelite (*n*) calcium tungstate, $CaWO_4$. Usually massive or granular (p.45). Found in pneumatolytic and hydrothermal deposits (p.145).
fluorite, fluorspar (*n*) calcium fluoride, CaF_2. Crystals are common, usually cubes. Occurs in veins (p.145) and replacement deposits (p.145) and in pegmatites (p.79).
fluorspar (*n*) = fluorite (↑).
halite, rock salt (*n*) sodium chloride (common salt), NaCl. Crystals are common. Colourless or white when pure. Halite occurs as beds which are produced by the evaporation (p.156), of sea water. It can flow under high pressure; salt domes (p.131) then result which in their form resemble igneous intrusions (p.64).
rock salt = halite (↑).
sylvite (*n*) potassium chloride, KCl. It occurs with halite (↑), etc.

石膏(名)　成分為硫酸鈣($CaSO_4 \cdot 2H_2O$)。常見其晶體，一般為孿晶(第 41 頁)。無色，很軟。是一種蒸發岩(第 85 頁)。

透石膏(名)　一種結晶的(第 44 頁)石膏(↑)變種。

重晶石(名)　成分為硫酸鋇($BaSO_4$)。其晶體常見。常以脈石礦物(第 145 頁)產出。

重晶石(↑)英文亦拼寫為 barites。

硬石膏(名)　成分為硫酸鈣($CaSO_4$)。通常呈纖維狀、頁片狀(第 45 頁)或粒狀產出。是一種蒸發岩；與石膏(↑)及石鹽(↓)一起產於沉積岩中。

天青石(名)　成分為硫酸鍶($SrSO_4$)。常以板狀(第 45 頁)晶體(第 40 頁)產出，與重晶石(↑)相似；也成塊狀(第 45 頁)和纖維狀(第 45 頁)。色白或淡藍。天青石賦存於沉積岩(第 80 頁)和熱液礦脈(第 145 頁)中。它是鍶的來源。

天青石(↑)英文亦稱celestine。

黑鎢礦；鎢錳鐵礦(名)　成分為鐵和錳的鎢酸鹽($(Fe, Mn)WO_4$)。以板狀(第 45 頁)或柱狀晶體(第 45 頁)產出；也呈塊狀(第 45 頁)。黑鎢礦見於氣成(第 63 頁)礦脈中和熱液礦脈(第 145 頁)中。

白鎢礦；重石礦(名)　成分為鎢酸鈣($CaWO_4$)。常為塊狀或粒狀(第 45 頁)。見於氣成和熱液礦床(第 145 頁)中。

螢石，氟石(名)　成分為氟化鈣(CaF_2)。其晶體很常見，通常為立方體。見於礦脈(第 145 頁)中，亦見於交代礦床(第 145 頁)和偉晶岩(第 79 頁)中。

氟石(名)　同螢石(↑)。

石鹽，岩鹽(名)　成分為氯化鈉(NaCl)(食鹽)。常見其晶體。純淨的石鹽為無色或白色。石鹽呈層狀產出，可由海水的蒸發(第 156 頁)製得。在高壓下可流動，因而能形成鹽丘(第 131 頁)，其形狀與火成侵入體(第 64 頁)相似。

岩鹽　同石鹽(↑)。

鉀鹽(名)　成分為氯化鉀(KCl)。與石鹽(↑)等一起產出。

barytes 重晶石

fluorite 螢石

silicate structures most silicate minerals are built up from SiO_4 tetrahedra (p.41) in which four oxygen atoms (p.152) are arranged round a silicon atom. These tetrahedra can be joined together to form larger groups: rings, chains (single and double), layers (sheets), and frameworks in three dimensions. It is also possible for some of the silicon atoms in the tetrahedra to be replaced by aluminium atoms, which have a similar ionic radius (p.15). Various metals can be fitted into the octahedral (p.41) spaces between the SiO_4 tetrahedra; e.g. magnesium, titanium, iron, aluminium, sodium, and potassium. (Aluminium can thus be included in the structure in two different ways.) A great many structures and compositions (p.15) are therefore possible and the actual compositions of silicate minerals can be highly complicated.

矽酸鹽構造 矽酸鹽礦物大多數是由 SiO_4 四面體(第 41 頁)構成的，四面體中的四個氧原子(第 152 頁)排列在一個矽原子周圍。這些四面體可以連成更大的原子團，包括：環、鏈(單鏈和雙鏈)、層(片)和三維空間框架。四面體中的某些矽原子也可以被離子半徑(第 15 頁)相近的鋁原子所置換。各種金屬例如鎂、鈦、鐵、鋁、鈉和鉀都可以放入 SiO_4 四面體之間的八面體(第 41 頁)的空間中(因而鋁可以兩種方式進入構造中)，從而有可能出現許多種構造和成分(第 15 頁)，也因此可以使其礦物的實際成分極為複雜。

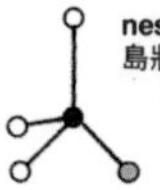

nesosilicate
島狀矽酸鹽
SiO_4^{4-}

sorosilicate
群島狀矽酸鹽
SiO_7^{6-}

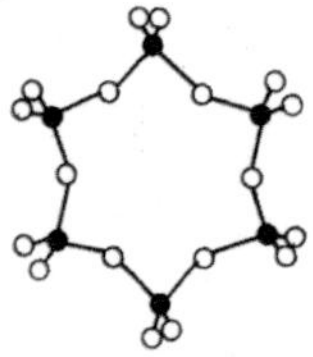

$SiO_6O_{18}^{12-}$

cyclosilicate
環狀矽酸鹽
ring structure
環狀構造

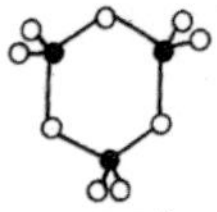

$Si_3O_9^{6-}$

nesosilicates (*n.pl.*) silicates (p.16) containing separate SiO_4 ions. These are the simplest silicate structures. The olivines, zircon, kyanite, sillimanite, and andalusite are all nesosilicates. The structure of olivine (p.58) can, for example, be thought of as a collection of SiO_4 groups with Mg^{2+} (magnesium) ions in the holes between them, each Mg^{2+} ion being surrounded by six oxygen atoms.

(孤)島狀矽酸鹽類(名、複) 所含 SiO_4 離子不相連的矽酸鹽(第 16 頁)。這是最簡單的矽酸鹽構造。橄欖石、鋯英石、藍晶石、矽線石和紅柱石都是孤島狀矽酸鹽。例如，橄欖石(第 58 頁)的構造可以看成是由一些 SiO_4 原子團的聚合，這些原子團之間的空穴中有 Mg^{2+}(鎂)離子起聯系作用，即每一個 Mg^{2+} 離子都被六個氧原子所環繞。

sorosilicates (*n.pl.*) silicates (p.16) in which two SiO_4 tetrahedra (p.41) share one oxygen atom between them. The unit is thus Si_2O_7. Melilite, $Ca_2MgSi_2O_7$, is an example.

群島狀矽酸鹽類；儔矽酸鹽類(名、複) 兩個 SiO_4 四面體(第 41 頁)共用兩者之間的一個氧原子的矽酸鹽(第 16 頁)。因此其單元為 Si_2O_7。黃長石($Ca_2MgSi_2O_7$)即其一例。

cyclosilicates (*n.pl.*) silicates with ring structures in which two oxygen atoms of each SiO_4 tetrahedron are shared with other SiO_4 tetrahedra. The general formula is thus $(SiO_3)^{2n-}$. The rings can have three, four, or six $SiO_4{}^n$ units. The structure with six SiO_4 units occurs in beryl, $Be_3Al_2Si_6O_{18}$, and cordierite (p.60). Ring structures are not, however, common.

環狀矽酸鹽類(名、複) 具環狀構造的矽酸鹽，其中每個 SiO_4 四面體有兩個氧原子與其他 SiO_4 四面體共用。因此其通式為 $(SiO_3)^{2n-}$。每一環可以有三個、四個或六個 SiO_4^n 單元。具有六個 SiO_4 單元的構造見於綠柱石($Be_3Al_2Si_6O_{18}$)和堇青石(第 60 頁)中。但環狀構造不常見。

ring silicates = cyclosilicates (↑).

環矽酸鹽類 同環狀矽酸鹽類(↑)。

inosilicates (*n.pl.*) silicates with a single chain structure. Two oxygen atoms of each SiO_4 unit are shared with other SiO_4 tetrahedra (p.41). Single or simple chains have the general formula $(SiO_3)_n$. The pyroxenes (p.57) are of this type. Double chains are also possible. These have the general formula $(Si_4O_{11})_n$. The amphiboles (p.57) are of this type. The two types of chain structure account for the different cleavages (p.44) of the pyroxenes and amphiboles.

鏈狀矽酸鹽類(名、複) 具單鏈構造的矽酸鹽。每個 SiO_4 單元有兩個氧原子與其他 SiO_4 四面體(第 41 頁)共用。單鏈型的通式為 $(SiO_3)_n$，輝石(第 57 頁)即屬此種類型。也可能有雙鏈型，通式為 $(Si_4O_{11})_n$。角閃石(第 57 頁)即屬此類型。這兩種鏈狀構造的差別是輝石和角閃石解理(第 44 頁)不相同的原因。

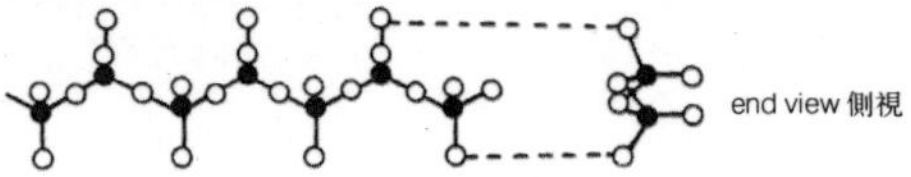

single chain 單鏈 $(SiO_3^{2-})_n$

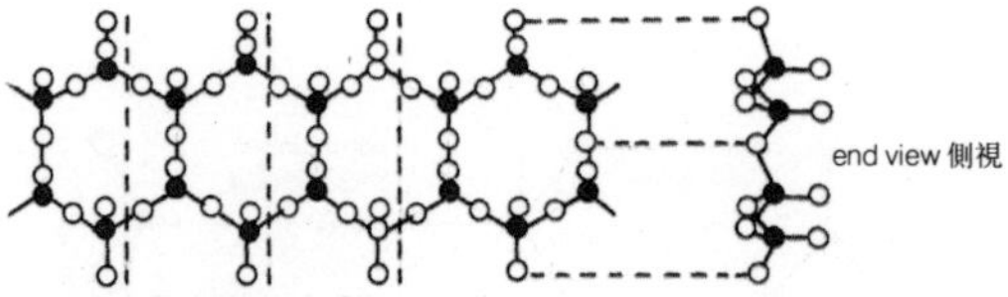

double chain 雙鏈 $(Si_4O_{11}^{6-})_n$

phyllosilicates (*n.pl.*) in these silicate structures (p.53) layers or sheets are built up of SiO_4 tetrahedra (p.41) in which three oxygen atoms of each SiO_4 unit are shared with those next to it. The most important type of layer structure in mineralogy is one in which one or two layers of this kind are combined with layers of hydroxyl (OH) groups and with magnesium or aluminium (Mg or Al) atoms. The micas (↓) are of this type. Talc (p.61), the clay minerals (p.61), and chlorite (p.61) also have layer structures.

sheet silicates = phyllosilicates (↑).

layer lattice silicates = phyllosilicates (↑).

tectosilicates (*n.pl.*) silicate structures (p.53) in which there is a framework or network of silicon and oxygen atoms. Some of the silicon atoms in minerals are replaced by aluminium atoms. Positive ions (p.15) such as Na^+ and Ca^{2+} are then also present. The most important minerals with framework structures are the feldspars (p.56).

層狀矽酸鹽類(名、複) 這種矽酸鹽構造(第 53 頁)中的層或片是由 SiO_4 四面體(第 41 頁)構成，其中每個 SiO_4 單元有三個氧原子與其緊鄰的單元共用。那些具有一層或二層 SiO_4 四面體層與氫氧(OH)原子團、鎂(Mg)或鋁(Al)原子組成的層相結合的類型是礦物學上最重要的層狀結構類型。雲母類(↓)即屬此類型。滑石(第 61 頁)、黏土礦物(第 61 頁)和綠泥石(第 61 頁)也具層狀構造。

片狀矽酸鹽類 同層狀矽酸鹽類(↑)。

層狀格子矽酸鹽類 同狀矽酸鹽類(↑)。

架狀矽酸鹽類 (名、複) 構造中具有矽和氧原子組成的柜架或網格的矽酸鹽構造(第 53 頁)。其礦物中某些矽原子被鋁原子所置換。而且也存在諸如 Na^+ 和 Ca^{2+} 這些正離子(第 15 頁)。長石(第 56 頁)是最重要的具架狀構造礦物。

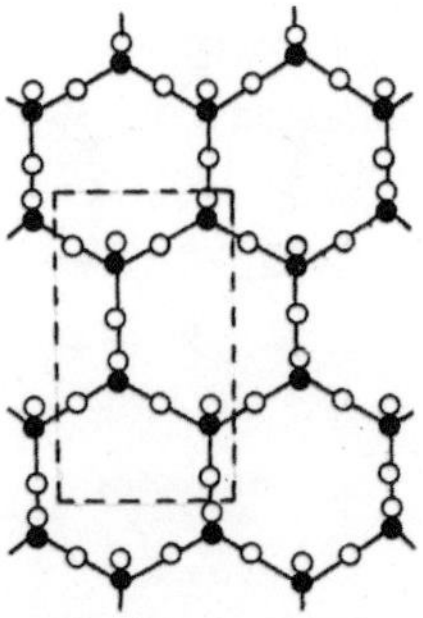

phyllosilicates 層狀矽酸鹽 $(Si_4O_{10}^{4-})_n$

quartz 石英

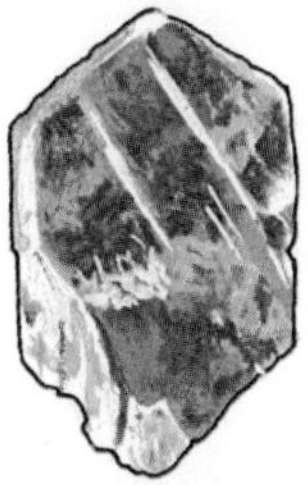

mica 雲母
muscovite 白雲母

quartz (*n*) a hard, glass-like mineral; chemical composition: silica, SiO_2. The structure of quartz is a framework of SiO_4 tetrahedra (p.41).

石英(名) 一種堅硬的玻璃狀礦物；化學成分為二氧化矽(SiO_2)。石英的構造是一個由 SiO_4 四面體(第 41 頁)組成的框架。

alpha quartz or **low quartz** found in a great variety of igneous (p.62) and metamorphic rocks (p.90), as veins (p.145), and in sandstones (p.87).

α 石英或低溫石英 見於多種火成岩(第 62 頁)和變質岩(第 90 頁)中，呈脈狀(第 145 頁)，亦見於砂岩(第 87 頁)中。

beta quartz or **high quartz** formed at 573°C.

β 石英或高溫石英 在 573°C 時形成的石英。

tridymite a form of quartz (↑) stable above 870°C.

鱗石英 在 870°C 以上穩定的石英(↑)變體。

cristobalite a form of quartz (↑) stable above 1470°C.

方英石 在 1470°C 以上穩定的石英(↑)變體。

coesite a polymorph (p.46) of silica formed at very high pressures.

柯石英 於極高壓力下形成的一種二氧化矽的多形晶(第 46 頁)。

chalcedony a cryptocrystalline (p.44) variety of silica.

石髓；玉髓 二氧化矽的一種隱晶質(第 44 頁)變種。

micas (*n.pl.*) a group of rock-forming minerals made up of layers of SiO_4 tetrahedra (p.41); chemically they are silicates of aluminium and potassium with hydroxyl (OH) groups; magnesium and iron are present in the dark micas. Most of the micas are monoclinic (p.43); all show perfect cleavage (p.44).

雲母族(名、複) 由 SiO_4 四面體(第 41 頁)層組成的一類造岩礦物；其化學成分為鋁和鉀的矽酸鹽，含氫氧根(OH)；暗色雲母中存在鎂和鐵。雲母大多屬單斜晶系(第 43 頁)；全都顯示完全解理(第 44 頁)。

muscovite (*n*) 'white mica': $KAl_2(AlSi_3)O_{10}(OH)_2$. Colourless or light in colour. Found in acid igneous rocks (p.74), metamorphic rocks (p.90), and sedimentary rocks (p.80).

白雲母(名) "白色的雲母" $KAl_2(AlSi_3)O_{10}(OH)_2$。無色或淺色。見於酸性火成岩(第 74 頁)、變質岩(第 90)和沉積岩(第 80 頁)中。

biotite (*n*) brown or black mica (↑) containing iron and magnesium: $K(Mg,Fe^{2+})_3(Al,Fe^{3+})Si_3O_{10}(OH)_2$. Common in igneous (p.62) and metamorphic rocks (p.90).

黑雲母(名) 含鐵和鎂的棕色或黑色的雲母(↑)：$K(Mg, Fe^{2+})_3(Al, Fe^{3+})Si_3O_{10}(OH)_2$。常見於火成岩(第 62 頁)和變質岩(第 90 頁)中。

phlogopite (*n*) a brown or black mica (↑) containing iron and magnesium: $K(Mg,Fe^{2+})_3(AlSi_3)O_{10}(F,OH_2)$. It is common in igneous rocks (p.62).

金雲母(名) 含鐵和鎂的棕色或黑色雲母(↑)：$K(Mg, Fe^{2+})_3(AlSi_3)O_{10}(F, OH_2)$。常見於火成岩(第 62 頁)中。

lepidolite (*n*) a light mica (↑) containing lithium, $K(Mg,Li,Al)_3(Al,Si_3)O_{10}(OH,F)_2$. Violet in colour. Lepidolite occurs in pegmatites (p.79).

鋰雲母；鱗雲母(名) 一種含鋰的淺色雲母(↑)，$K(Mg, Li, Al)_3(AlSi_3)O_{10}(OH, F)_2$。紫色，鋰雲母見於偉晶岩(第 79 頁)中。

glauconite (*n*) a green mineral closely related to the micas (↑). It occurs in marine sediments (p.80).

海綠石(名) 一種與雲母(↑)關係密切的綠色礦物。產於海相沉積物(第 80 頁)中。

ferromagnesian minerals a general term for silicate minerals that contain more than small amounts of iron or magnesium, or both. The ferromagnesian minerals include olivine (p.58), augite (p.57), hornblende (p.57), and biotite (↑).

鎂鐵礦物類 含很少量鐵、鎂(二者或其中之一)矽酸鹽物的通稱。鎂鐵礦物包括橄欖石(第 58 頁)、普通輝石(第 57 頁)。普通角閃石(第 57 頁)和黑雲母(↑)。

feldspars, felspars (*n.pl.*) silicates (p.16) of sodium, potassium, calcium, and barium. Their basic structure is a network of SiO_4 tetrahedra (p.41). All feldspars are triclinic or monoclinic (p.43). They occur in rocks of all types.

alkali feldspars feldspars containing sodium silicate, $NaAlSi_3O_8$ ('soda feldspar') or potassium silicate, $KAlSi_3O_8$ ('potash feldspar'), or the two together. The group includes orthoclase (↓), adularia (↓), sanidine (↓), and microcline (↓).

orthoclase (*n*) an alkali feldspar (↑) $KAlSi_3O_8$. Monoclinic (p.43). Twin crystals (p.41) are common. A characteristic mineral of granites (p.76).

adularia (*n*) a low-temperature form of $KAlSi_3O_8$ found in veins (p.145), etc.

sanidine (*n*) a high-temperature form of $KAlSi_3O_8$ found in lavas (p.70) and dykes (p.67). On slow cooling it changes to orthoclase (↑).

microcline (*n*) the triclinic (p.43) form of $KAlSi_3O_8$; sodium is usually present as well. Microcline shows characteristic complex twinning (p.41).

potash feldspar potassium feldspar, $KAlSi_3O_8$; an alkali feldspar (↑).

soda feldspar sodium feldspar, $NaAlSi_3O_8$; an alkali feldspar (↑).

plagioclase feldspar a series of triclinic (p.43) feldspars (↑) with albite, $NaAlSi_3O_8$, and anorthite, $CaAl_2Si_2O_8$, as the end-members (p.46). Lamellar twinning (p.41) is common. Plagioclase is found in igneous rocks (p.62) and metamorphic rocks (p.90).

albite (*n*) *see* **plagioclase feldspar** (↑).

oligoclase (*n*) plagioclase feldspar (↑) with 70–90% albite (↑).

andesine (*n*) plagioclase feldspar (↑) with 50–70% albite (↑).

labradorite (*n*) plagioclase feldspar (↑) with 30–50% albite (↑).

bytownite (*n*) plagioclase feldspar (↑) with 10–30% albite (↑).

anorthite (*n*) *see* **plagioclase feldspar** (↑).

perthite (*n*) an intergrowth (p.73) of two feldspars. **perthitic** (*adj*).

長石類(名、複)　鈉、鉀、鈣和鋇的矽酸鹽(第16頁)。其基本構造為 SiO_4 四面體(第41頁)組成的網架。一切長石均屬三斜晶系或單斜晶系(第43頁)。長石存在於所有岩石類型中。

鹼性長石　含矽酸鈉($NaAlSi_3O_8$)("鈉長石")或矽酸鉀($KAlSi_3O_8$)("鉀長石")或兩者的長石。這一族包括正長石(↓)、冰長石(↓)、透長石(↓)和微斜長石(↓)。

正長石(名)　一種鹼性長石(↑)($KAlSi_3O_8$)。屬單斜晶系(第43頁)。其雙晶(第41頁)常見。是花崗岩(第76頁)的一種特徵礦物。

冰長石(名)　$KAlSi_3O_8$ 的一種低温變體，見於脈體(第145頁)等之中。

透長石(名)　$KAlSi_3O_8$ 的一種高温變體，見於溶岩(第70頁)和岩牆(第67頁)中。緩慢冷却時可變為正長石(↑)。

微斜長石(名)　$KAlSi_3O_8$ 的三斜(第43頁)變體，通常也含鈉。微斜長石顯示特徵性的複合雙晶(第41頁)。

鉀長石　含鉀的長石($KAlSi_3O_8$)，一種鹼性長石(↑)。

鈉長石　含鈉的長石，($NaAlSi_3O_8$)，一種鹼性長石(↑)。

斜長石　以鈉長石($NaAlSi_3O_8$)和鈣長石($CaAl_2Si_2O_8$)作為端員(第46頁)的三斜(第43頁)長石(↑)系列。常見其聚片雙晶(第41頁)。斜長石見於火成岩(第62頁)和變質岩(第90頁)中。

鈉長石(名)　見**斜長石**(↑)。

奧長石；更長石；鈉灰長石(名)　含70–90%鈉長石(↑)的一種斜長石(↑)。

中長石(名)　含50–70%鈉長石(↑)的一種斜長石(↑)。

拉長石(名)　含30–50%鈉長石(↑)的一種斜長石(↑)。

培長石；倍長石(名)　含10–30%鈉長石(↑)的一種斜長石(↑)。

鈣長石(名)　見**斜長石**(↑)。

條紋長石(名)　兩種長石的共生混合體(第73頁)。(形容詞為 perthitic)

feldspar 長石
microcline feldspar
微斜長石

pyroxenes (*n.pl.*) inosilicates (p.54) with a single-chain structure of SiO_4 tetrahedra (pp.41, 53). Most pyroxenes are monoclinic (p.43). All have cleavage (p.44) at 90°, unlike the amphiboles (↓), which have a cleavage at 124°. Pyroxenes are found in basic and ultrabasic igneous rocks (p.75) and in metamorphic rocks (p.90).

clinopyroxenes (*n.pl.*) a general name for monoclinic (p.43) pyroxenes (↑). The most common of these is augite, $(Ca,Mg,Fe)(Si,Al)_2O_6$. Others include diopside, $CaMgSi_2O_6$, and the alkali-pyroxenes, which contain Na, Fe, Al, or Li.

orthopyroxenes (*n.pl.*) a general name for orthorhombic (p.43) pyroxenes (↑). They form a solid-solution series (p.46) with enstatite, $MgSiO_3$, and ferrosilite, $FeSiO_3$, as the end-members (p.46).

augite (*n*) *see* **clinopyroxenes** (↑).

diopside (*n*) *see* **clinopyroxenes** (↑).

alkali pyroxenes *see* **clinopyroxenes** (↑).

enstatite (*n*) *see* **orthopyroxenes** (↑).

ferrosilite (*n*) *see* **orthopyroxenes** (↑).

amphiboles (*n.pl.*) inosilicates (p.54) with a double-chain structure of SiO_4 tetrahedra (p.41). Most amphiboles are monoclinic (p.43); a few are orthorhombic (p.43). All have a cleavage (p.44) at 124°, unlike the pyroxenes (↑), which have a cleavage at 90°. Amphiboles are found in plutonic (p.64) and other igneous rocks (p.62), in metamorphic rocks (p.90), and in sediments (p.80). Fibrous (p.45) forms of amphibole are included in the asbestos group of minerals (p.61). The monoclinic amphiboles include *hornblende*, an important rock-forming mineral, $(Ca,Na)_{2-3}(Mg,Fe^{2+},Fe^{3+}Al)_5(Al,Si)_8O_{22}(OH)_2$, *tremolite* and *actinolite*, which are found in metamorphic rocks, and *glaucophane* and *riebeckite*, which are found in igneous rocks.

pyriboles (*n,pl.*) a general term for pyroxenes (↑) and amphiboles (↑) together.

hornblende (*n*) *see* **amphiboles** (↑).

tremolite (*n*) *see* **amphiboles** (↑).

actinolite (*n*) *see* **amphiboles** (↑).

glaucophane (*n*) *see* **amphiboles** (↑).

riebeckite (*n*) *see* **amphiboles** (↑).

輝石類(名、複)　具有 SiO_4 四面體(第 41、53 頁)單鏈構造的鏈狀矽酸鹽(第 54 頁)。大多數輝石屬單斜(第 43 頁)晶系。所有輝石都有成 90° 相交的解理(第 44 頁)，而角閃石(↓)的解理成 124° 相交。輝石見於基性火成岩、超基性火成岩(第 75 頁)和變質岩(第 90 頁)中。

單斜輝石類(名、複)　屬單斜晶系(第 43 頁)的輝石(↑)的總稱。其中最常見是普通輝石 $(Ca, Mg, Fe)(Si, Al)_2O_6$。其餘的包括透輝石 $(CaMgSi_2O_6)$ 和含有 Na, Fe, Al 或 Li 的鹼性輝石。

斜方輝石類(名、複)　屬斜方晶系(第 43 頁)的輝石(↑)的總稱。此類輝石以頑火輝石 $(MgSiO_3)$ 和鐵輝石 $(FeSiO_3)$ 為兩端員(第 46 頁)組成固溶體系列(第 46 頁)。

普通輝石(名)　**見單斜輝石**(↑)。

透輝石(名)　**見單斜輝石**(↑)。

鹼性輝石　**見單斜方石**(↑)。

頑火輝石(名)　**見斜方輝石**(↑)。

鐵輝石(名)　**見斜方輝石**(↑)。

角閃石類(名、複)　具 SiO_4 四面體(第 41 頁)雙鏈構造的鏈狀矽酸鹽(第 54 頁)。大多數角閃石屬單斜(第 43 頁)晶系，只有少數屬斜方(第 43 頁)晶系。所有角閃石都有交角為 124° 的解理(第 44 頁)，這和輝石(↑)的解理交角為 90° 是不同的。角閃石見於深成岩(第 64 頁)和其他火成岩(第 62 頁)、變質岩(第 90 頁)和沉積物(第 80 頁)中。纖維狀(第 45 頁)角閃石歸入石棉族礦物(第 61 頁)中。單斜角閃石包括普通角閃石，這是一種重要的造岩礦物，$(Ca, Na)_{2-3}(Mg, Fe^{2+}, Fe^{3+}, Al)_5(Al, Si)_8O_{22}(OH)_2$，見於變質岩中的透閃石和陽起石，以及見於火成岩中的藍閃石和鈉閃石。

輝閃石類(名、複)　輝石類(↑)和角閃石類(↑)的總稱。

普通角閃石(名)　**見角閃石類**(↑)。

透閃石(名)　**見角閃石類**(↑)。

陽起石(名)　**見角閃石類**(↑)。

藍閃石(名)　**見角閃石類**(↑)。

鈉閃石(名)　**見角閃石類**(↑)。

augite 普通輝石

hornblende
普通角閃石

olivines (*n.pl.*) a group of magnesium and iron silicate minerals built up of isolated SiO_4 tetrahedra (pp.41, 53) joined by cations (p.15). All olivines are orthorhombic (p.43). Crystals are green or green brown with a glassy lustre (p.47) and smoothly curved fracture (p.44). The olivines form a solid-solution series (p.46) with forsterite, Mg_2SiO_4, and fayalite, Fe_2SiO_4, as the end-members (p.46). Olivines are important minerals in basic and ultrabasic igneous rocks (pp.74, 75).
forsterite (*n*) *see* **olivines** (↑).
fayalite (*n*) *see* **olivines** (↑).
garnets (*n.pl.*) a group of silicate minerals with the general formula $R_3^{2+}R_2^{3+}(SiO_4)_3$, where R^{2+} is Ca, Mg, Fe^{2+}, or Mn and R^{3+} is Al, Cr, or Fe^{3+}. In the garnets SiO_4 tetrahedra (p.41) are packed together with the R^{2+} and R^{3+} ions between them. Garnets are generally cubic (p.43). They have high refractive indices (p.159) and are strongly coloured (red, green, or black). They have no cleavage (p.44). Garnets occur in igneous (p.62) and metamorphic rocks (p.90), e.g. in schists (p.97).
pyrope (*n*) a red garnet (↑), $Mg_3Al_2(SiO_4)_3$.
almandine (*n*) a red garnet (↑), $Fe_3Al_2(SiO_4)_3$.
spessartite (*n*) a red garnet (↑), $Mn_3Al_2(SiO_4)_3$.
grossularite (*n*) a green garnet (↑), $Ca_3Al_2(SiO_4)_3$.
andradite (*n*) a red or green garnet (↑), $Ca_3Fe_2(SiO_4)_3$.
uvarovite (*n*) a green garnet (↑), $Ca_3Cr_2(SiO_4)_3$.
feldspathoids (*n.pl.*) sodium and potassium silicate minerals (p.44) with framework structures (p.122). The feldspathoids are closely related to the feldspars (p.56) but are chemically undersaturated (p.74), i.e. they are never present at the same time as quartz (p.55). Most feldspathoids are cubic or hexagonal (p.43).
nepheline (*n*) a feldspathoid (↑), $NaAlSiO_4$.
leucite (*n*) a feldspathoid (↑), $KAlSi_2O_6$.
sodalite (*n*) a complex feldspathoid (↑).
nosean (*n*) a complex feldspathoid (↑).
haüyne (*n*) a complex feldspathoid (↑).
cancrinite (*n*) a complex feldspathoid (↑).

橄欖石類（名、複） 陽離子（第 15 頁）將孤立的 SiO_4 四面體（第 41、53 頁）相連而構成的鎂和鐵矽酸鹽礦物族。一切橄欖石均屬斜方（第 43 頁）晶系。晶體呈綠色或綠棕色，具玻璃光澤（第 47 頁）和圓滑彎曲的斷口（第 44 頁）。橄欖石以鎂橄欖石（Mg_2SiO_4）和鐵橄欖石（Fe_2SiO_4）為兩端員（第 46 頁）組成固溶體系列（第 46 頁）。橄欖石是基性火成岩和超基性火成岩（第 74、75 頁）中的重要礦物。

鎂橄欖石（名） 見**橄欖石**（↑）。
鐵橄欖石（名） 見**橄欖石**（↑）。
石榴石類（名、複） 通式為 $R_3^{2+}R_2^{3+}(SiO_4)_3$ 的一族矽酸鹽礦物，其中 R^{2+} 代表 Ca、Mg、Fe^{2+}、Mn，R^{3+} 代表 Al、Cr、Fe^{3+}。在石榴石中，SiO_4 四面體（第 41 頁）堆積在一起，四面體之間有 R^{2+} 和 R^{3+} 離子。石榴石一般屬立方（第 43 頁）晶系，其折射率（第 159 頁）高，顏色很深（紅、綠或黑色）。不具解理（第 44 頁）。石榴石產於火成岩（第 62 頁）和變質岩（第 90 頁），例如片岩（第 97 頁）中。

鎂鋁榴石（名） 紅石榴石（↑），$Mg_3Al_2(SiO_4)_3$。
鐵鋁榴石（名） 紅石榴石（↑），$Fe_3Al_2(SiO_4)_3$。
錳鋁榴石（名） 紅石榴石（↑），$Mn_3Al_2(SiO_4)_3$。
鈣鋁榴石（名） 綠石榴石（↑），$Ca_3Al_2(SiO_4)_3$。

鈣鐵榴石（名） 紅或綠的石榴石（↑），$Ca_3Fe_2(SiO_4)_3$。
鈣鉻榴石（名） 綠石榴石（↑），$Ca_3Cr_2(SiO_4)_3$。
副長石類；似長石類（名、複） 鈉和鉀的矽酸鹽礦物（第 44 頁），具框架構造（第 122 頁）。副長石雖與長石（第 56 頁）關係密切，但在化學上是不飽和的（第 74 頁），即它們絕不會和石英（第 55 頁）同時存在。副長石大多屬立方或六方晶系（第 43 頁）。

霞石（名） 一種副長石（↑），$NaAlSiO_4$。
白榴石（名） 一種副長石（↑），$KAlSi_2O_6$。
方鈉石（名） 一種複雜的副長石（↑）。
黝方石（名） 一種複雜的副長石（↑）。
藍方石（名） 一種複雜的副長石（↑）。
鈣霞石（名） 一種複雜的副長石（↑）。

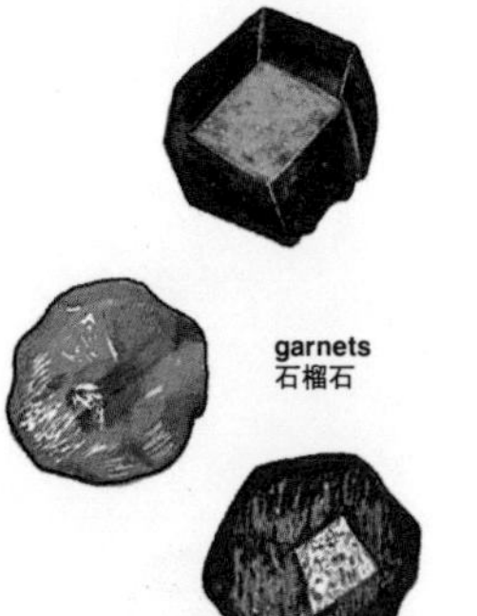
garnets
石榴石

aluminium silicate minerals there are three common minerals with the chemical composition Al_2SiO_5: andalusite, sillimanite, and kyanite (↓). They are found in metamorphic rocks (p.90).

andalusite (*n*) aluminium silicate, Al_2SiO_5. Orthorhombic (p.43); crystals are prismatic (p.45). Andalusite is formed at high temperature and low stress (p.122) and is typical of thermal metamorphism (p.90), especially of argillaceous rocks (p.85). *See also* **sillimanite** (↓), **kyanite** (↓).

sillimanite (*n*) aluminium silicate, Al_2SiO_5. Orthorhombic (p.43). Commonly occurs as fibrous masses (p.45). Sillimanite is formed at high temperatures and is found in regionally metamorphosed rocks of high grade (pp.90, 91). It gives its name to the sillimanite zone (p.91). *See also* **andalusite** (↑), **kyanite** (↓).

kyanite (*n*) aluminium silicate, Al_2SiO_5. Triclinic (p.43). Commonly occurs as blue crystals. Kyanite does not change under stress and is characteristic of metamorphic rocks of intermediate grade (pp.90, 91). It gives its name to the kyanite zone (p.91). *See also* **andalusite** (↑), **sillimanite** (↑).

staurolite (*n*) a complex iron aluminium silicate. Monoclinic (p.43), though it appears to be orthorhombic. Staurolite is found in regionally metamorphosed argillaceous rocks (p.85) of medium grade (p.91). It gives its name to the staurolite zone (p.91).

sphene (*n*) a calcium titanium silicate, $CaTiSiO_5$. Twin crystals (p.41) are common; it also occurs in massive form (p.45). Sphene occurs as an accessory mineral (p.75) in acid igneous rocks (p.74) and in contact-metamorphosed limestones (pp.86, 90).

titanite (*n*) = sphene (↑).

zircon (*n*) zirconium silicate, $ZrSiO_4$. Occurs as prismatic crystals (p.45). Zircon is found in acid igneous rocks (p.74), in metamorphic rocks (p.90), and in sediments (p.80).

topaz (*n*) an aluminium silicate, $Al_2SiO_4(OH,F)_2$. It is found in acid igneous rocks (p.74), in tin veins (p.145), and in contact zones (p.92).

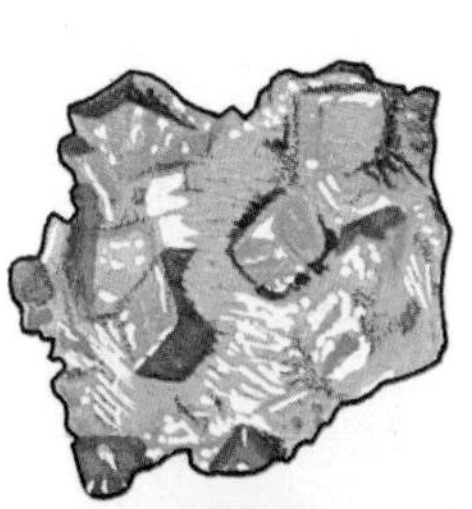

andalusite
紅柱石

sillimanite
矽線石

矽酸鋁礦物類　化學成分為 Al_2SiO_5 的常見礦物有三種：紅柱石、矽線石和藍晶石（↓）。這些礦物存在於變質岩（第 90 頁）中。

紅柱石；赤柱石（名）　成分為矽酸鋁（Al_2SiO_5），屬斜方晶系（第 43 頁）；晶體呈柱狀（第 45 頁）。紅柱石在高温及低應力（第 122 頁）下生成，尤其是由泥質岩石（第 85 頁）經熱力變質作用（第 90 頁）形成的典型產物。參見**矽線石**（↓）、**藍晶石**（↓）。

矽線石（名）　成分為矽酸鋁（Al_2SiO_5），屬斜方晶系（第 43 頁）。常以纖維狀塊體（第 45 頁）存在。矽線石在高温下生成並存在於高級區域變質岩中（第 90、91 頁）。矽線石帶（第 91 頁）以矽線石命名。參見**紅柱石**（↑）、**藍晶石**（↓）。

藍晶石（名）　成分為矽酸鋁（Al_2SiO_5），屬三斜晶系（第 43 頁）。常呈藍色晶體出現。藍晶石在應力下不變化，是中級變質岩（第 90、91 頁）的特徵。藍晶石帶（第 91 頁）以藍晶石命名。參見**紅柱石**（↑）、**矽線石**（↑）。

十字石（名）　一種成分複雜的矽酸鋁鐵鹽，屬單斜晶系（第 43 頁），外表似斜方晶系。十字石見於中級區域變質（第 91 頁）的泥質岩石（第 85 頁）中。十字石帶（第 91 頁）以十字石命名。

榍石（名）榍成分為鈣鈦的矽酸鹽（$CaTiSiO_5$）。常見其雙晶（第 41 頁）；但也呈塊狀（第 45 頁）產出。榍石在酸性火成岩（第 74 頁）及接觸變質的石灰岩（第 86、90 頁）中以副礦物（第 75 頁）形式存在。

榍石（↑）的另一英文名稱為 Titanite。

鋯石（名）　成分為矽酸鋯（$ZrSiO_4$）。以柱狀晶體（第 45 頁）產出。鋯石見於酸性火成岩（第 74 頁）、變質岩（第 90 頁）和沉積岩（第 80 頁）中。

黃晶；黃玉；黃石英（名）　成分為鋁的矽酸鹽（$Al_2SiO_4(OH,F)_2$）。見於酸性火成岩（第 74 頁）、錫礦脈（第 145 頁）和接觸帶（第 92 頁）中。

cordierite (*n*) an aluminium iron magnesium silicate. The structure is built up of SiO_4 units in a six-membered ring (p.53). Orthorhombic (p.43); usually granular or massive (p.45); blue in colour and glassy. Cordierite is found in rocks that have been contact metamorphosed (p.90) and in thermally metamorphosed argillaceous rocks (p.85).

堇青石（名） 成分為鋁鐵鎂的矽酸鹽。由六個 SiO_4 單元組成的六方環（第 53 頁）構造。屬斜方晶系（第 43 頁）；常為粒狀或塊狀（第 45 頁）；呈藍色，有玻璃光澤。堇青石見於曾受接觸變質（第 90 頁）的岩石中，也見於熱力變質的泥質岩石（第 85 頁）中。

beryl (*n*) a beryllium aluminium silicate, $Be_3Al_2Si_6O_{18}$. The structure is a six-membered ring (p.53). Hexagonal (p.43); crystals are common. Very hard; used for jewellery. Beryl occurs as an accessory mineral in acid igneous rocks (p.74) and in metamorphic rocks (p.90).

綠柱石；綠玉（名） 成分為鈹鋁的矽酸鹽（$Be_3Al_2Si_6O_{18}$），六方環（第 53 頁）構造，屬六方晶系（第 43 頁）；其晶體常見，質堅硬，用於製作寶石。綠柱石以副礦物產於酸性火成岩（第 74 頁）和變質岩（第 90 頁）中。

beryl 綠柱石

tourmaline (*n*) a complex borosilicate containing sodium, magnesium or iron, manganese, and lithium. Its structure is of the six-membered ring type (p.53). Hexagonal (p.43); needle-shaped crystals are common, often in groups. Tourmaline occurs as an accessory mineral (p.75) in igneous (p.62) and metamorphic rocks (p.90) and in sediments (p.80).

電氣石（名） 成分為複雜的含鈉、鎂或鐵、錳和鋰的硼矽酸鹽。六方環型（第 53 頁）構造。屬六方晶系（第 43 頁）；針狀晶體很普遍，常成群出現。電氣石以副礦物（第 75 頁）產於火成岩（第 62 頁）及變質岩（第 90 頁）和沉積岩（第 80 頁）中。

idocrase (*n*) a complex calcium aluminium magnesium silicate (p.16) formed by the contact metamorphism (p.90) of impure limestones (p.86).

符山石（名） 成分為複雜的鈣鋁鎂的矽酸鹽（第 16 頁），由不純石灰岩（第 86 頁）經接觸變質作用（第 90 頁）形成。

vesuvianite (*n*) = idocrase (↑).

維蘇威石（名） 同符山石（↑）。

epidotes (*n.pl.*) silicate minerals with SiO_4 and Si_2O_7 units. Most epidotes contain calcium and/or aluminium, iron, and manganese. Some are orthorhombic, others are monoclinic (p.43). The epidotes include zoisite, a calcium aluminium silicate, and epidote (in the narrower sense), a calcium iron aluminium silicate. Zoisite is orthorhombic (p.43) but there is also a monoclinic form, *clinozoisite*. Epidote is monoclinic (p.43). Epidotes are found in metamorphic rocks (p.90) of lower grade (p.91), in igneous rocks (p.62), and in sandstones (p.87).

綠簾石類（名、複） 具 SiO_4 和 Si_2O_7 單元的矽酸鹽礦物。大多數綠簾石含鈣或鋁、鐵、錳或兼含。綠簾石屬斜方晶系，有些則屬單斜晶系（第 43 頁）。綠簾石類包括黝簾石（係鈣鋁的矽酸鹽）和綠簾石（狹義的，係鈣鐵鋁的矽酸鹽）。黝簾石屬斜方晶系（第 43 頁），但也有單斜變體（斜黝簾石）。綠簾石屬單斜晶系（第 43 頁）。綠簾石類見於較低級（第 91 頁）的變質岩（第 90 頁）中，以及火成岩（第 62 頁）和砂岩（第 87 頁）中。

piedmontite (*n*) an epidote (↑) containing manganese.

紅簾石（名） 含錳的綠簾石（↑）。

allanite (*n*) an epidote (↑) containing cerium.

褐簾石（名） 含鈰的綠簾石（↑）。

orthite (*n*) = allanite (↑).

褐簾石（↑）亦稱 orthite。

zoisite (*n*) *see* **epidotes** (↑).

黝簾石（名） 見**綠簾石類**（↑）。

clinozoisite (*n*) *see* **epidotes** (↑).

斜黝簾石（名） 見**綠簾石類**（↑）。

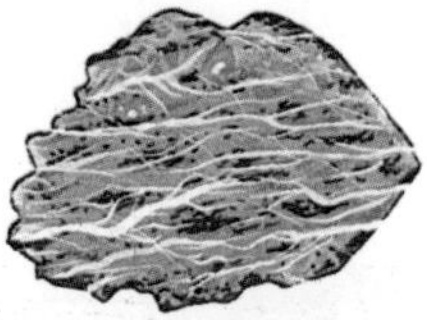

serpentine 蛇紋石

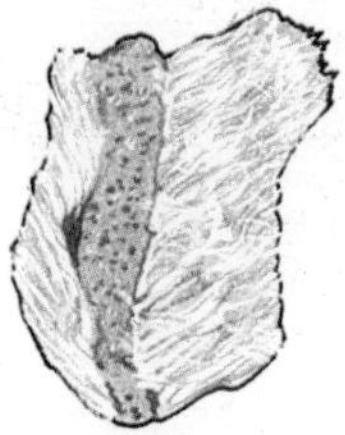

asbestos 石棉

serpentine (*n*) a hydrous (p.157) magnesium silicate, green or nearly black, usually with fine lines or spotted. It occurs in metamorphosed (p.90) basic and ultrabasic igneous rocks (pp.74–5) by the alteration of olivines (p.58) and pyroxenes (p.57).

talc (*n*) a hydrated magnesium silicate. It is very soft and is usually massive (p.45). Talc occurs in metamorphosed (p.90) basic rocks (p.74) with serpentine (↑).

asbestos (*n*) (1) the fibrous (p.45) forms of various silicate minerals, including certain amphiboles (p.57) and serpentine (↑); (2) (in mineralogy) the fibrous (p.45) forms of amphiboles only.

clay minerals a group of silicate minerals (p.16) with sheet or layer structures (p.54) formed by the weathering (p.20) of other silicate minerals. Clay minerals occur as very fine particles and can readily take up water.

kaolinite (*n*) a group of clay minerals (↑) with the formula $Al_4Si_4O_{10}(OH)_8$. They are formed by the decomposition of feldspars (p.56).

illite (*n*) a group of clay minerals (↑) formed by the decomposition of certain silicate minerals.

montmorillonite (*n*) a group of clay minerals formed by the alteration of certain silicate minerals.

smectite = montmorillonite (↑).

fullers' earth a fine earthy material containing montmorillonite (↑).

vermiculite (*n*) a group of clay minerals related to chlorite and montmorillonite (↑). Vermiculite is formed by the alteration of micas, especially biotite, and other minerals.

chlorite (*n*) a group of minerals with a phyllosilicate structure (p.54) related to talc (↑). They occur in igneous rocks (p.62), where they are formed by alteration of ferromagnesian minerals (p.55), and in metamorphic rocks (p.90).

zeolites (*n*) a group of hydrous (p.157) aluminium silicate minerals containing sodium, potassium, calcium, and barium. These positive ions (p.15) can easily be replaced. Zeolites occur in volcanic rocks (p.68) and in hydrothermal veins (p.145).

蛇紋石（名） 成分為水合（第 157 頁）矽酸鎂，綠色或近黑色，通常帶細條紋或斑點。產於變質（第 90 頁）基性火成岩和超基性火成岩（第 74-75 頁）中，是由其中的橄欖石類（第 58 頁）和輝石類（第 57 頁）蝕變而成。

滑石（名） 成分為水合矽酸鎂。質極軟，通常呈塊狀（第 45 頁）。滑石與蛇紋石（↑）共存於變質（第 90 頁）基性岩（第 74 頁）中。

石棉（名） (1)各種矽酸鹽礦物，包括某些角閃石類（第 57 頁）和蛇紋石（↑）的纖維狀（第 45 頁）變體；(2)（礦物學上）僅指角閃石類的纖維狀（第 45 頁）變體。

黏土礦物 由其他矽酸鹽礦物風化（第 20 頁）而成、具片狀或層狀構造（第 54 頁）的一族矽酸鹽礦物（第 16 頁）。黏土礦物以極細顆粒產出，易吸水。

高嶺土（名） 分子式為 $Al_4Si_4O_{10}(OH)_8$ 的一族黏土礦物（↑），由長石類礦物（第 56 頁）分解形成。

伊萊石；水白雲母（名） 由某些矽酸鹽礦物分解形成的一族黏土礦物（↑）。

蒙脱石（名） 由某些矽酸鹽礦物蝕變形成的一族黏土礦物。

膠嶺石 同蒙脱石（↑）。

漂白土 一種含蒙脱石（↑）的細土狀物質。

蛭石（名） 和綠泥石和蒙脱石（↑）有關的一族黏土礦物。蛭石是由雲母類，特別是黑雲母及其他礦物蝕變形成的。

綠泥石（名） 所具的層狀矽酸鹽構造（第 54 頁）與滑石（↑）有關的一族礦物，產於火成岩（第 62 頁）中，由其中的鎂鐵礦物（第 55 頁）蝕變形成，也出現在變質岩（第 90 頁）中。

沸石類（名、複） 一族含鈉、鉀、鈣和鋇的含水（第 157 頁）鋁矽酸鹽礦物。所含的這些正離子（第 15 頁）易被置換。沸石產於火山岩（第 68 頁）和熱液脈（第 145 頁）中。

rock (*n*) in geology, any natural material formed of minerals (p.44), or less commonly of a single mineral, whether solid or not. Rocks are divided into three main classes: igneous (↓), sedimentary (p.80), and metamorphic (p.90).

petrology (*n*) the study of rocks: their origin, their occurrence, and what they are made of. **petrological** (*adj*).

petrography (*n*) the description of rocks and their grouping into classes. A branch of petrology (↑). **petrographic** (*adj*).

igneous (*adj*) igneous rocks are, with certain exceptions, the crystalline (p.40) or glassy rocks that have solidified from magma (↓): the lavas that have been poured out at the surface of the Earth and the rocks that have solidified at various depths in the crust (p.9).

petrogenesis (*n*) the origin of rocks and the ways in which they are formed. The word is generally used only in igneous petrology (↑).

magma (*n*) rock material in a molten state. Magmas consist of silicates (p.16), water, and gases at high temperatures. **magmatic** (*adj*).

magmatism (*n*) the formation and movement of magma (↑) and of igneous rocks (↑) in the Earth's crust (p.9).

magmatic differentiation the separation of a magma (↑) into two or more parts, called *fractions*, with differing compositions (p.15). The separate fractions may then crystallize out (p.40) as rocks of different chemical composition (p.15) containing different minerals (p.44).

fraction (*n*) *see* magmatic differentiation (↑).

hybrid (*n, adj*) a hybrid rock is an igneous rock (↑) that has been produced by the mixing together of two magmas (↑) of different compositions or by the assimilation (↓) of other rocks by a magma. **hybridization** (*n*).

enrichment (*n*) a process in which the amount of one mineral or chemical element in a rock is increased in relation to the rest.

assimilation (*n*) the process of taking material (usually solid rock) into an igneous rock (↑) by melting it.

岩石(名)　地質學上指由幾種礦物(第 44 頁)或少數由一種礦物形成的任何天然物質，而不論其是否為固體。岩石分三大類：火成(↓)岩、沉積(第 80 頁)岩和變質(第 90 頁)岩。

岩石學(名)　研究岩石成因、產狀和組成的學科。(形容詞為 petrological)

岩相學；岩類學(名)　岩石的描述和分類的學科。岩石學(↑)的一門分科。(形容詞為 petrographic)

火成的(形)　除某些例外，火成的是描述從岩漿(↓)凝固出來的結晶質(第 40 頁)或玻璃質岩石：即溢出地球表面的熔岩和在地殼(第 9 頁)不同深度凝固出來的岩石。

岩石成因論；岩理學(名)　岩石的成因和其形成的方式。這個詞通常僅用于火成岩石學(↑)。

岩漿(名)　熔融狀態的岩石物質。岩漿是在高温下由矽酸鹽類(第 16 頁)、水和氣體組成的。(形容詞為 magmatic)

岩漿作用(名)　岩漿(↑)和火成岩(↑)在地殼(第 9 頁)內的形成與運動。

岩漿分異(作用)　一種岩漿(↑)分離為兩個或多個具有不同成分(第 15 頁)的部分，稱作"分餾部分"。其後各分離部分可結晶出(第 40 頁)化學成分(第 15 頁)不同的含不同礦物(第 44 頁)的各種岩石。

分餾部分(名)　見岩漿分異(↑)。

混染岩(的)(名、形)　混染岩是一種火成岩(↑)，由成分不同的兩種岩漿(↑)混合一起而成或由岩漿將別的岩石同化(↓)而成。(名詞為 hybridization)

富集作用(名)　岩石中某一種礦物或化學元素的數量較其餘部分相對增加的過程。

同化作用(名)　將物質(通常為固結的岩石)熔化於火成岩(↑)中的過程。

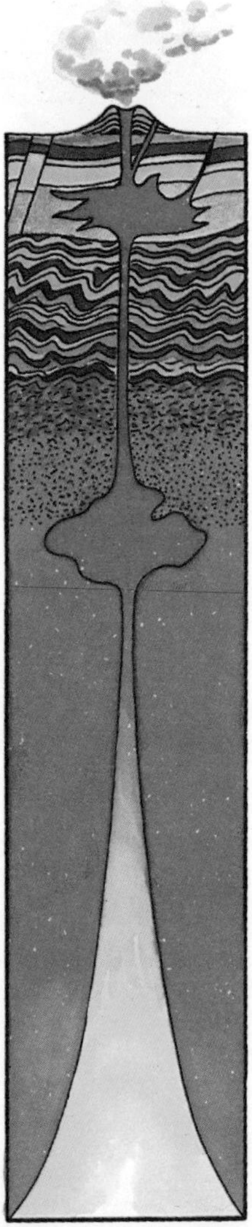

ascent of magma 岩漿的上升

pneumatolysis (*n*) changes produced by hot gases (e.g. fluorine, hydrofluoric acid) given off by a magma in the later stages of cooling and solidification. **pneumatolytic** (*adj*).

氣成（名） 由岩漿冷却和凝固晚期釋放出的熱氣體（例如氟、氫氟酸）所產生的變化。（形容詞為 Pneumatolytic）

hydrothermal (*adj*) caused by the action of water at high temperatures during the formation of igneous rocks. *See also p.145.*

熱液的（形） 火成岩形成過程中由高温水作用所引起的。參見第 145 頁。

diffusion (*n*) the spreading out of ions (p.15), atoms (p.152), or molecules (p.15) into a liquid, a gas, or a porous (p.84) solid (such as a rock). The process of diffusion tends to distribute a substance more evenly throughout the system. **diffuse** (*v*), **diffused** (*adj*).

擴散（名） 離子（第 15 頁）、原子（第 152 頁）或分子（第 15 頁）分散入一種液體、氣體或多孔（第 84 頁）固體（例如岩石）中。擴散過程促使一種物質更均勻地分佈於整個系統中。（動詞為 diffuse，形容詞為 diffused）

granitization (*n*) a word used for a process by which solid rocks were thought to be changed into granite (p.76) by the action of liquids or gases without first being turned into magma (↑).

花崗岩化（名） 顯現帶狀的固體沉積岩（↑）時不需首先轉變為岩漿（↑）而是通過液體或氣體的作用變成花崗岩（第 76 頁）的一種過程。

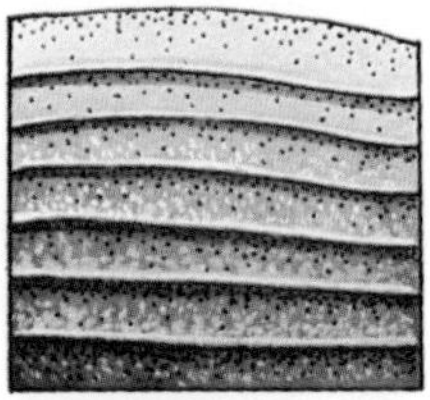
layering 層狀構造

layering, banding (*n*) terms used to describe bodies of igneous rock (p.62) that show bands when seen in a vertical section (p.147). 'Layering' is a better term than 'banding'. Layering may be the result of gravity separation (↓) or of other physical processes.

層狀構造，帶狀構造（名） 描述在垂直剖面（第 147 頁）上顯示帶狀火成岩（第 62 頁）體所用的術語。使用"層狀構造"這一術語比使用"帶狀構造"這一術語為佳。層狀構造可以是重力分離（↓）或其他物理作用的結果。

banding (*n*) = layering (↑).

帶狀構造（名） 同層狀構造（↑）。

gravity separation it has generally been thought that if the crystals (p.40) that are first to form in a magma (↑) are heavier or lighter than the liquid they will sink or rise through it and will gather together at a lower or higher level in the magma chamber (p.64). This process is called *gravity separation*. If the movement of the crystals is downward the term 'crystal settling' can be used.

重力分離 一般認為：假如岩漿（↑）內首先形成的一些晶體（第 40 頁）重於或輕於液體，它們便會在液體中下沉或上浮，並會在岩漿房（第 64 頁）內一個較低的或較高的層位上聚集。此作用稱為重力分離。假如是一些晶體向下運動，則可用"晶體沉降"這個術語。

crystal settling *see* **gravity separation** (↑).

晶體沉降 見**重力分離**（↑）。

cumulate (*n*) an igneous rock that is thought to have formed by crystal settling (↑).

堆積岩（名） 由晶體沉降（↑）而成的一種火成岩。

compositional zoning differences in the temperature and pressure in different parts of a magma chamber (p.64) can result in variation in the composition of the magma. The magma erupted will then vary in composition with time.

成分分帶 岩漿房（第 64 頁）不同部分的温度差異和壓力差異，可導致岩漿成分變化，因而所噴發的岩漿，其成分隨時間而發生變化。

devitrification (*n*) the formation of crystals (p.40) in a glass (p.77), which in time may become completely crystalline. **devitrified** (*adj*).

去玻作用（名） 在天然玻璃（第 77 頁）中形成晶體（第 40 頁），最後會使天然玻璃變成完全的結晶質的作用。（形容詞為 devitrified）

igneous intrusion a body of igneous rock (p.62) that has been put into place among rocks that were already there before it crystallized (p.44). **intrusive** (*adj*), **intrude** (*v*).

emplacement (*n*) any process by which an igneous rock is put into place. The word 'emplacement' can be used without suggesting any particular method of emplacement. **emplace** (*v*).

magma chamber a space below the Earth's surface containing magma (p.62).

injection (*n*) the intrusion (↑) of magma into rocks that are already in place. **inject** (*v*).

hypabyssal (*adj*) at no great depth. The word is used to describe bodies of igneous rock, such as dykes (p.67) and sills (p.66), that have formed at greater depths than volcanic rocks (p.68) but not at such great depth as plutonic rocks (↓).

plutonic rocks those igneous rocks (p.62) that have formed from magma (p.62) or by chemical alteration – metasomatism (p.90) – at great depth in the Earth's crust.

pluton (*n*) a body of plutonic rock (↑). The term 'pluton' is now often used to mean a granitic (p.76) rock body of roughly circular shape in plan that has been emplaced (↑) at a relatively low temperature.

火成侵入體　一個火成岩體(第 62 頁)定位在其結晶(第 44 頁)之前早已存在的岩石中間。(動詞為 intrude，形容詞為 intrusive)

侵位(名)　一種火成岩定位的任何過程。使用"侵位"這一詞時無需指出任何特定的定位方式。(動詞為 emplace)

岩漿房　地下儲藏岩漿(第 62 頁)的空間。

貫入(名)　岩漿侵入(↑)已在位的岩石中間。(動詞為 inject)

半深的；淺成的(形)　處於不很深之處的。這個詞用於描述一些火成岩體，例如岩牆(第 67 頁)和岩床(第 66 頁)，它們形成於比火山岩(第 68 頁)更深之處，但又不及深成岩(↓)那麼深。

深成岩類　在地殼很深處由岩漿(第 62 頁)或經化學蝕變作用(交代作用(第 90 頁))所形成的火成岩(第 62 頁)。

深成岩體(名)　深成岩(↑)的岩體。"深成岩體"這一術語現常指在平面上近似圓形，在較低溫度下侵位(↑)的花崗質(第 76 頁)岩體。

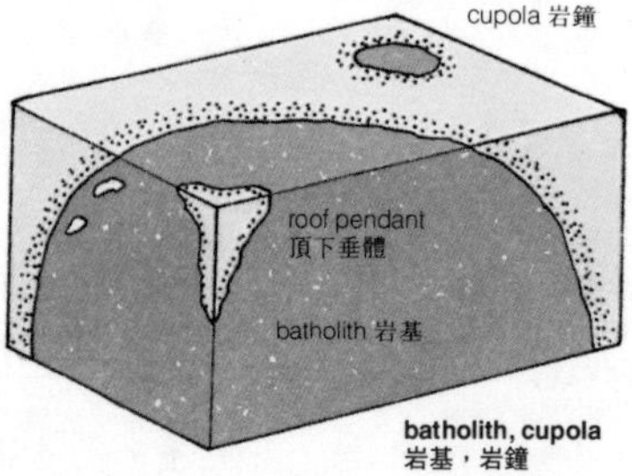

batholith, cupola
岩基，岩鐘

batholith, bathylith (*n*) a pluton (↑) with a large outcrop (p.122) and no visible base. Usually a granite (p.76). **batholithic** (*adj*).

cupola (*n*) an igneous (p.62) rock-body, round in plan, which is joined below to a larger body of igneous rock, such as a batholith (↑).

岩基(名)　具有巨大露頭(第 122 頁)且不見底的深成岩體(↑)。通常為花崗岩(第 76 頁)。(形容詞為 batholithic)

岩鐘(名)　火成(第 62 頁)岩體，平面上呈圓形，它下面連接一個較大的火成岩體，如岩基(↑)。

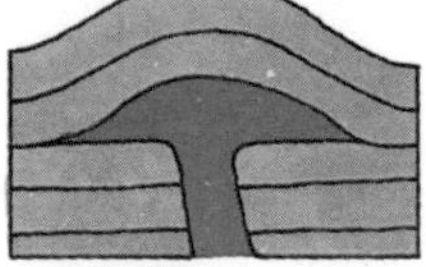

laccolith 岩蓋

laccolith (*n*) an igneous intrusion (↑) with a flat floor and a rounded roof which has pushed up the sediments (p.80) above it into the shape of a dome.

岩蓋(名)　具平底和圓形頂蓋的火成侵入體(↑)，頂蓋將其上的沉積物(第 80 頁)向上推而成穹隆狀。

stock (*n*) an igneous intrusion (↑) like a batholith (↑) but smaller (less than 100 km^2 in area) and with a more or less circular shape in plan.

岩株(名)　類似岩基(↑)，但比岩基小(面積小於 100 km^2)，在平面上近似圓形的火成侵入體(↑)。

boss (*n*) an igneous intrusion (↑) that is circular in plan and has steeply dipping contacts (p.148) with the country-rock (↓).

岩瘤(名)　平面上呈圓形，與原岩(↓)呈陡傾斜接觸(第 148 頁)的火成侵入體(↑)。

multiple intrusion an igneous intrusion (↑) that has been put into place in more than one injection (↑) of magma (p.62). The material is usually all of the same kind. Multiple intrusions are usually sills or dykes (pp.66, 67).

多次侵入體；重複侵入體　岩漿(第 62 頁)不只一次貫入(↑)已定位的火成侵入體(↑)。各次貫入的物質通常都相同。多次侵入體通常都是一些岩床或岩牆(第 66、67 頁)。

country-rock (*n*) the rock or rocks into which an igneous intrusion (↑) or a mineral vein (p.145) is emplaced.

原岩；圍岩(名)　火成侵入體(↑)或礦脉(第 145 頁)定位於其中的岩石。

xenolith (*n*) a piece of 'foreign' rock that is enclosed by an igneous rock body, e.g. a piece of country-rock broken off from the wall of the intrusion. **xenolithic** (*adj*).

捕擄岩(名)　為火成岩體所封閉的一片"外來"岩石。例如從侵入體的一壁崩落出來的一片圍岩。(形容詞為 xenolithic)

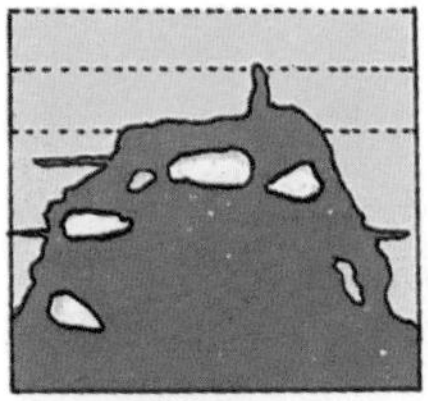

stoping 頂蝕作用

stoping, magmatic stoping (*n*) a way in which an igneous rock (p.62) may be emplaced (↑) by forcing its way into joints (p.21) in the country-rock (↑) and pushing out blocks of it, which then sink into the magma (p.62) to be assimilated (p.62).

頂蝕作用，岩漿頂蝕作用(名)　火成岩(第 62 頁)通過在圍岩(↑)節理(第 21 頁)中擠出通道和剝落圍岩岩塊(↑)而進行侵位(↑)的一種方式，剝出的岩塊隨後下沉到岩漿(第 62 頁)中而被同化(第 62 頁)。

subjacent (*adj*) without a known floor. The word is used to describe large igneous intrusions (↑) for which there is no sign of a base.

深成的；深淵的(形)　無底的。這個詞用於描述沒有底界跡象的大的火成侵入體(↑)。

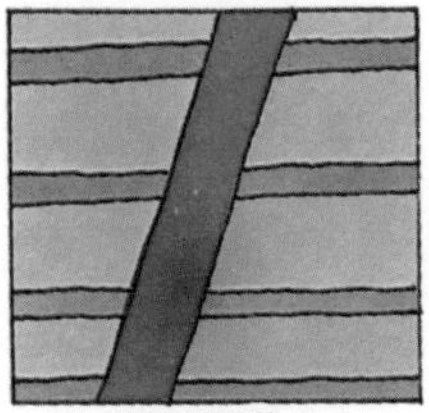

cross-cutting 橫切的

cross-cutting (*adj*) cutting across other, earlier rocks.

橫切的(形)　橫割別的、較早期的岩石。

discordant (*adj*) a word used to describe an igneous rock that cuts across the bedding (p.80) or the foliation (p.95) of the rocks into which it is intruded. *See also* **concordant** (↓).

不整合的(形)　這個詞用於描述橫割岩石的層理(第 80 頁)或葉理(第 95 頁)並侵入該岩石的火成岩。參見**整合的**(↓)。

concordant (*adj*) a word applied to an igneous rock that is parallel to the bedding (p.80) or the foliation (p.95) of the rocks into which it is intruded. *See also* **discordant** (↑).

整合的(形)　這個詞用於描述平行岩石的層理(第 80 頁)或葉理(第 95 頁)並侵入該岩石的火成岩。參見**不整合的**(↑)。

schlieren (*n.pl.German*) Long-drawn-out lines or areas in an igneous rock. Schlieren are usually of different composition (p.15) from the rest of the rock.

析離體；異離體(名)　在火成岩中呈拉長的線或面。析離體的成分(第 15 頁)通常與所在岩石其餘部分不同。

apophysis (*n*) (*apophyses*) a vein (p.145) or branch of an igneous rock (p.62) that is joined to a larger body.

tongue (*n*) = apophysis (↑).

lit-par-lit (*French*) 'Bed by bed'. *Lit-par-lit* intrusion (p.64) is the injection (p.64) of magma (p.62) along bedding-planes (p.80) or foliation-planes (p.95) to give a rock with thin, closely spaced layers of igneous material, e.g. a foliated gneiss (p.97).

chilled margin, chilled zone a border of fine-grained rock at the edge of an igneous intrusion (p.64). The fine grain is the result of rapid cooling of the magma (p.62) by the country-rock (p.65).

岩枝（名） 與較大火成岩體相連的火成岩脈（第145頁）或分枝。（複數為 apophyses）

岩舌（名） 同岩枝（↑）。

間層的 "一層又一層"。間層侵入體（第64頁）是岩漿（第62頁）沿着層面（第80頁）或葉理面（第95頁）貫入（第64頁），使岩石帶有薄而密集的火成物質層。例如葉片狀的片麻岩（第97頁）。

冷凝邊，冷凝帶 火成侵入體（第64頁）邊緣的一條細粒岩石邊。細晶粒是由靠近圍岩（第65頁）的岩漿（第62頁）急速冷却形成的。

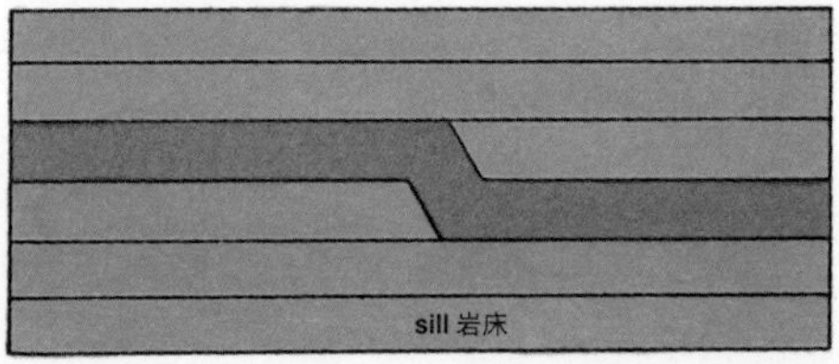

sill (*n*) an igneous intrusion (p.64) in the form of a concordant (p.65) sheet; usually more or less horizontal.

lopolith (*n*) a large igneous intrusion (p.64) shaped like a nearly flat dish. A lopolith is generally concordant (p.65) and its width is about ten to twenty times its thickness.

phacolith (*n*) an igneous intrusion (p.64) in folded sedimentary rocks (p.80) that is convex upwards and concave below.

岩床（名） 一種呈整合（第65頁）席狀的火成侵入體（第64頁）；通常近似水平。

岩盆（名） 一種大的火成侵入體（第64頁），形狀像一個近於扁平的盤，岩盆一般是整合的（第65頁），其寬度約為厚度的10–20倍。

岩鞍（名） 在褶皺的沉積岩層（第80頁）中的一種向上凸和向下凹的火成侵入體（第64頁）。

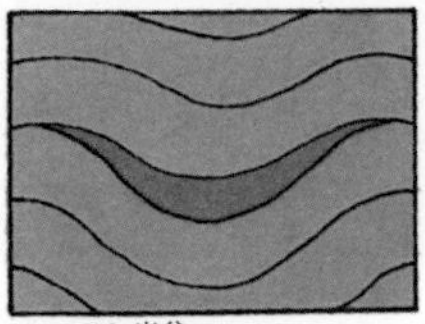
lopolith 岩盆

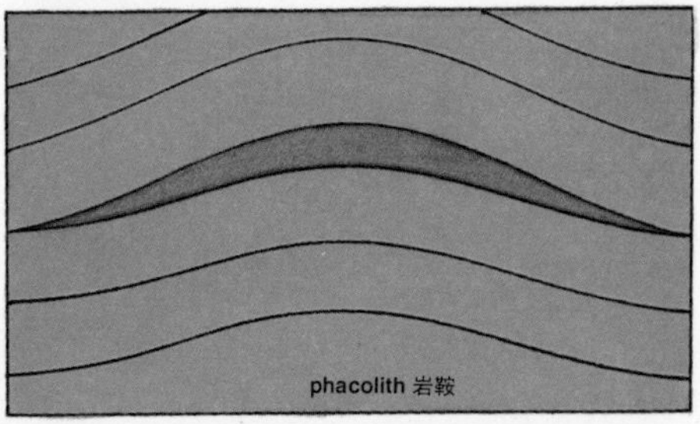

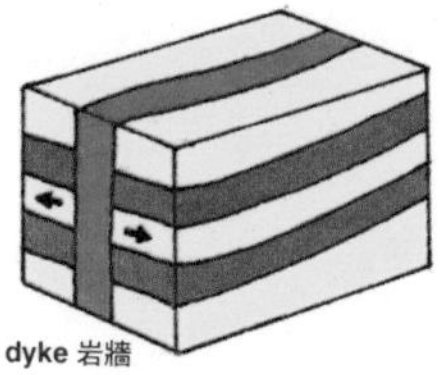

dyke 岩牆

dyke (*n*) an igneous intrusion (p.64) like a wall that is discordant (p.65) to the country-rock (p.65).

岩牆(名) 似牆狀的一種火成侵入體(第 64 頁)，與圍岩(第 65 頁)不整合(第 65 頁)。

dike (*n*) American spelling of dyke (↑).

岩牆(↑)的美語拼寫為 dike。

dyke-swarm (*n*) a group of dykes (↑) injected at about the same time and either parallel to each other or in lines meeting at a point.

岩牆群(名) 大致於同時注入的一群岩牆(↑)，它們或彼此平行，或成排相交於一點。

ring-dyke (*n*) a dyke (↑) with an outcrop (p.122) in the shape of a curve or a circle.

環狀岩牆(名) 具有弧形或環形露頭(第 122 頁)的一種岩牆(↑)。

ring-complex (*n*) a group of igneous intrusions (p.64) occurring together that have an outcrop (p.122) shaped like a ring, e.g. a complex of cone-sheets (↓) or ring-dykes (↑).

環狀雜岩(名) 產在一起的一群火成侵入體(第 64 頁)，它們具有形似環狀的露頭(第 122 頁)。例如錐狀岩席(↓)或環狀岩牆(↑)的雜岩。

cone-sheet (*n*) a dyke (↑) that has an outcrop in the shape of a curve and is in three dimensions shaped like a cone, dipping inwards towards a point below the Earth's surface.

錐狀岩席(名) 一種岩牆(↑)，具有弧狀露頭和在三維空間上形似錐體，在地下朝着一點向內傾斜。

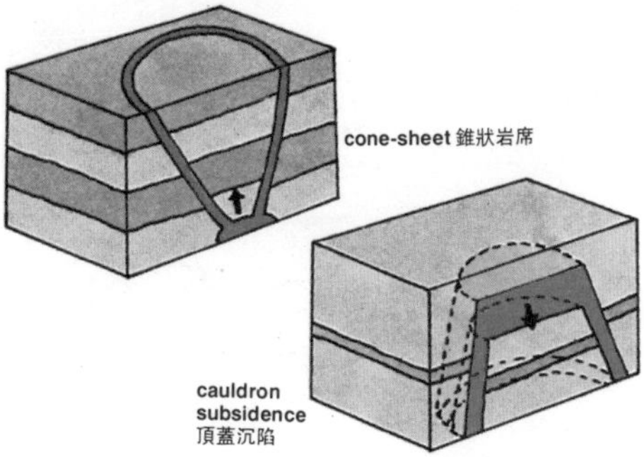

cone-sheet 錐狀岩席

cauldron subsidence 頂蓋沉陷

cauldron subsidence the descent of a large mass of rock shaped like a drum with magma (p.62) coming up from below round its sides and over it. Ring-complexes (↑) could be formed in this way.

頂蓋沉陷；火山口沉陷 鼓狀大岩塊的降落，伴有岩漿(第 62 頁)繞其四周從下面上升，並覆蓋在上面。環狀雜岩(↑)可能由這種方式形成。

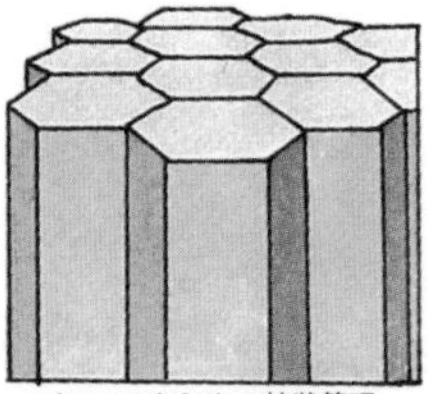

columnar jointing 柱狀節理

columnar jointing, columnar structure a regular form of jointing (p.21) that produces many-sided pillars or columns in igneous rocks (p.62). The joints are perpendicular to the contact (p.148) between the intrusion and the country-rock (p.65). Columnar jointing is seen especially in lava flows (p.70) and sills (↑).

柱狀節理，柱狀構造 一種規則的節理(第 21 頁)形式，在火成岩(第 62 頁)中它顯示六面柱或柱狀。節理垂直於侵入體和圍岩(第 65 頁)之間的接觸面(第 148 頁)。柱狀節理主要見於熔岩流(第 70 頁)和岩床(↑)中。

minor intrusions dykes (↑) and sills (↑).

小侵入體 指岩牆(↑)和岩床(↑)。

volcano (*n*) (1) a hole (vent) or fissure (p.21) in the Earth's crust (p.9) from which molten lava (p.70), pyroclastic materials (↓), and gases come out; (2) a hill or mountain, usually shaped like a cone, built of the materials that have come out of such a vent. **volcanic** (*adj*), **volcanism, vulcanism** (*n*).

火山(名)　(1)地殼(第 9 頁)中噴湧出熔岩(第 70 頁)、火成碎屑物(↓)和氣體的孔洞(裂口)或裂隙(第 21 頁)；(2)由此裂口湧出的物質所構成，通常似錐狀的山丘。(形容詞為 volcanic，名詞 volcanism，vulcanism 意為 火山作用)

eruption (*n*) the sending out of volcanic material from a volcanic vent (↓) or fissure (p.21) at the Earth's surface. **erupt** (*v*), **eruptive** (*adj*).

噴發(名)　從地球表面的一個火山口(↓)或裂隙(第 21 頁)發射出火山物質。(動詞為 erupt，形容詞為 eruptive)

central eruption a volcanic eruption (↑) from a circular vent (↓), i.e. an eruption of the ordinary kind.

中心噴發　從圓形火山口(↓)的一種火山噴發(↑)，即正常型式的噴發。

fissure eruption a volcanic eruption (↑) from a crack or fissure (p.21) in the Earth's crust (p.9).

裂隙噴發　從地殼(第 9 頁)中的一個裂隙(第 21 頁)或裂縫的一種火山噴發(↑)。

volcanic vent the hole or pipe through which a volcano sends out lava (p.70) and other igneous material (p.62) such as volcanic ash (↓).

火山口　一座火山賴以發射出熔岩(第 70 頁)和其他火成物質(第 62 頁)(如火山灰(↓))的孔洞或管道。

volcanic conduit = volcanic vent (↑).

火山道　同火山口(↑)。

volcanic cone the heap of volcanic material – solidified lava (p.70), volcanic ash (↓), etc. – that forms round and above a volcanic vent (↑).

火山錐　堆積在火山口(↑)四周及其上方的火山物質，這些物質包括固結的熔岩(第 70 頁)、火山灰(↓)等等。

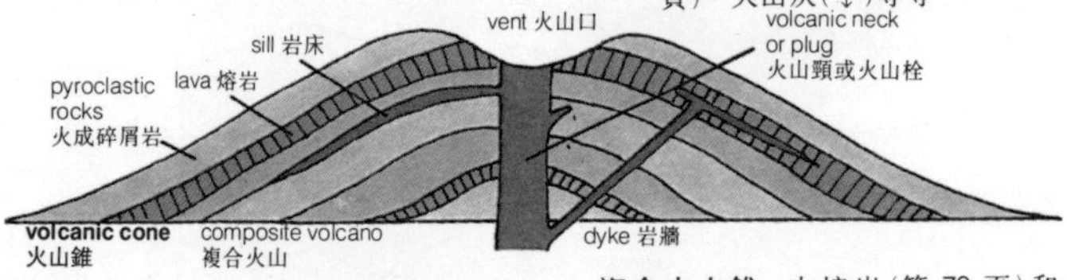

composite cone a volcanic cone (↑) formed of lavas (p.70) and pyroclastic rocks (↓).

複合火山錐　由熔岩(第 70 頁)和火成碎屑岩(↓)構成的火山錐(↑)。

stratovolcano (*n*) = composite cone (↑).

成層火山(名)　同複合火山錐(↑)。

cinder cone a volcanic cone (↑) formed of pyroclastic rocks (↓).

火山渣錐　由火成碎屑岩(↓)構成的火山錐(↑)。

shield volcano a volcanic cone (↑) formed by the eruption (↑) of large quantities of lava (p.70) of a type that flows easily. Shield volcanoes are large and their sides have gentle slopes.

盾形火山　由大量易流型熔岩(第 70 頁)噴發(↑)形成的火山錐(↑)。盾形火山巨大，其山坡坡度平緩。

volcanic plug solidified magma (p.62) in a volcanic vent (↑).

火山栓　火山口(↑)中固結的岩漿(第 62 頁)。

volcanic neck (1) a volcanic plug; (2) the pipe that joins a magma chamber to a volcanic vent (↑).

火山頸　(1)指火山栓；(2)指連接岩漿房與火山口(↑)的管道。

caldera (*n*) a large hollow, more or less circular in plan, formed by the falling in or the explosion of a volcano (↑).

破火山口(名)　平面上近似圓形的一個大凹地，係由火山(↑)陷落或爆發所形成。

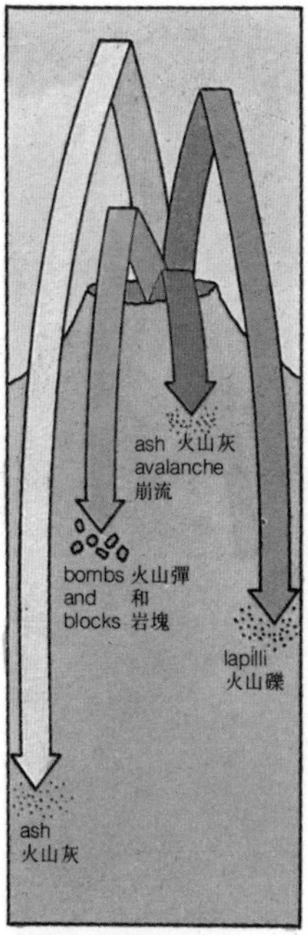

ejectamenta 噴出物

diatreme (*n*) a volcanic vent (↑) that has cut through bedded rocks as a result of an explosive eruption (↑).

extrusive (*adj*) extrusive rocks are those rocks that have flowed out as magma (p.62) at the Earth's surface – volcanic rocks.

pyroclastic (*adj*) describes rocks formed of pieces of material that have been thrown into the air by volcanic action.

ejectamenta (*n.pl.*) solid material thrown out from a volcanic vent (↑).

volcanic dust very fine material in the form of particles less than 0.06 mm in diameter blown out by a volcano.

volcanic ash material in the form of small fragments from 0.06 to 4 mm in diameter blown out by a volcano.

tuff (*n*) a consolidated (p.84) volcanic ash (↑).

lapilli (*n.pl.*) small fragments from 4 to 32 mm in diameter blown out by a volcano.

volcanic bomb an irregular or long, rounded block of lava more than 32 mm in diameter thrown out by a volcano.

volcanic block a large mass of rock thrown out by a volcano.

volcanic agglomerate a pyroclastic rock (↑) made up of fragments that are 20 to 30 mm in diameter or larger.

agglomerate (*n*) = volcanic agglomerate (↑).

volcanic breccia a rock consisting of large angular pieces of volcanic rock in a matrix (p.73) of fine-grained pyroclastic (↑) material.

pumice (*n*) a light-coloured, glassy, vesicular (p.73) rock of acid (p.74) composition. Pumice is formed when gases pass through newly erupted rhyolitic (p.76) lava (p.70).

scoria (*n.pl.*) pieces of rough, highly vesicular (p.73) lava (p.70) that have been thrown out from a volcano or have been formed by the cooling of the surface of molten lava by the air. *See also p.73.*

spatter cone a tower-shaped heap of lava (p.70) built up layer by layer by eruptions of lava one after the other from a small opening on the side of a volcano.

火山爆發口(道)(名) 由火山爆裂噴發(↑)造成切穿層狀岩石的一種火山口(↑)。

噴出的(形) 噴出岩是指那些由流出地面的岩漿(第 62 頁)所形成的岩石——火山岩。

火成碎屑的(形) 描述由火山活動拋到空中的物質碎片所組成的岩石。

噴出物(名、複) 從火山口(↑)拋出的固體物質。

火山塵 火山所噴出的極細小物質，顆粒直徑小於 0.06 mm。

火山灰 火山所噴出的細小碎塊物質，直徑為 0.06–4 mm。

凝灰岩(名) 固結的(第 84 頁)火山灰(↑)。

火山礫(名、複) 火山所噴出的細小碎塊，直徑為 4–32 mm。

火山彈 火山拋出的不規則的或長、圓形塊體，直徑大於 32 mm。

火山塊 火山拋出的大岩塊。

火山集塊岩 直徑 20–30 mm 或更大的碎塊組成的一種火成碎屑岩(↑)。

集塊岩(名) 同火山集塊岩(↑)。

火山角礫岩 由細粒的火成碎屑(↑)物為填質(第 73 頁)及大塊稜角狀火山岩碎片組成的一種岩石。

浮岩(名) 一種淺色、玻璃質、多孔狀(第 73 頁)的含酸性(第 74 頁)成分的岩石。浮岩是氣體穿過新噴出的流紋質(第 76 頁)熔岩(第 70 頁)時形成的。

火山渣(名、複) 粗糙而氣孔極多的(第 73 頁)熔岩(第 70 頁)碎片，它為火山所拋射出或是由於接觸空氣使熔融的熔岩表面冷却而形成。參見第 73 頁。

熔岩滴錐；寄生熔岩錐 一個塔狀的熔岩(第 70 頁)堆積，是在火山一側的一個細孔洞陸續噴出的熔岩一層一層地堆構成的。

lava (*n*) (1) the molten material that is thrown out from a volcano; magma (p.62) that reaches the Earth's surface; (2) the rock that forms when the molten material solidifies.

lava flow (1) lava poured out at the Earth's surface from a vent (p.68) or fissure (p.68); (2) the solid rock formed from lava poured out in this way.

áá (*n*) a lava (↑) flow with a rough surface.

pahoehoe (*n*) a lava (↑) flow with a surface like rope and a glassy outer skin.

lava tube the surface of a lava flow (↑) may cool and solidify while liquid lava continues to flow beneath it in a lava tube. When the flow of lava stops the tube may be left empty.

lava tunnel a lava tube (↑) that is open at both ends.

fire fountain a stream of very hot lava (↑) and gas coming from a hole in the ground. The lava is basic (p.74) and flows readily. Fire fountains may reach heights of several hundred metres. A line of fire fountains may form a *fire curtain* (↓).

fire curtain a *line of fire fountains* (↑).

nuée ardente (*n. French*) (*nuées ardentes*) a white-hot cloud of gas and volcanic ash (p.69) from a volcano. A *nuée ardente* can cause great damage.

ignimbrite (*n*) a volcanic rock (p.68) formed from a *nuée ardente* (↑); a welded tuff (p.69).

phreatic eruption a volcanic eruption that takes place when water under the ground is rapidly and violently turned to steam.

lahar (*n*) a mud flow of ash, etc. and water on the sides of a volcano.

pillow lava a lava (rock) formed by an eruption under water (typically under the sea). Pillow lavas have characteristic rounded masses that are shaped like pillows.

fumarole (*n*) a hole in the ground in a volcanic area from which steam and other gases (volatiles, p.18) come out. Fumaroles are characteristic of the later stages of volcanic activity.

geyser (*n*) a spring in a volcanic area from which hot water and steam come out from time to time.

熔岩(名)　(1)火山拋射出的熔融物質：到達地球表面的岩漿(第 62 頁)；(2)熔融物質固結形成的岩石。

熔岩流　(1)從火山口(第 68 頁)或裂隙(第 68 頁)流出地面的熔岩；(2)以這種方式流出的熔岩所形成的固體岩石。

塊熔岩(阿阿熔岩)(名)　表面粗糙的熔岩(↑)流。

繩狀熔岩(名)　具繩狀表面和玻璃質外皮的熔岩(↑)流。

熔岩管　在熔岩流(↑)表面冷却和固結的同時液態熔岩在其下方的熔岩管內繼續流動。當熔岩流走後，所留的空管。

熔岩燧道　兩端開口的熔岩管(↑)。

火噴泉　地下孔洞流出的熾熱熔岩(↑)和氣體流。熔岩是基性(第 74 頁)的，易於流動。火噴泉可以高達幾百米。一排火噴泉則形成火簾(↓)。

火簾　一排的噴泉(↑)。

熾熱火山雲(名)　火山發出的白熱氣體和火山灰(第 69 頁)。熾熱火山雲可造成巨大災害。(複數為 *nuées ardentes*)

熔結凝灰岩(名)　由熾熱火山雲(↑)形成的火山岩(第 68 頁)；一種熔凝的灰岩(第 69 頁)。

蒸氣噴發　地下水迅速而猛烈地轉變為蒸氣時所發生的一種火山噴發。

火山泥流物(名)　火山山坡的火山灰等物質和水混成的泥流。

枕狀熔岩　水下噴發(一般在海底)形成的熔岩(岩石)。枕狀熔岩具枕頭狀的特有圓塊。

噴氣孔(名)　火山地區地下一個噴發出蒸氣和其他氣體(揮發物，第 18 頁)的孔洞。噴氣孔是火山活動較後期的特徵。

間歇噴泉(名)　火山地區不時地噴發出熱水和蒸氣的一種泉。

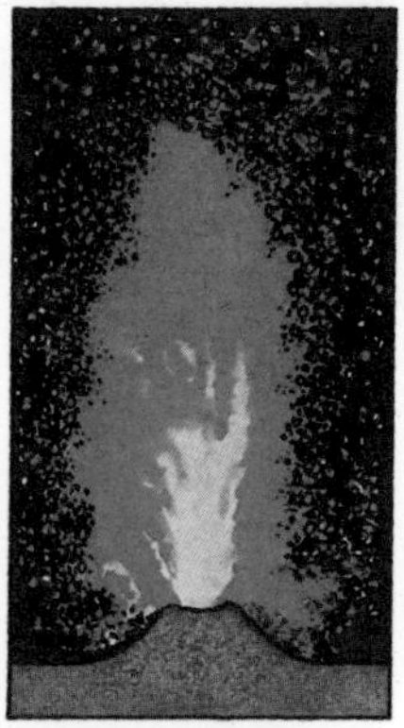

lava fountain 熔岩噴泉

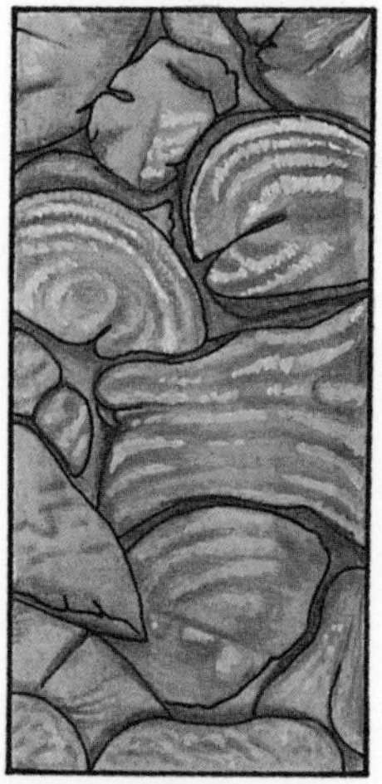

pillow lava 枕狀熔岩

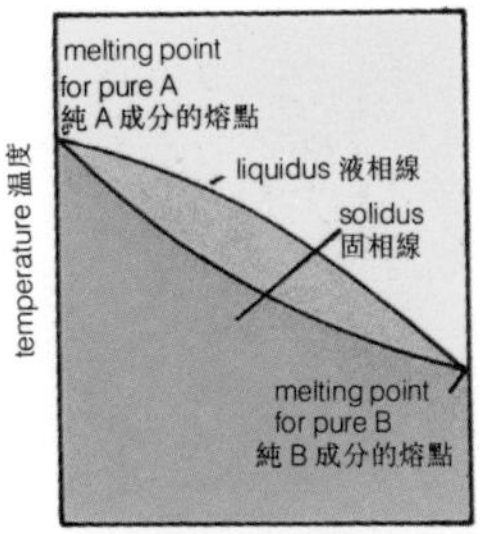

temperature – composition
溫度 - 組成圖

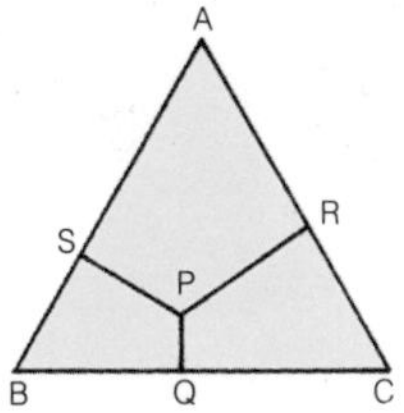

represents the percentage
omponent A present in
nposition P; PR the
centage of component B;
PS the percentage of
nponent C

PQ 代表 A 成分在
P 組成中的百分比；
PR 代表 B 成分在
該點的百分比；
PS 為 C 成分在
該點的百分比

triangular diagram for system on three components
三元體系的三角形圖解

temperature–composition diagram a figure drawn to show the relationship between the composition (p.15) of a mixture and temperature. Diagrams of this kind can be used to show the crystallization (p.40) of a magma (p.62) consisting of two components (p.19).

liquidus (*n*) a line joining points on a temperature–composition diagram (↑) of a two-component system at which the liquid phase (p.19) contains as much of a solid phase or component (p.19) as can dissolve (p.155) in it (i.e. it is *saturated*). If the temperature of the liquid falls, crystallization (p.40) will begin when the liquidus temperature is reached that corresponds to the composition of the melt.

solidus (*n*) a line on a temperature–composition diagram that joins points above which solid and liquid are in equilibrium (p.155) and below which only the solid phase exists.

eutectic point a point on a temperature – composition diagram at which two components crystallize together; the lowest temperature at which a mixture of given components (p.19) will melt provided that they do not form solid solutions (p.46).

incongruent melting a solid may melt to form another solid phase (p.19) and a liquid, neither of which has the same chemical composition (p.15) as the original solid. This is called *incongruent melting*. For example, orthoclase (p.56) melts to form leucite (p.58) and a liquid that contains more silica (p.16) than orthoclase.

exsolution (*n*) unmixing. The appearance of two mineral phases (p.19) in the solid state when a solid solution (p.46) is cooled slowly to a certain temperature (called the *exsolution temperature*). One mineral then separates out. **exsolved** (*adj*).

reaction series a group of minerals that are formed one after the other by chemical reaction (p.17) during the cooling of a magma (p.62).

triangular diagram a figure or drawing showing the chemical composition of a system (e.g. a magma) made up of three components (p.19). Each side represents a two-component system.

ternary diagram = triangular diagram (↑).

溫度 - 組成圖解 用於表示混合物成分(第 15 頁)與溫度之間的關係的一種圖形。這種圖解可用於表示二元(第 19 頁)岩漿(第 62 頁)的結晶作用(第 40 頁)。

液相曲綫(名) 二元體系溫度 - 組成圖解(↑)上連接各點的一條曲綫。此二元體系的液相(第 19 頁)中所溶解(第 155 頁)的固相或組分(第 19 頁)達最大限度，即體系是飽和的。在液體溫度下降到液相綫溫度時，開始出現結晶(第 40 頁)，而此溫度與熔體成分的溫度相一致。

固相曲綫 溫度 - 組成圖解上連接各點的一條曲綫，在此綫上方區域固體和液體處於平衡(第 155 頁)，在下方區域則只存在固相。

共晶點；低共熔點 溫度 - 組成圖解上兩種組分一起結晶的點；多組分(第 19 頁)混合物中各組分不形成固溶體(第 46 頁)時，此混合物開始熔化的最低溫度。

異元熔融；不一致熔融 一種固體熔化形成另一種固相(第 19 頁)和一種液體，當兩者的化學成分(第 15 頁)和原先的固體不一樣時，稱之為異元熔融。例如，正長石(第 56 頁)熔化形成白榴石(第 58 頁)和一種比正長石含更多二氧化矽(第 16 頁)的液體。

出溶作用(名) 不混溶。固溶體(第 46 頁)在緩慢冷却到一定溫度(稱為出溶溫度)時，該固態物質中出現兩種礦物相(第 19 頁)，因而分離出一種礦物的作用。(形容詞為 exsolved)

反應系列(名) 岩漿(第 62 頁)冷却期間，因化學反應(第 17 頁)而陸續形成的一組礦物。

三角形圖解 用以表示三元(第 19 頁)體系(例如一種岩漿)的化學成分的一種圖形。三角形的各邊分別代表一種二元體系。

三元相圖 同三角形圖解(↑)。

grain (*n*) (1) one of the mineral particles or individual crystals (p.40) that make up a rock; (2) the texture (fine or coarse) of a rock. **grained** (*adj*).

grain boundary the surface that separates two grains (↑) that are next to each other in a rock.

texture (*n*) the relationships between the grains of minerals that make up a rock. **textural** (*adj*).

phaneritic (*adj*) a rock texture (↑) with crystals that are large enough to be seen by the eye without using a lens.

aphanitic (*adj*) a rock texture (↑) in which the crystals are too small to be seen by the eye without using a lens.

phanerocrystalline (*adj*) a rock texture (↑) with crystals that can be seen by the eye without using a lens.

microcrystalline (*adj*) a rock texture (↑) with crystals that are so small that they can be seen only by using a microscope (p.147).

cryptocrystalline (*adj*) a rock texture (↑) with crystals (p.40) that are so small that they cannot be seen without a powerful microscope (p.147); a finer texture than microcrystalline (↑).

holocrystalline (*adj*) wholly made up of crystals, without any glassy (↓) material.

glassy (*adj*) without any crystals: like glass in appearance. *See also* **glass** (p.77).

hyaline (*adj*) = glass-like in appearance (↑).

crystalline texture (*adj*) a rock texture (↑) in which the mineral grains are firmly locked together.

decussate (*adj*) a rock texture (↑) with long narrow crystals arranged in an irregular way.

idiomorphic (*adj*) a rock texture (↑) in which the mineral grains (↑) show more or less complete crystal forms.

panidiomorphic (*adj*) a rock texture (↑) in which most of the mineral grains (↑) are idiomorphic (↑).

hypidiomorphic (*adj*) a rock texture (↑) in which the mineral grains (↑) show some crystal form.

allotriomorphic (*adj*) a rock texture in which the mineral grains (↑) do not show crystal form (p.40).

晶粒；顆粒(名)　(1)礦物顆粒中之一顆或組成岩石的各單個晶體(第 40 頁)；(2)岩石的結構(如細粒的或粗粒的)。(形容詞為 grained)

晶粒間界　岩石中分開兩顆相鄰晶粒(↑)的面。

結構(名)　構成岩石的各種礦物顆粒之間的關係。(形容詞為 textural)

顯晶的(形)　指一種岩石結構(↑)所具的晶體足夠大，不需用放大鏡，只憑肉眼便可看見。

非顯晶的；隱晶的(形)　指一種岩石結構(↑)的晶體太小，只憑肉眼不使用放大鏡便無法看見。

顯晶質的(形)　指一種岩石結構(↑)所具的晶體，不需用放大鏡，只憑肉眼便可見。

微晶的(形)　指一種岩石結構(↑)所具的晶體太小，要使用顯微鏡(第 147 頁)才能看見。

隱晶結構的(形)　指一種岩石結構(↑)所具的晶體(第 40 頁)太小，必須用高倍率顯微鏡(第 147 頁)才可見：一種比微晶(↑)更細小的結構。

全晶質的(形)　指整個岩石由晶體構成，沒有任何玻璃質(↓)物質。

玻璃質的(形)　無任何晶體，外表像玻璃。參見**火山玻璃**(第 77 頁)。

玻璃狀的(形)　即外表像玻璃(↑)。

晶質結構的(形)　指一種岩石結構(↑)中的礦物顆粒穩固地交接在一起。

交錯的(形)　指一種岩石結構(↑)具有不規則排列的窄長晶體。

自形的(形)　指一種岩石結構(↑)中的礦物顆粒(↑)顯現近似完整的晶形。

全自形的(形)　指一種岩石結構(↑)中的大部分礦物顆粒(↑)是自形的(↑)。

半自形的(形)　指一種岩石結構(↑)中的礦物顆粒(↑)顯現部分晶形。

他形的(形)　指一種岩石結構中的礦物顆粒(↑)不顯現晶形(第 40 頁)。

glassy texture (obsidian)
玻璃質結構(黑曜岩)

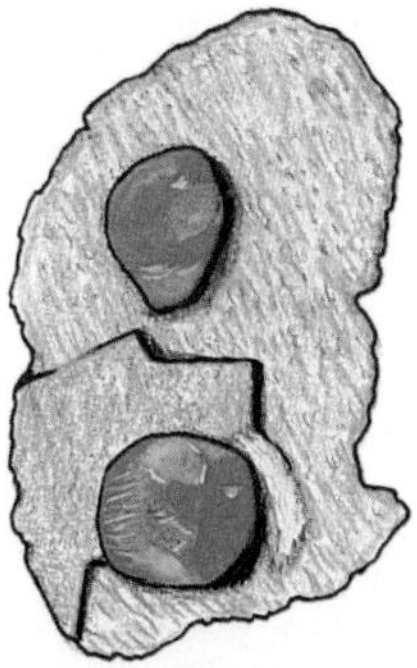

phenocryst 斑晶
garnet in feldspar 長石中的石榴石

porphyritic (*adj*) a rock texture (↑) with large crystals or phenocrysts (↓) in a fine-grained groundmass (↓).

phenocryst (*n*) a larger crystal in a groundmass (↓) of finer grain (↑). *See also* **porphyritic** (↑).

groundmass (*n*) the material of finer grain (↑) that forms the main part of an igneous rock with larger crystals.

matrix (*n*) material of finer grain (↑) between larger grains.

equigranular (*adj*) a rock texture (↑) in which all the mineral grains (↑) are of about the same size.

xenocryst (*n*) a crystal (p.40) that was not formed in the rock in which it is found.

spherulite (*n*) a small round mass of crystals formed when a volcanic (p.68) glass is devitrified (p.63). **spherulitic** (*adj*).

amygdaloidal (*adj*) a texture (↑) shown by lavas (p.70) in which holes, more or less round in shape, were formed when gas was given off by the magma. These holes are called *amygdales* or *amygdules*. They may later be filled with a mineral; some geologists use the word only for rocks in which the amygdales have been filled in this way. *See also* **scoriaceous** (↓).

vesicle (*n*) a more or less round hole in a lava (p.70) which has been formed by gas from the magma (p.62). **vesicular** (*adj*). *See also* **amygdaloidal** (↑).

scoriaceous (*adj*) a word used for an amygdaloidal (↑) texture (↑) in which the holes (amygdales) are empty.

ophitic (*adj*) a rock texture in which larger crystals of pyroxene (p.57) contain euhedral or subhedral (p.45) crystals of plagioclase (p.56). *See also* **poikilitic** (↓), **symplectic** (↓).

poikilitic (*adj*) a texture in which small crystals, facing in various directions, are inside a larger crystal. *See also* **ophitic** (↑), **symplectic** (↓).

intergrowth (*n*) a relationship in which crystals of different minerals are firmly locked together.

symplectic (*adj*) a rock texture in which there is an intergrowth (↑) of two different minerals, e.g. ophitic texture (↑), poikilitic texture (↑).

斑狀的(形)　指一種岩石結構(↑)的細粒基質(↓)中具有大的晶體或斑晶(↓)。

斑晶(名)　較細晶粒(↑)的基質(↓)中的較大晶體。參見**斑狀的**(↑)。

基質(名)　晶粒(↑)較細的物質，它構成一種含有較大晶體的火成岩的主要部分。

填質(名)　在較大顆粒之間的較細顆粒(↑)物質。

等粒狀(形)　指一種岩石結構(↑)中所具的全部礦物顆粒(↑)大小均大致相同。

捕擄晶(名)　見於某一岩石中但非形成於該岩石中的一種晶體(第 40 頁)。

球粒(名)　由火山(第 68 頁)玻璃脫玻(第 63 頁)所形成的晶體的小圓塊。(形容詞為 spherulitic)

杏仁狀的(形)　描述熔岩(第 70 頁)所呈現的一種結構(↑)，係由氣體從岩漿釋放時所形成的一些近似圓形的孔洞(稱為"杏仁孔"或"小杏仁孔")，而後可能為某種礦物所充填：某些地質學家規限該詞用於指杏仁孔已按此方式充填的岩石。參見**渣狀的**(↓)。

氣孔(名)　熔岩(第 70 頁)中近似圓形的孔洞，係由岩漿(第 62 頁)中的氣體所形成。(形容詞為 vesticular)。參見**杏仁狀的**(↑)。

渣狀的(形)　形容杏仁狀(↑)結構(↑)用的詞，此結構中的孔(杏仁孔)是空的。

輝綠結構的(形)　指岩石結構中較大的輝石(第 57 頁)晶體含有斜長石(第 56 頁)的自形或半自形(第 45 頁)晶體。參見**嵌晶狀的**(↓)、**後成合晶的**(↓)。

嵌晶狀的(形)　指結構中一個較大的晶體里面，有許多面向不同方向的小晶體。參見**輝綠結構的**(↑)、**後成合晶的**(↓)。

連晶(名)　不同礦物晶體牢固地交接在一起的一種關係。

後成合晶的(形)　指有兩種不同礦物的連晶(↑)的一種岩石結構。例如輝綠結構(↑)、嵌晶結構(↑)。

saturation (*n*) the degree to which a rock is saturated. In petrology (p.62) an *oversaturated* rock is one that contains free silica, SiO_2; that is, silica that is not chemically combined. The free silica may appear as quartz (p.55). An *undersaturated* rock contains only minerals that do not appear when free silica is present (such as olivine (p.58)). **saturated** (*adj*).

飽和(名) 岩石所飽和的程度。在岩石學(第 62 頁)上，"過飽和"岩石是指含有游離二氧化矽(SiO_2)的岩石，即所含二氧化矽未與其他物質化學結合的岩石。游離二氧化矽可成為石英(第 55 頁)出現。"不飽和"岩石所含礦物(如橄欖石(第 58 頁))，在有二氧化矽存在時絕不會出現。(形容詞為 saturated)

oversaturated (*adj*) *see* **saturation** (↑).

過飽和的(形) 參見**飽和**(↑)。

undersaturated (*adj*) *see* **saturation** (↑).

不飽和的(形) 參見**飽和**(↑)。

acid (*adj*) in petrology an acid igneous rock is one that has 10 per cent or more free quartz; granite and rhyolite (p.76) are examples. The use of the word 'acid' in this way has its origin in an earlier view of silicates as compounds of silica (SiO_2) with oxides of metals. Silica was then thought of as playing the part of an acid. *See also* **basic** (↓).

酸性的(形) 在岩石學上，酸性火成岩是指含 10% 或更多游離石英的岩石：花崗岩和流紋岩(第 76 頁)為其例。"acid"這個詞的這種用法起源於早年的觀點：將矽酸鹽視為金屬氧化物與二氧化矽(SiO_2)的化合物，視二氧化矽為起酸的作用。參見**基性的**(↓)。

basic (*adj*) in petrology a basic igneous rock is one that contains no quartz and has feldspars (p.56) with more calcium than sodium; basalt (p.77) and gabbro (p.76) are examples. Basic rocks contain 45 to 55 per cent silica. *See also* **acid** (↑).

基性的(形) 在岩石學上，基性火成岩是指長石(第 56 頁)不含石英而含鈣多於鈉的火成岩：例如玄武岩(第 77 頁)和輝長岩(第 76 頁)。基性岩含 45–55% 的二氧化矽。參見**酸性的**(↑)。

intermediate (*adj*) an intermediate igneous rock is one with less than 10 per cent quartz and with plagioclase feldspar containing 50–70% albite (p.55) or alkali feldspar (p.56), or both. *See also* **acid** (↑) *and* **basic** (↑).

中性的(形) 中性火成岩是指含石英少於 10% 並含 50–70% 鈉長石(第 55 頁)的斜長石或鹼性長石(第 56 頁)、或兩者均含的火成岩。參見**酸性的**(↑)和**基性的**(↑)。

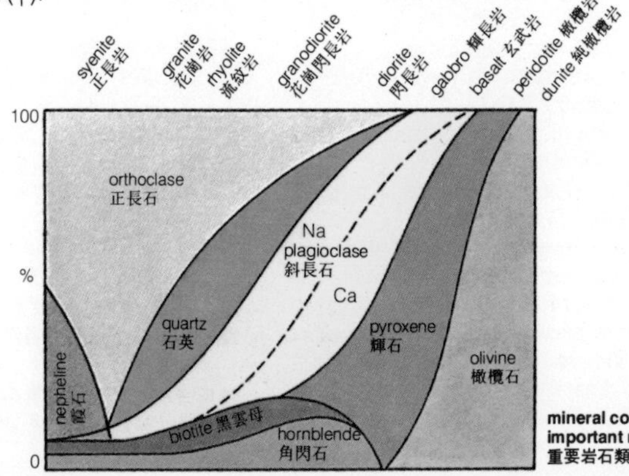

mineral composition of important rock types
重要岩石類型的礦物成分

ultrabasic, ultramafic (*adj*) ultrabasic or ultramafic igneous rocks are those composed essentially of ferromagnesian minerals (p.55); quartz, feldspars, and feldspathoids (pp.55, 56, 58) are absent.

超基性的，超鎂鐵質的(形)　超基性或超鎂鐵質火成岩基本上是由鎂鐵質礦物(第55頁)組成的火成岩：其中缺石英、長石及副長石(第55、56、58頁)。

ultramafic, *see* **ultrabasic** (↑).

超鎂鐵質的　見**超基性的**(↑)。

alkaline (*adj*) alkaline igneous rocks are those in which the feldspar (p.56) is mainly sodium-bearing or potassium-bearing. ('Alkaline' here does not mean the opposite of 'acid'.) The word is also used with the same meaning for minerals. *See also* **calc-alkaline** (↓) *and* **alkali** (p.16).

鹼性的(形)　鹼性火成岩是指岩中的長石(第56頁)主要含鈉或含鉀的火成岩(在此"鹼性的"這一詞並非"酸性的"反義詞)。該詞用於礦物方面也具有同一意思。參見**鈣鹼性**(↓)和**鹼性**(第16頁)。

calc-alkaline (*adj*) calc-alkaline igneous rocks are those in which the feldspar (p.56) is rich in calcium. The word is also used for minerals that are rich in calcium. *See also* **alkaline** (↑).

鈣鹼性(形)　鈣鹼性火成岩是指岩中的長石(第56頁)富含鈣的火成岩。該詞也可用於指富鈣的礦物。參見**鹼性的**(↑)。

essential mineral a mineral (p.44) that must be present in an igneous rock (p.62), if only in small amounts, for it to be given a particular name, e.g. quartz and alkali feldspar are essential minerals in granite. *See also* **accessory mineral** (↓).

主要礦物　某種火成岩(第62頁)中必有的(即使只含少量礦物(第44頁)，岩石將依據它命名。例如石英和鹼性長石是花崗岩中的主要礦物。參見**副礦物**(↓)。

accessory mineral a mineral (p.44) whose presence or absence does not affect the name given to an igneous rock (p.62). *See also* **essential mineral** (↑).

副礦物；附生礦物　一種存在與否俱不影響火成岩(第62頁)定名的礦物(第44頁)。參見**主要礦物**(↑)。

monomineralic (*adj*) consisting of only one mineral.

單礦物的(形)　只由一種礦物組成的。

mafic (*adj*) a word used for the ferromagnesian minerals (p.55) in igneous rocks. *See also* **felsic** (↓).

鎂鐵質的(形)　描述火成岩中鎂鐵礦物(第55頁)用的詞。參見**長英質的**(↓)。

felsic (*adj*) a word used for the light-coloured minerals in igneous rocks: feldspars, feldspathoids, quartz, etc. *See also* **mafic** (↑).

長英質的(形)　描述火成岩中的淡色礦物(如長石、副長石、石英等)用的詞。參見**鎂鐵質的**(↑)。

basalt, andesite, granite: basalt is melanocratic, granite is leucocratic
玄武岩、安山岩、花崗岩： 玄武岩是暗色的，花崗岩是淡色的

leucocratic (*adj*) describes igneous rocks consisting mainly of felsic (↑) minerals. *See also* **melanocratic** (↓).

淡色的(形)　描述主要含長英質(↑)礦物的火成岩。參見**暗色的**(↓)。

melanocratic (*adj*) describes igneous rocks consisting mainly of dark minerals. *See also* **leucocratic** (↑).

暗色的(形)　描述主要含深色礦物的火成岩。參見**淡色的**(↑)。

petrographic province an area in which a group of igneous rocks formed at about the same time are similar in petrographic (p.62) character and have a common origin, e.g. the British–Icelandic Tertiary igneous province.

岩區　大約同時形成、岩石(第62頁)特徵彼此相似並具有共同成因的一組火成岩賦存地區。例如英國、冰島之間的第三紀火成岩區。

granite (*n*) a coarse-grained acid plutonic (p.64) rock consisting essentially of quartz (p.55), alkali feldspar (p.56), and mica (p.55). The silica content is about 70 per cent. Granite is the most common intrusive rock. **granitic** (*adj*).

花崗岩（名） 主要由石英（第 55 頁）、鹼性長石（第 56 頁）和雲母（第 55 頁）組成粗粒酸性深成（第 64 頁）岩。二氧化矽含量約 70%。花崗岩是最普通的侵入岩。（形容詞為 granitic）

granite
花崗岩

granodiorite (*n*) a coarse-grained acid plutonic rock (p.64) containing quartz (p.55), plagioclase (p.56), orthoclase (p.56), and ferromagnesian minerals (p.55), chiefly biotite (p.55) and hornblende (p.57). **granodioritic** (*adj*).

花崗閃長岩（名） 一種粗粒酸性深成岩（第 64 頁），含石英（第 55 頁）、斜長石（第 56 頁）、正長石（第 56 頁）以及主要為黑雲母（第 55 頁）與普通角閃石（第 57 頁）的鎂鐵質礦物（第 55 頁）。（形容詞為 granodioritic）

syenite (*n*) a coarse-grained intermediate (p.74) igneous rock consisting essentially of alkali feldspar (p.56) with hornblende (p.57), biotite, or other mafic minerals. Most syenites are plutonic (p.64). **syenitic** (*adj*).

正長岩（名） 一種粗粒中性（第 74 頁）火成岩，主要由鹼性長石（第 56 頁）及普通角閃石（第 57 頁）、黑雲母或其他鎂鐵質礦物組成。正長岩大多數是深成（第 64 頁）岩。（形容詞為 syenitic）

diorite (*n*) a coarse-grained plutonic rock of intermediate composition (p.74) consisting essentially of intermediate plagioclase (oligoclase-andesine) (p.56) hornblende, and biotite or pyroxene as ferromagnesian minerals (p.55). A diorite may contain up to 10 per cent quartz (p.55) and some alkali feldspar (p.56). **dioritic** (*adj*).

閃長岩（名） 一種中性成分（第 74 頁）的粗粒深成岩，主要由中性斜長石（奧長石 - 中長石）（第 56 頁）以及鎂鐵質礦物（第 55 頁）如角閃石和黑雲母或輝石組成。閃長岩可含高達 10% 的石英（第 55 頁）和一些鹼性長石（第 56 頁）。（形容詞為 dioritic）

gabbro (*n*) a coarse-grained plutonic rock of basic composition (p.74) consisting of calcic plagioclase (labradorite–andesine) (p.56) and pyroxene (p.57). Many gabbros also contain olivine. **gabbroic** (*adj*).

輝長石（名） 一種基性成分（第 74 頁）的粗粒深成岩，由鈣質斜長石（拉長石 - 中長石）（第 56 頁）和輝石（第 57 頁）組成。許多輝長岩也含橄欖石。（形容詞為 gabbroic）

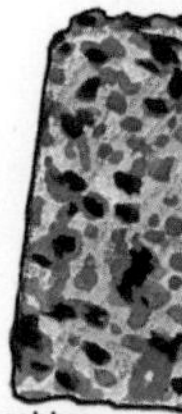
gabbro
輝長岩

rhyolite (*n*) an acid (p.74) volcanic rock, fine-grained or glassy (p.72), approximately equivalent to granite (↑). **rhyolitic** (*adj*).

流紋岩（名） 一種酸性（第 74 頁）的細粒或玻璃質（第 72 頁）火山岩，成分大致相當於花崗岩（↑）。（形容詞為 rhyolitic）

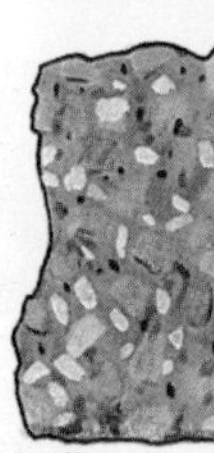
rhyolite
流紋岩

dacite (*n*) a fine-grained rock equivalent to a granodiorite (↑).

英安岩（名） 一種相當於花崗閃長岩（↑）的細粒岩石。

andesite (*n*) a fine-grained intermediate rock (p.74) composed essentially of plagioclase (oligoclase–andesine) (p.56) with biotite (p.55), hornblende (p.57), or pyroxene (p.57). Andesites are more or less equivalent in chemical and mineralogical composition to diorites (↑). Andesites occur as extrusive rocks (p.69) and as dykes and sills (pp.66, 67). **andesitic** (*adj*).

安山岩（名） 一種細粒的中性岩（第 74 頁），主要由斜長石（奧長石 - 中長石）（第 56 頁）、黑雲母（第 55 頁）、普通角閃石（第 57 頁）或輝石類（第 57 頁）組成。安山岩的化學成分和礦物成分大致相當於閃長岩（↑）。安山岩以噴出岩（第 69 頁）和以岩牆、岩床（第 66 及第 67 頁）形式產出。（形容詞為 andesitic）

microgranite (*n*) a medium-grained acid igneous rock (p.62) similar in chemical and mineral composition to a granite (↑).

微花崗岩（名） 一種中粒的酸性火成岩（第 62 頁）、化學和礦物成分類似花崗岩（↑）。

basalt (*n*) a fine-grained basic igneous rock (p.74) consisting essentially of calcic plagioclase (p.56) and pyroxene (p.57), usually augite, with or without olivine (p.58). Basalt is more or less equivalent to gabbro (↑). Basalts occur mainly as lavas (p.70). They may be porphyritic, vesicular, or amygdaloidal (p.73). **basaltic** (*adj*).

玄武岩（名） 一種細粒的基性火成岩（第 74 頁），主要由鈣質斜長石（第 56 頁）和輝石（第 57 頁）（通常為普通輝石）組成，可有或無橄欖石（第 58 頁）。玄武岩大致相當於輝長岩（↑）。玄武岩主要以熔岩（第 70 頁）形式產出。它們可以是斑狀的、多孔的或杏仁狀的（第 73 頁）。（形容詞為 basaltic）

plateau basalt a basalt (↑) that has been poured out in large quantities from a fissure eruption (p.68). Lava-flows of this kind form *plateaux* – wide areas of land at a high level.

高原玄武岩 由裂隙噴發中（第 68 頁）大量溢出來的玄武岩（↑）。這類熔岩流構成"高原"，即具高海拔的大片地域。

tholeiite (*n*) a basalt (↑) in which the plagioclase (p.56) is labradorite or bytownite and the pyroxene (p.57) is an Mg/Fe variety. **tholeiitic** (*adj*).

拉斑玄武岩（名） 所含斜長石（第 56 頁）為拉長石或倍長石，而輝石（第 57 頁）則為 Mg/Fe 變種的玄武岩（↑）。（形容詞為 tholeiitic）

pitchstone (*n*) a glassy acid igneous rock (p.74) with a dull lustre (p.47) like pitch (a black sticky material that is used to fill the cracks between the boards of boats). Pitchstone occurs as dykes and sills (p.66).

松脂岩；瀝青岩（名） 具瀝青（填補船板裂縫用的一種黑色黏性物質）狀，光澤（第 47 頁）暗淡的玻璃質酸性火成岩（第 74 頁）。松脂岩以岩牆和岩床（第 66 頁）形式產出。

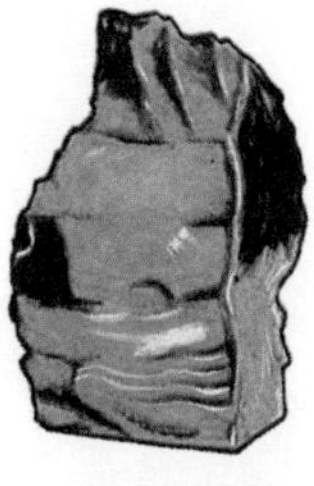

obsidian 黑曜岩

obsidian (*n*) a glassy volcanic rock with a composition equivalent to granite or rhyolite (↑). It is black in colour and has a conchoidal fracture (p.44).

黑曜岩（名） 成分相當於花崗岩或流紋岩（↑）的玻璃質火山岩，呈黑色並具貝殼狀斷口（第 44 頁）。

felsite (*n*) a fine-grained acid igneous rock (p.74), light in colour. Quartz and feldspar (p.56) are the chief minerals. Felsites occur as dykes and veins (pp.67, 145). **felsitic** (*adj*).

霏細岩；長英岩（名） 淺色、細粒的酸性火成岩（第 74 頁）。石英和長石（第 56 頁）是主要礦物。霏細岩呈岩牆和岩脈（第 67、145 頁）產出。（形容詞為 felsitic）

trachyte (*n*) a fine-grained intermediate rock (p.74) composed of alkali feldspar (p.56) and ferromagnesian minerals (p.55). **trachytic** (*adj*).

粗面岩（名） 由鹼性長石（第 56 頁）和鎂鐵質礦物（第 55 頁）組成的細粒中性岩石（第 74 頁）。（形容詞為 trachytic）

phonolite (*n*) an undersaturated (p.74) trachyte (↑) containing the feldspathoid nepheline (p.58), alkali feldspar (usually sanidine) (p.56), and a ferromagnesian mineral (p.55). Phonolites are the fine-grained equivalents of nepheline-syenites (↑). **phonolitic** (*adj*).

響岩（名） 含副長石類霞石（第 58 頁）、鹼性長石（通常為透長石）（第 56 頁）和鎂鐵礦物（第 55 頁）的不飽和（第 74 頁）粗面岩（↑）。響岩相當於細粒的霞石正長岩（↑）。（形容詞為 phonolitic）

MORB short for mid-ocean ridge basalt; a basalt (↑) erupted at a mid-oceanic ridge (p.35).

MORB 洋中脊玄武岩的英文縮寫：在洋中脊（第 35 頁）噴發出來的玄武岩（↑）。

glass (*n*) an amorphous (p.44) rock that contains no crystalline (p.44) material, e.g. obsidian (↑). Glasses are formed when a magma (p.62) is cooled very rapidly.

火山玻璃（名） 不含結晶（第 44 頁）物質的一種非結晶質（第 44 頁）岩石。例如黑曜岩（↑）。火山玻璃是岩漿（第 62 頁）極快速冷却時形成的。

dolerite (*n*) a medium-grained hypabyssal rock (p.64) with calcic plagioclase (p.56) and pyroxene (p.57) as essential minerals. Mineralogically equivalent to gabbro (p.76) and basalt (p.77). Dolerites are very common as minor intrusions. Ophitic texture (p.73) is common. **doleritic** (*adj*).

diabase (*n*) a word used by American geologists for dolerite (↑).

monzonite (*n*) a coarse-grained igneous rock between syenite (p.76) and gabbro (p.76), with more or less equal amounts of potassium feldspar (p.56) and plagioclase (oligoclase–andesine) (p.56). Pyroxene, biotite, or hornblende (p.57) may also be present. Monzonites range from acid to basic (p.74) in composition. **monzonitic** (*adj*).

syenodiorite (*n*) = monzonite (↑).

lamprophyre (*n*) a dark-coloured basic igneous rock with phenocrysts (p.73) of dark mica, augite, hornblende, or olivine (p.58) in a groundmass (p.73) of feldspar, usually alkali feldspar (p.56). Lamprophyres occur typically as dykes (p.67). **lamprophyric** (*adj*).

porphyry (*n*) a word used, generally with the name of a mineral, to describe rocks with phenocrysts (p.73) in a fine-grained groundmass, i.e. rocks with a porphyritic texture (p.73). Porphyries are usually hypabyssal rocks (p.64).

ophiolite assemblage, ophiolite complex the occurrence together of deep-sea sediments (p.80), basaltic pillow lavas (p.70), basaltic dykes (p.67), gabbros (p.76), and peridotites (↓). These assemblages may be fragments of oceanic crust (p.9). The word 'ophiolite' was first used for basic and ultrabasic rocks (pp.74, 75) made up largely of the mineral serpentine (p.61), with garnet (p.58) and other minerals, occurring as lavas (p.70) and minor intrusions (p.67).

spilite (*n*) an altered basalt (p.77) in which albite replaces plagioclase feldspar (p.56), and chlorite (p.61) replaces augite and olivine (p.58). Spilites occur as pillow-lavas (p.70) with geosynclinal (p.132) sediments. **spilitic** (*adj*).

粗玄岩（名） 主要礦物為鈣質斜長石（第 56 頁）和輝石（第 57 頁）的中粒淺成岩（第 64 頁）。礦物學上相當於輝長岩（第 76 頁）和玄武岩（第 77 頁）。粗玄岩常呈小的侵入體。一般為輝綠結構（第 73 頁）。（形容詞為 doleritic）

dolerite 粒玄岩

輝綠岩（名） 美國地質學家稱粗玄岩（↑）的用語。

二長岩（名） 介於正長岩（第 76 頁）和輝長岩（第 76 頁）之間的一種粗粒火成岩，所含的鉀長石（第 56 頁）和斜長石（奧長石 - 中長石）（第 56 頁）量大致相等。也可能含有輝石類、雲母或普通角閃石（第 57 頁）。二長岩的成分在酸性與基性（第 74 頁）之間變化。（形容詞為 monzonitic）

正長閃長岩（名） 同二長岩（↑）。

煌斑岩（名） 一種暗色的基性火成岩，在長石基質（第 73 頁），（通常為鹼性長石（第 56 頁））中含暗色雲母、普通輝石、普通角閃石或橄欖石（第 58 頁）的斑晶（第 73 頁）。煌斑岩一般呈岩牆（第 67 頁）產出。（形容詞為 lamprophyric）

斑岩（名） 描述細粒基質中含斑晶（第 73 頁）岩石用的一個詞（一般連同礦物名稱使用），即具斑狀結構（第 73 頁）的岩石。斑岩通常為淺成岩（第 64 頁）。

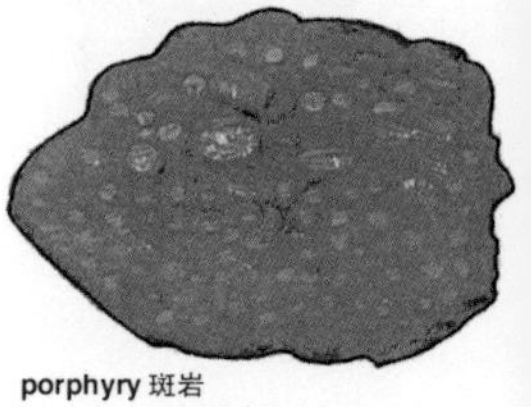

porphyry 斑岩

蛇綠岩，蛇綠岩組合 指深海沉積物（第 80 頁）、玄武岩質枕狀熔岩（第 70 頁）、玄武岩質岩牆（第 67 頁）、輝長岩（第 76 頁）和橄欖岩（↓）相伴產出。這些組合可能就是大洋地殼（第 9 頁）的各種碎塊。"蛇綠岩"這個詞最初是使用於那些主要是由礦物蛇紋石（第 61 頁）伴隨石榴石（第 58 頁）和其他礦物構成、呈熔岩（第 70 頁）和小侵入體（第 67 頁）產出的基性和超基性岩（第 74、75 頁）。

細碧岩（名） 一種蝕變的玄武岩（第 77 頁），其中鈉長石交代斜長石（第 56 頁），綠泥石（第 61 頁）交代普通輝石和橄欖石（第 58 頁）。細碧岩呈枕狀熔岩（第 70 頁）與地槽（第 132 頁）沉積物一起產出。（形容詞為 splitic）

peridotite (*n*) a dark-coloured coarse-grained ultrabasic (p.75) igneous rock consisting largely of olivine (p.58), with or without pyroxene or hornblende (p.57). Quartz and feldspar (p.56) are absent. **peridotitic** (*adj*).

dunite (*n*) an ultrabasic igneous rock composed almost entirely of olivine (p.58).

pyroxenite (*n*) a dark-coloured coarse-grained ultrabasic (p.75) igneous rock consisting essentially of one or more pyroxenes (p.57).

anorthosite (*n*) a coarse-grained plutonic rock consisting largely of plagioclase feldspar (andesine to labradorite) (p.56); the other minerals present are those found in gabbros (p.76). **anorthositic** (*adj*).

hornblendite (*n*) a coarse-grained igneous rock consisting largely of hornblende (p.57).

perknite (*n*) the perknites are ultrabasic rocks (p.75) consisting largely of ferromagnesian minerals (p.55) with the exception of olivine (p.58). They include pyroxenites (↑) and hornblendites (↑).

pegmatite (*n*) an igneous rock of coarse or very coarse grain. The crystals in a pegmatite may be up to a metre or even more in length. Granite-pegmatites, which are usually simply called pegmatites, consist of quartz and feldspar (p.56). Coarse-grained varieties of other plutonic rocks are called by the name of the equivalent rock, e.g. gabbro-pegmatite. Pegmatites occur as dykes (p.67) and veins (p.145). They are formed towards the end of the crystallization of a magma. **pegmatitic** (*adj*).

橄欖岩（名） 主要由橄欖石（第 58 頁）組成，可具或不具輝石或普通角閃石（第 57 頁）的暗色粗粒超基性（第 75 頁）火成岩。不含石英和長石（第 56 頁）。（形容詞為 peridotitic）

純橄欖岩（名） 幾乎全由橄欖石（第 58 頁）組成的超基性火成岩。

輝石岩（名） 主要由一種或多種輝石（第 57 頁）組成的暗色粗粒超基性（第 75 頁）火成岩。

斜長岩（名） 主要由斜長石（中長石乃至拉長石）（第 56 頁）組成的粗粒深成岩；其餘都是那些見於輝長岩（第 76 頁）中的礦物。（形容詞為 anorthositic）

角閃石岩（名） 主要由普通角閃石（第 57 頁）組成的粗粒火成岩。

輝閃岩類（名） 輝閃岩類是一些超基性岩（第 75 頁），主要由不包括橄欖石（第 58 頁）在內的鎂鐵質礦物（第 55 頁）組成，包括輝石岩類（↑）和角閃石岩類（↑）。

偉晶岩（名） 一種粗粒或極粗粒的火成岩。偉晶岩中的晶體，可長達一米或更長。由石英和長石（第 56 頁）組成的花崗偉晶岩通常簡稱為偉晶岩。其他一些深成岩的粗粒品種都是以相應的岩石命名的。例如輝長偉晶岩。偉晶岩呈岩牆（第 67 頁）和岩脈（第 145 頁）產出。它們是岩漿結晶趨近末尾時形成的。（形容詞為 pegmatitic）

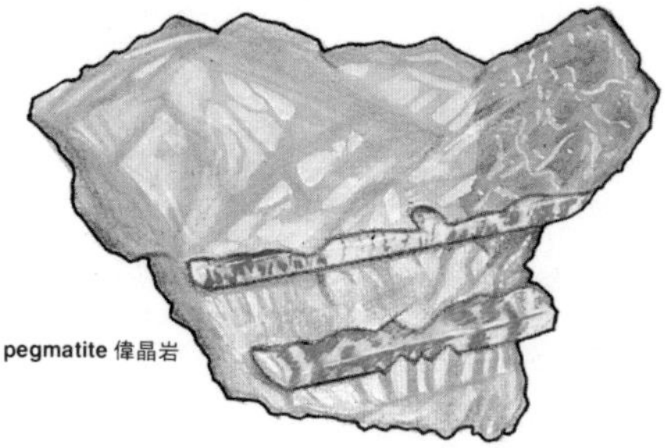

sediment (*n*) material that has been deposited (↓) in water (e.g. on the sea floor or on the bed of a lake), having settled after being in suspension (p.24); any material that has been obtained from earlier rocks by denudation (p.32). In the wider sense, sediment includes material deposited by ice and the wind or chemically precipitated (p.18) in water, together with material from plants and animals. **sedimentary** (*adj*).

沉積物（名） 水域（例如海床或湖底）中經懸移（第 24 頁）後沉降下的沉積（↓）物質；因剝蝕作用（第 32 頁）而從較老岩石獲得的任何物質。廣義而言，沉積物包括冰成和風成沉積物質或水中化學沉澱的（第 18 頁）物質以及來自植物和動物的物質。（形容詞為 sedimentary）

sedimentation (*n*) the process of forming sediments (↑).

沉積物形成作用（名） 形成沉積物（↑）的過程。

sedimentology (*n*) the study of sediments (↑) and their formation.

沉積學（名） 研究沉積物（↑）及其形成的學科。

deposit (*n*) anything that is laid down; a sediment (↑); minerals precipitated (p.18) from solution (p.159) in veins (p.145) and ore bodies (p.145). **deposited** (*adj*), **deposit** (*v*).

沉積；礦床（名） 沉澱下的任何東西；沉積物（↑）；礦脈（第 145 頁）或礦體（第 145 頁）中從溶液（第 159 頁）中沉澱（第 18 頁）下的礦物。（形容詞為 deposited，動詞為 deposit）

deposition (*n*) the laying down of material that may later form a rock; sedimentation (↑); the precipitation (p.18) of minerals from solution (p.159).

沉積作用（名） 指過後可形成岩石的物質的沉澱；沉積物形成作用（↑）；礦物自溶液（第 159 頁）中沉澱（第 18 頁）。

bed (*n*) a layer of sedimentary (↑) rock that is marked off above and below by surfaces that can be seen and is made up of material that is the same in all parts. **bedding** (*n*).

層（名） 沉積（↑）岩的一層，其組成物質處處相同，其上、下為可見的面所劃分。（名詞 bedding 意為層理）

stratum (*n*) (*strata*) = bed (↑).

地層（名） 同層（↑）。（複數為 strata）

bedding-plane (*n*) a surface in a sedimentary (↑) rock that is parallel to the original surface on which the sediment was deposited.

層面（名） 沉積（↑）岩中和原始沉積面平行的一個面。

lamina (*n*) (*laminae*) a thin layer (less than 10 mm thick) in a sedimentary rock that can be distinguished from the material above and below it. **laminated** (*adj*).

紋層（名） 沉積岩中與其上、下的物質有異的一個薄層（厚度小於 10 mm）。（複數為 laminae，形容詞為 laminated）

lamination (*n*) the presence or formation of laminae (↑).

紋理（名） 紋層（↑）的出現或紋層的形成。

stratification (*n*) the presence of layers or beds (↑) in a sedimentary rock (↑).

層理（名） 沉積岩（↑）中存在的各個層（↑）。

stratified (*adj*) formed of layers or beds (↑).

成層的；分層的（形） 岩層（↑）的形成。

unstratified (*adj*) not stratified (↑).

不成層的（形） 非成層的（↑）。

fissile (*adj*) easily divided up along closely spaced parallel surfaces.

易剝裂的（形） 易於沿一些密集平行面分開的。

sorting (*n*) the degree to which the particles that make up a sediment (↑) or other material are alike in some respect, e.g. their size or shape. **sorted** (*adj*).

分選（名） 沉積物（↑）中所含顆粒或其他物質在某些方面（例如其大小和形狀）的相似程度。（形容詞為 sorted）

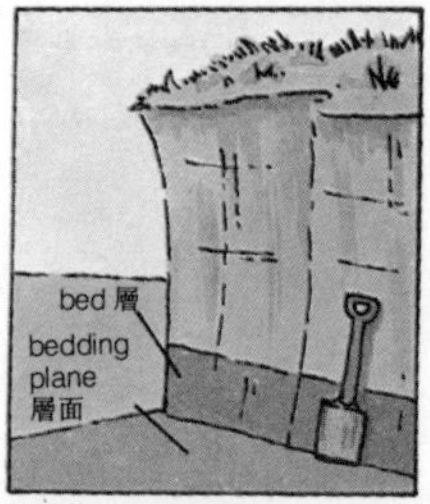

bedding plane 層面

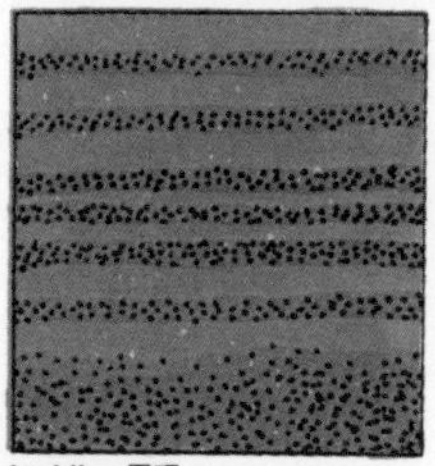
bedding 層理

environments of the seas and oceans
海洋的環境

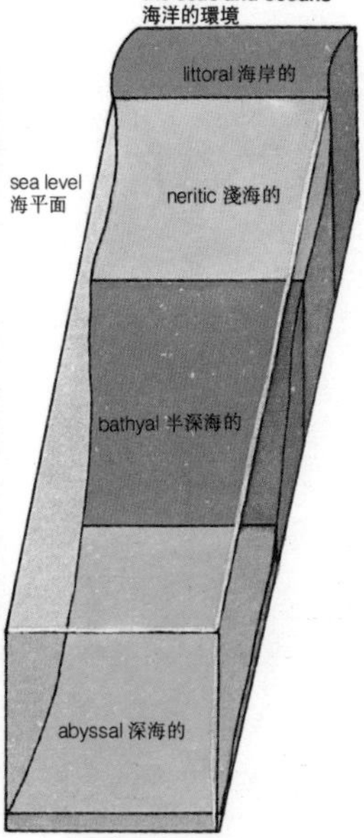

sedimentary environment the conditions under which a sediment (↑) is deposited (↑); e.g. the depth and temperature of the water, the strength and direction of the currents. These can vary widely, and they affect the texture, composition, and structure of the sediments that are formed.

沉降環境 沉積物(↑)沉積(↑)的條件，例如水的深度和溫度、水流的強度和方向。這些條件變化範圍寬廣，影響所形成的沉積物的結構、成分和構造。

depositional environment = sedimentary environment (↑).

沉積環境 同沉降環境(↑)。

marine environments these include the *littoral* (p.37) *environment*, the *neritic* (p.34) *environment*, the *bathyal* (p.34) *environment*, and the *abyssal* (p.34) *environment*.

海洋環境 包括濱岸(第 37 頁)環境、淺海(第 34 頁)環境、半深海(第 34 頁)環境和深海(第 34 頁)環境。

paralic (*adj*) relating to sediments deposited on the coast in shallow water. Marine (p.34) and non-marine sediments may both be formed here. *See also* **limnic** (↓).

近海的(形) 與沉積在淺水海岸的沉積物有關的。海成(第 34 頁)和非海成沉積物均可在此處形成。參見**湖沼的**(↓)。

terrestrial (*adj*) of the land, formed on the land; not marine (p.34).

陸地的(形) 屬於陸地、在陸地上形成的；不是海成(第 34 頁)的。

limnic (*adj*) relating to sediments deposited in fresh-water lakes. *See also* **lacustrine** (↓).

湖沼的(形) 與沉積在淡水湖中的沉積物有關的。參見**湖泊的**(↓)。

lacustrine (*adj*) of, or formed in or by, a lake. *See also* **limnic** (↑).

湖泊的(形) 屬於湖的，或在湖中或湖岸形成的。參見**湖沼的**(↑)。

paludal (*adj*) of low wet land (swamp, marsh).

沼澤的(形) 屬於低濕陸地的(沼澤、草沼)。

lagoonal (*adj*) relating to areas of shallow water on the coast, usually between a barrier island (p.38) and the mainland.

瀉湖的(形) 與海岸的淺水區有關的，通常介於堤島(第 38 頁)和大陸之間。

fluvial (*adj*) relating to rivers.

河成的(形) 與河流有關的。

estuarine (*adj*) relating to the part of a river nearest to the sea, where tides, i.e. the regular rise and fall of the sea, occur and where fresh and salt water mix. The sediments deposited in estuaries are fine silts, clays, and muds.

河口灣的(形) 與河流最靠近海部分有關的，該處有潮汐(即海水的規則漲落)出現，淡水與鹹水混合。沉積在一些河口灣中的沉積物是一些細小的粉砂、黏土和泥。

deltaic (*adj*) relating to a delta (p.26). Sediments laid down in a delta typically show cross bedding (p.82).

三角洲的(形) 與三角洲(第 26 頁)有關的。在三角洲內的沉積物一般都顯示交錯層理(第 82 頁)。

anaerobic (*adj*) without oxygen. Sediments formed under anaerobic conditions are typically black muds rich in organic (p.17) material and sulphides (p.16).

厭氧的(形) 形容無氧的。在厭氧條件下形成的沉積物一般都是富含有機(第 17 頁)質和硫化物(第 16 頁)的黑泥。

allothigenous, allothigenic (*adj*) refers to the part of a sediment (↑) that has been transported (p.21) from some other place to the area in which it was deposited (↑).

他生的(形) 指從別處搬運來(第 21 頁)並沉積(↑)於本地區的那一部分沉積物(↑)。

allogenic (*adj*) = allothigenous (↑).

外源的(形) 同他生的(↑)。

graded bedding bedding (p.80) in which the largest particles are at the bottom of a unit and the smallest particles are at the top.
cross bedding original bedding (p.80) in which the bedding-planes (p.80) are at an angle to the main surface on which the sediments were deposited (p.80). **cross bedded** (*adj*).
current bedding = cross bedding (↑).

遞變層理；粒級層理 最大顆粒在沉積單元底部，最小的顆粒在頂部的層理(第 80 頁)。

交錯層理 一些層理面(第 80 頁)與沉積物的主要沉積(第 80 頁)界面呈某一交角的原生層理(第 80 頁)。(形容詞為 cross bedded)

水流層理；波狀層理 同交錯層理(↑)。

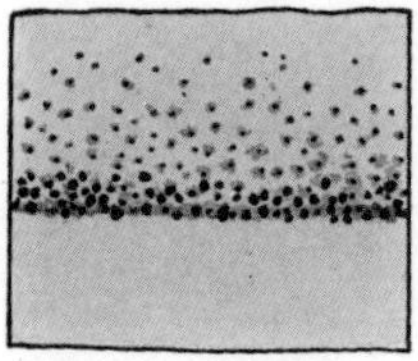

graded bedding 遞變層理

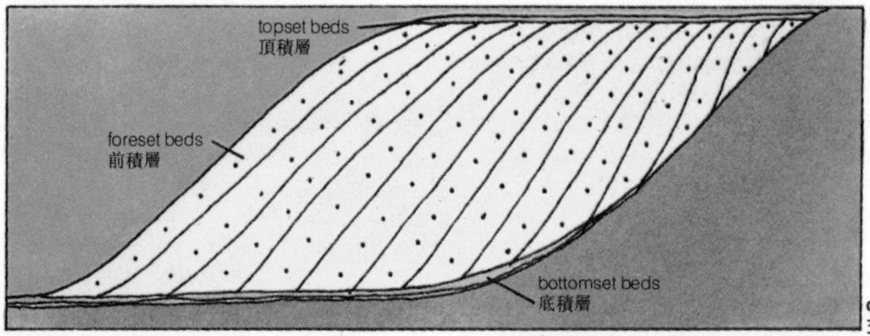

cross-(current-)-bedding
交錯（水流）層理

dune bedding cross bedding (↑) of relatively large size. **dune-bedded** (*adj*).
topset beds horizontal strata (p.80) at the top of sediments deposited (p.80) in a delta (p.26).
bottomset beds horizontal strata (p.80) at the bottom of sediments deposited (p.80) in a delta (p.26).
foreset beds steeply sloping beds (p.80) in sediments deposited (p.80) in a delta (p.26).
washout (*n*) a gap in a bed (p.80) that has been filled with later sediments (p.80). Formed when a stream flows across sediments soon after they have been deposited. The sediments in the course of the stream will then be removed and the space will later be filled by other material.
scour-and-fill the process of cutting a channel in a sediment and filling it in again. *See also* **washout** (↑).
convolute bedding a structure in which the laminae (p.80) of a sediment are bent into folds, which are cut off by the beds above.
palaeocurrent (*n*) a flow of water that took place at some time in the geological past while a sediment was being deposited (p.80).

沙丘層理 比較大型的交錯層理(↑)。(形容詞為 dune-bedded)

頂積層 三角洲(第 26 頁)沉積物(第 80 頁)頂部的水平岩層(第 80 頁)。

底積層 三角洲(第 26 頁)沉積物(第 80 頁)底部的水平岩層(第 80 頁)。

前積層 三角洲(第 26 頁)沉積物(第 80 頁)內部的陡斜層(第 80 頁)。

沖蝕溝(名) 岩層(第 80 頁)中為後來的沉積物(第 80 頁)所填充的缺口。是沉積物沉積之後不久即被水流流過而形成的。河道中的沉積物於是被搬走，而後其空間又為別種物質填充。

沖淤作用 在沉積物中沖刷出一條溝又再把它填充的過程。參見**沖蝕溝**(↑)。

旋捲層理 沉積物的紋層(第 80 頁)撓曲成褶皺並被上覆岩層切斷的一種構造。

古水流(名) 發生在過去某一地質時期的一種水流，其時某種沉積物正在沉積(第 80 頁)。

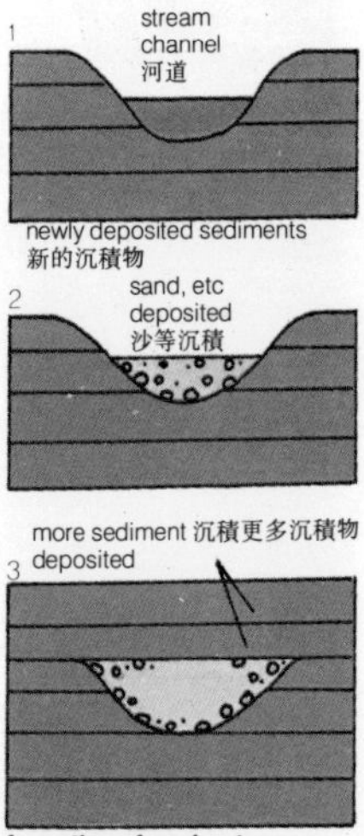

formation of washout
沖蝕溝的形成

angular (*adj*) with sharp edges and corners; showing little sign of wear.

稜角狀的（形） 具尖銳稜和角的；很少磨損痕跡的。

subangular (*adj*) with edges and corners rounded off to some degree; showing signs of wear. *See also* **angular** (↑).

近稜角狀的（形） 具磨圓到一定程度的稜和角的；顯示磨損痕跡的。參見**稜角狀的**（↑）。

subrounded (*adj*) with edges and corners rounded to smooth curves but with the original shape of the grain still to be seen.

近圓形的（形） 指具磨圓成平滑曲線的稜和角，但仍能看出顆粒的原有形狀。

rounded (*adj*) with edges and corners rounded to smooth curves.

圓形的（形） 指具磨圓成平滑曲綫的稜和角的。

well-rounded (*adj*) worn to a completely smooth curved shape without any sign of the original shape of the particle.

渾圓的（形） 指磨成完全平滑曲綫形狀，無任何顆粒原有形狀的痕跡。

sedimentary structures structures formed while a sediment (p.80) was being deposited (p.80) or very soon after it was deposited.

沉積構造 沉積物（第 80 頁）沉積（第 80 頁）時或沉積後不久所形成的構造。

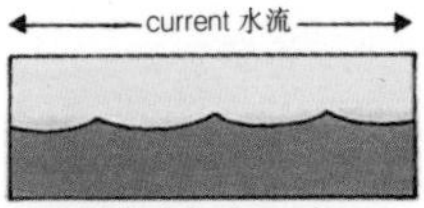

symmetrical ripple mark
對稱波痕

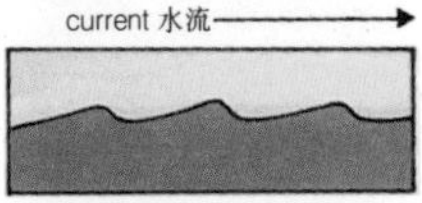

asymmetrical ripple mark
不對稱波痕

ripple mark wave-like marks formed by the movement of water or of the air over the surface of a newly deposited sediment. The two main types of ripple mark are *oscillation ripples*, which are symmetrical, and *current ripples*, which are not symmetrical.

波痕 水或空氣在新近沉積的沉積物表面上運動所形成的波狀痕跡。波痕的兩種主要類型是"擺動波痕"，即對稱的波痕；"流水波痕"，即不對稱的波痕。

swash marks patterns made in the sand of a beach by the movement of water after a wave has broken. These patterns are usually curved.

沖痕 波浪沖擊形成的水流運移，而在海灘沙中造成的圖形。這些圖形通常是弧形的。

crescent marks marks formed by the washing away by water of the sediment round a pebble, a shell, or other object on the beach as the sea flows back after each wave.

新月形痕 每次海浪沖擊後海水回流而在海灘上環繞礫石、貝殼或其他物體沖刷所形成的痕迹。

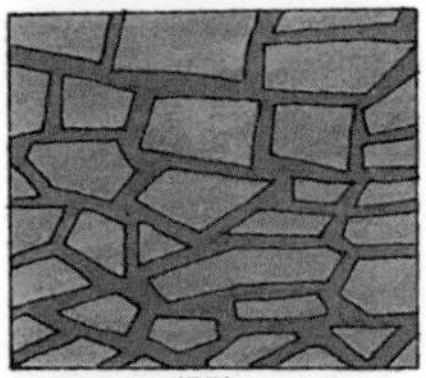

mud cracks 泥裂

rill marks branching patterns of small channels formed by water running down a beach.

流痕 由流下海灘的水所形成的一些細小河道的分枝圖樣。

mud cracks the cracks produced when wet mud dries in the air. A pattern of many-sided cracks results, typical of dried up shallow lakes.

泥裂 濕泥在空氣中乾涸所產生的多邊形裂隙圖樣，在乾涸的淺水湖中很典型。

flow cast a swelling formed when sediment is deposited on a soft material that can flow under pressure.

流動底模 沉積物沉積在一種能在壓力下流動的軟性物質下面所形成的突起印模。

load cast = flow cast (↑).

負荷印模 同流動底模（↑）。

sole marks
底痕

sole marks, sole markings marks on what was originally the under surface of a bed (p.80). The bed must be different lithologically (p.85) from the bed below it for sole marks to be seen. Types of sole mark include various marks produced by animals and objects moving on the surface on which the sediment was deposited (p.80).

底痕，底面印痕作用 一個層（第 80 頁）的原來底面上的痕跡。這個層必定在岩性上（第 85 頁）與其下伏層不同，這樣才能使底痕可見。底痕類型包括各種各樣痕跡，它們是由在沉積（第 80 頁）面上活動的動物及物體產生的。

compaction (*n*) the process in which a sediment is reduced in volume by pressure, e.g. by the weight of material above it. **compact** (*v*), **compacted** (*adj*).

壓實(名)　由於壓力(如上覆物質的重量)，使沉積物體積減少的過程。(動詞為 compact，形容詞為 compacted)

consolidation (*n*) the process of forming a solid mass from loose or liquid material, e.g. the formation of a firm rock from loose sediment. **consolidate** (*v*), **consolidated** (*adj*).

固結(名)　從鬆散物質或液體物質形成固體的過程。例如從疏鬆沉積物形成堅固岩石。(動詞為 consolidate，形容詞為 consolidated)

unconsolidated (*adj*) not consolidated (↑); loose.

未固結的(形)　非固結的(↑)；疏鬆的。

lithification (*n*) the processes by which unconsolidated sediments become consolidated (↑) into rocks. **lithify** (*v*), **lithified** (*adj*).

石化作用(名)　未固結沉積物轉變成固結(↑)岩石的過程。(動詞為 lithify，形容詞為 lithified)

lithifaction = lithification (↑).

石化作用(↑)的另一英文拼寫為 lithifaction。

induration (*n*) the process of making hard, e.g. by heat or by pressure. **indurate** (*v*), **indurated** (*adj*).

硬化(名)　變硬的過程。例如因受熱或壓力而變硬。(動詞為 indurate，形容詞為 indurated)

cement (*n*) material between the particles of a sedimentary rock that holds them together.

膠結物(名)　在沉積岩顆粒之間將顆粒保持在一起的物質。

cementation (*n*) the deposition of cement (↑) between the particles of a sediment to form a solid rock.

膠結作用(名)　膠結物(↑)在沉積物顆粒之間形成固體岩石的沉積作用。

impregnation (*n*) the filling-in of the pore spaces (↓) in a rock by minerals or the replacement (p.159) of pore material. Impregnation generally relates to an event that takes place after cementation (↑). **impregnate** (*v*), **impregnated** (*adj*).

浸染(名)　礦物在岩石孔隙(↓)內的填充或孔隙物質的交代(第 159 頁)。侵染一般與膠結作用(↑)後發生的地質事件有關。(動詞為 impregnate，形容詞為 impregnated)

pore space the space between the particles that make up a rock.

孔隙　岩石中顆粒之間的空間。

porosity (*n*) the proportion of empty space in a rock.

孔隙度(名)　岩石中空隙所佔的比例。

porous (*adj*) containing pore spaces (↑).

多孔隙的(形)　含孔隙(↑)的。

diagenesis (*n*) the changes in mineral composition and texture that take place in a sediment after it has been deposited, except those that occur at great depth. **diagenetic** (*adj*).

成岩作用(名)　沉積物沉積後，其礦物成分和結構所發生的變化，發生在深處的那些變化則除外。(形容詞為 diagenetic)

authigenic (*adj*) minerals (p.44) formed during diagenesis (↑) are called *authigenic*.

自生的(形)　指礦物(第 44 頁)在成岩作用(↑)期間形成的。

concretion (*n*) a mass of round or irregular shape in a sediment, formed either during deposition (p.80) or during diagenesis (↑).

固結作用(名)　沉積(第 80 頁)期間或成岩作用(↑)期間在沉積物中形成圓形或不規則形狀的塊體。

nodule (*n*) a concretion (↑) of rounded shape.

結核(名)　圓形的固結作用(↑)。

Neptunean dyke a sheet of sediment (p.80) filling a crack in another rock.

水成岩牆　填充於另一岩石裂隙中的薄板狀沉積物(第 80 頁)。

sandstone dyke a type of Neptunean dyke (↑).

砂岩岩牆　水成岩牆(↑)的一種類型。

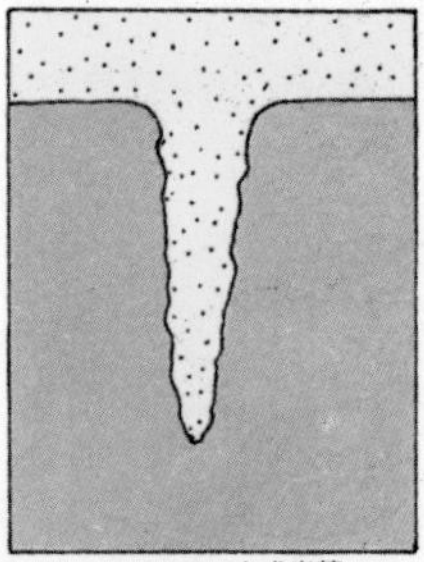

Neptunean dyke 水成岩牆

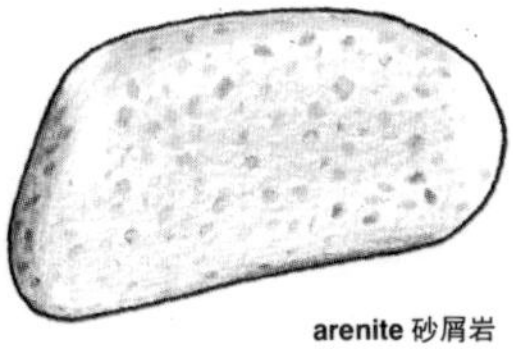

arenite 砂屑岩

lithology (*n*) the general character of a rock, or a sedimentary formation, more particularly as seen in exposures (p.122) and hand specimens (p.147). **lithological** (*adj*).

rudite (*n*) a sedimentary rock with an average grain size greater than 2 mm, e.g. conglomerates and breccias (p.87). **rudaceous** (*adj*).

psephite (*n*) = rudite (↑). **psephitic** (*adj*).

arenite (*n*) a sedimentary rock with a grain size from 1/16 to 2 mm; a sandstone. **arenaceous** (*adj*). *See also* **sandstone**, (p.87).

lutite (*n*) a sedimentary rock of any composition with particles between 1/256 and 1/16 mm in diameter; a rock composed of mud. **lutaceous** (*adj*).

argillite (*n*) a sedimentary rock with an average grain size less than 1/16 mm; clays, silts, mudstones, etc.; a hard mudstone. **argillaceous** (*adj*). *See also* **clay** (p.88).

clast (*n*) a piece of a sedimentary (p.80) rock formed by the breaking-up of a larger mass.

clastic (*adj*) describes sediments made up of fragments produced by the breaking-up of earlier rocks.

detrital (*adj*) = clastic (↑).

organic (*adj*) as applied to sediments, 'organic' refers to the remains of plants and animals; e.g. coral reefs and coal are organic sediments.

evaporite (*n*) a sediment formed by the *evaporation* (i.e. the drying up) of a body of water containing a chemical compound in solution, e.g. salt deposits, gypsum.

terrigenous (*adj*) applied to sediments deposited on the sea bed that contain material that has come from the land.

orthochemical (*adj*) refers to materials in sedimentary rocks that have been chemically precipitated (p.18) within the area in which the rock was deposited and show no sign of having been transported after their deposition.

allochemical (*adj*) refers to constituents of sedimentary rocks that have been chemically precipitated (p.18) within the area in which the rock was deposited but have since been transported elsewhere; e.g. shell fragments.

岩性(名) 岩石或沉積岩層的一般特徵，尤其是在露頭(第 122 頁)和手標本(第 147 頁)所見者。(形容詞為 Lithological)

礫屑岩(名) 平均粒徑大 2 mm 的沉積岩。例如礫岩和角礫岩(第 87 頁)。(形容詞為 rudaceous)

礫質岩(名) 同礫屑岩(↑)。(形容詞為 psephitic)

砂屑岩(名) 粒徑由 1/16 至 2 mm 的沉積岩；一種砂岩。(形容詞為 arenaceous)，參見**砂岩**(第 87 頁)。

泥屑岩(名) 顆粒直徑界於 1/256 至 1/16 mm 之間的任何成分的沉積岩；由泥組成的岩石。(形容詞為 lutaceous)

泥質岩(名) 平均粒徑小於 1/16 mm 的沉積岩；黏土、粉砂、泥岩等；一種硬質泥岩。(形容詞為 argillaceous)，參見**黏土**(第 88 頁)。

碎屑(名) 由較大塊岩石破碎而成的沉積(第 80 頁)岩石的碎片。

碎屑狀的(形) 描述由較早期岩石破碎而成的碎屑所組成的沉積物。

碎屑的(形) 同碎屑狀的(↑)。

有機的(形) 這個詞用於沉積物時是指與植物和動物遺體有關的。例如，珊瑚礁和煤是有機沉積物。

蒸發岩(名) 含某種化合物溶液的水體經蒸發作用(即乾涸)所形成的沉積物。例如鹽類的沉積物、石膏。

陸源的(形) 指沉積在海床上並含有來自大陸的物質的沉積物。

正源化學的(形) 指一些沉積岩內的物質是在該岩石沉積地區內經由化學沉澱(第 18 頁)的，且無跡象顯示它們沉積後遭受過搬運。

異源化學的(形) 指沉積岩的組分原先是在該岩石沉積地區內化學沉澱(第 18 頁)的，而此後則被搬往別處。例如殼碎片。

limestone (*n*) a sedimentary rock (p.80) containing more than 50 per cent of calcium carbonate, $CaCO_3$. Limestones can be of fresh-water or marine origin. The material of which they are composed may be chemically precipitated (p.18), organic (p.85), or detrital (p.85).

石灰岩(名) 含碳酸鈣($CaCO_3$)50%以上的沉積岩(第 80 頁)。石灰岩可以是淡水或海水成因，其組成物質可以是化學沉澱的(第 18 頁)、有機質的(第 85 頁)、或碎屑的(第 85 頁)。

calcareous (*adj*) containing calcium carbonate, $CaCO_3$.

鈣質的(形) 含碳酸鈣($CaCO_3$)的。

chalk (*n*) a very pure limestone, white in colour. The word is also used, with a capital C, as a stratigraphical name for a division of the Cretaceous (p.115).

白堊(名) 一種極純的白色石灰岩。此英文詞的首字母大寫時(Chalk)則作為白堊紀(第 115 頁)的一個地層劃分名稱(即上白堊紀)。

oolith (*n*) ooliths are small round particles that make up a sedimentary rock. In the mass they look like the roe (eggs) of a fish. Ooliths have a regular structure. They are usually calcareous (↑). **oolitic** (*adj*).

鮞石(名) 鮞石是構成一種沉積岩的一些細小圓粒，大體而言，象魚卵。鮞石通常是鈣質的(↑)，具有規則構造。(形容詞為 oolitic)

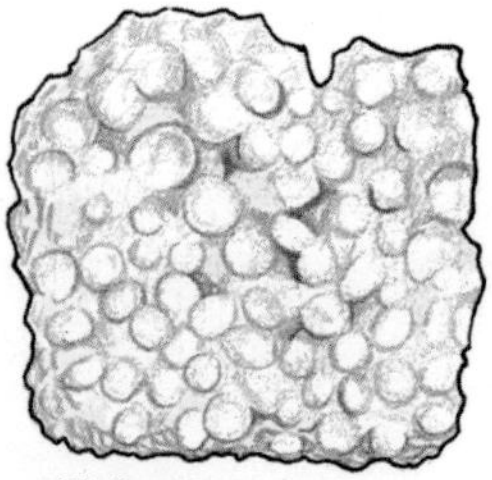

oolitic limestone
鮞狀石灰岩

oolite (*n*) an oolitic (↑) limestone (↑); a limestone composed of ooliths.

鮞狀岩(名) 鮞狀的(↑)石灰岩(↑)：由鮞石組成的石灰岩。

pisolith (*n*) a large oolith (↑), 3–6 mm in diameter.

豆石(名) 大的鮞石(↑)，直徑 3 至 6 mm。

pisolite (*n*) a coarse-grained oolite (↑) containing ooliths 3–6 mm in diameter.

豆狀岩(名) 粗粒的鮞狀岩(↑)，含直徑 3 至 6 mm 的鮞石。

micrite (*n*) very fine-grained calcite (p.51) forming the matrix (p.73) of a limestone (↑). *See also* **sparite** (↓).

泥晶石灰岩(名) 形成石灰岩(↑)基質(第 73 頁)的極細粒方解石(第 51 頁)。參見**亮晶石灰岩**(↓)。

sparite (*n*) coarse-grained calcite (p.51) forming the matrix (p.73) of a limestone (↑).

亮晶石灰岩(名) 形成石灰岩(↑)基質(第 73 頁)的粗粒方解石(第 51 頁)。

dolomitization (*n*) the conversion of calcium carbonate, $CaCO_3$, in a rock to dolomite, $CaMg(CO_3)_2$ (p.51). The process may be partial or complete. Dolomitization is usually caused by metasomatism (p.90). **dolomitized** (*adj*).

白雲石化作用(名) 岩石中的碳酸鈣($CaCO_3$)轉化成白雲石($CaMg(CO_3)_2$)(第 51 頁)的過程，它可以是局部轉化或完全轉化的過程。白雲石化作用通常是由交代作用(第 90 頁)所引起。(形容詞為 dolomitized)

dedolomitization (*n*) the conversion of dolomite, $CaMg(CO_3)_2$ (p.51), to calcite, $CaCO_3$, in a rock. This may be caused by thermal metamorphism (p.90) or diagenesis (p.84).

去白雲石化作用(名) 岩石中的白雲石($CaMg(CO_3)_2$)(第 51 頁)轉化為方解石($CaCO_3$)的過程。熱力變質作用(第 90 頁)或成岩作用(第 84 頁)可引起此種作用。

chert (*n*) a dense, hard, siliceous rock (↓); cryptocrystalline (p.44) silica, SiO_2; a form of quartz (p.55). Chert occurs as beds (bedded chert) and as nodules (p.84) in limestones (↑) and shales (p.88).

燧石(名) 一種緻密堅硬的矽質岩(↓)；隱晶結構的(第 44 頁)二氧化矽(SiO_2)；石英(第 55 頁)的一種形式。在石灰岩(↑)和頁岩(第 88 頁)中，燧石呈層(層狀燧石)和呈結核(第 84 頁)產出。

flint (*n*) a variety of chert (↑) occurring as irregular nodules (p.84), especially in chalk (↑).

打火石(名) 燧石(↑)的一種變種，呈不規則結核(第 84 頁)產出，尤其是在白堊(↑)中。

sandstone (*n*) a clastic (p.85) arenaceous rock (p.85) consisting of fragments from 1/16 to 2 mm in diameter. The fragments are rounded to subrounded (p.83) in shape. Most sandstones are composed mainly of quartz grains (p.55).

砂岩(名) 由直徑 1/16 至 2 mm 碎屑所組成的碎屑狀(第 85 頁)砂屑岩(第 85 頁)。碎屑形狀為圓形到近圓形(第 83 頁)。大部分砂岩主要由石英顆粒(第 55 頁)組成。

grit (*n*) an arenaceous rock (p.85) in which the particles are angular to subangular (p.83). *See also* **sandstone** (↑).

粗砂岩(名) 顆粒為稜角狀到近稜角狀(第 83 頁)的砂屑岩(第 85 頁)。參見**砂岩**(↑)。

quartzite (*n*) a quartz sandstone (↑) with a quartz (p.55) cement (p.84).

石英岩(名) 含石英(第 55 頁)膠結物(第 84 頁)的石英砂岩(↑)。

orthoquartzite = quartzite (↑).

正石英岩 同石英岩(↑)。

siliceous (*adj*) containing silica, SiO_2; containing large amounts of quartz (p.55).

矽質的(形) 含二氧化矽(SiO_2)的；含大量的石英(第 55 頁)的。

arkose (*n*) an arenaceous rock (p.85) composed mainly of fragments of quartz (p.55) and feldspar (p.56). Commonly red or pink in colour.

長石砂岩(名) 主要由石英(第 55 頁)和長石(第 56 頁)的碎屑組成的砂屑岩(第 85 頁)，常為紅色或粉紅色。

greywacke, graywacke (*n*) a dark-coloured arenaceous rock (p.85) consisting of angular and subangular (p.83) fragments of various sizes from fine to coarse; an impure sandstone (↑). Greywackes are formed in mobile belts (p.132); they show sedimentary structures (p.83).

雜砂岩(名) 粒度從細到粗的碎屑組成的稜角狀和近稜角狀(第 83 頁)、暗色砂屑岩(第 85 頁)；一種不純的砂岩(↑)。雜砂岩形成於活動帶(第 132 頁)，顯示沉積構造(第 83 頁)。

greensand (*n*) a sandstone (↑) containing glauconite (p.55).

綠砂岩(名) 含海綠石(第 55 頁)的砂岩(↑)。

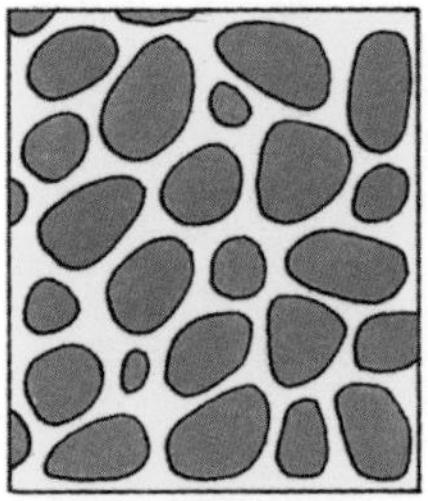

conglomerate 礫岩

conglomerate (*n*) a rudaceous rock (p.85) composed of rounded or subrounded (p.83) fragments. The fragments may be of any rock and may be from a few millimetres to several centimetres in diameter.

礫岩(名) 由圓形或近圓形(第 83 頁)碎屑組成的一種礫屑岩(第 85 頁)。這些碎屑可以是任何岩石，其直徑可由幾毫米到數厘米。

breccia (*n*) a clastic (p.85) sedimentary rock composed of angular (p.83) fragments mixed with finer material. *See also* **conglomerate** (↑).

角礫岩(名) 由稜角狀(第 83 頁)碎屑與較細物質混合組成的碎屑狀(第 85 頁)沉積岩。參見**礫岩**(↑)。

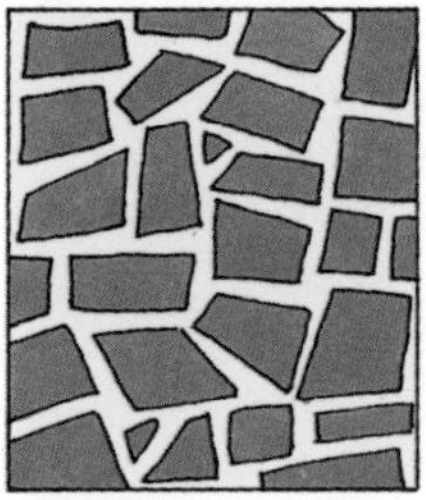

breccia 角礫岩

sand (*n*) detrital (p.85) material consisting of particles from 1/16 to 2 mm in diameter. *See also* **sandstone** (↑).

砂(名) 由粒徑 1/16 至 2 mm 的顆粒組成的碎屑(第 85 頁)物質。參見**砂岩**(↑)。

gravel (*n*) (1) a sediment with grains from 2 to 4 mm across; (2) loose detrital (p.85) material of a range of sizes.

礫石(名) (1)含有橫寬 2 至 4 mm 顆粒的一種沉積物；(2)顆粒大小不同的疏鬆碎屑(第 85 頁)物質。

pebble (*n*) a smooth rounded (p.83) piece of rock between 4 mm and 64 mm in diameter.

小漂礫；卵石(名) 直徑為 4 至 64 mm 的光滑圓形(第 83 頁)岩石碎塊。

cobble (*n*) a rounded (p.83) piece of rock between 64 mm and 256 mm in diameter.

中礫(名) 直徑 64 至 256 mm 的圓形(第 83 頁)岩石碎塊。

boulder (*n*) a large rounded piece of rock more than 256 mm in diameter.

漂礫(名) 直徑大於 256 mm 的大而圓的岩石碎塊。

clay (*n*) an argillaceous rock (p.85) with particles less than 1/256 mm in diameter. Clays are plastic when wet and show no bedding (p.80).

shale (*n*) an argillaceous rock (p.85) with particles less than 1/256 mm in diameter showing well-marked bedding. Shale is fissile (p.80).

mudstone (*n*) an argillaceous rock (p.85) like a shale (↑) but not fissile (p.80).

marl (*n*) a calcareous (p.86) mudstone (↑).

silt (*n*) a sediment with particles from 1/16 to 1/256 mm in diameter.

siltstone (*n*) a consolidated silt (↑); an argillaceous rock (p.85) like a mudstone (↑) but with particles of silt (↑) grade.

ferruginous (*adj*) containing iron. Ferruginous sediments may contain large amounts of iron compounds, e.g. bedded siderites and sedimentary haematite (p.48).

red beds a general term for sedimentary rocks that contain red iron compounds. Red beds are usually formed under very dry continental conditions.

黏土(名) 粒徑 1/256 mm 以下的泥質岩(第 85 頁)。黏土潤濕時具有塑性，不顯示層理(第 80 頁)。

頁岩(名) 粒徑 1/256 mm 以下並顯示明顯層理的泥質岩(第 85 頁)。頁岩易剝裂(第 80 頁)。

泥岩(名) 似頁岩(↑)但不易剝裂(第 80 頁)的泥質岩(第 85 頁)。

泥灰岩(名) 含鈣質的(第 86 頁)泥岩(↑)。

粉砂(名) 粒徑由 1/16 至 1/256 mm 的沉積物。

粉砂岩(名) 固結的粉砂(↑)；一種似泥岩(↑)而具粉砂(↑)粒級顆粒的泥質岩(第 85 頁)。

鐵質的(形) 含鐵的。鐵質沉積物可含大量的鐵化合物。例如層狀的菱鐵礦和沉積的赤鐵礦(第 48 頁)。

紅層 含紅色鐵化合物沉積岩的一般術語。紅層通常是在極乾旱的大陸性條件下形成的。

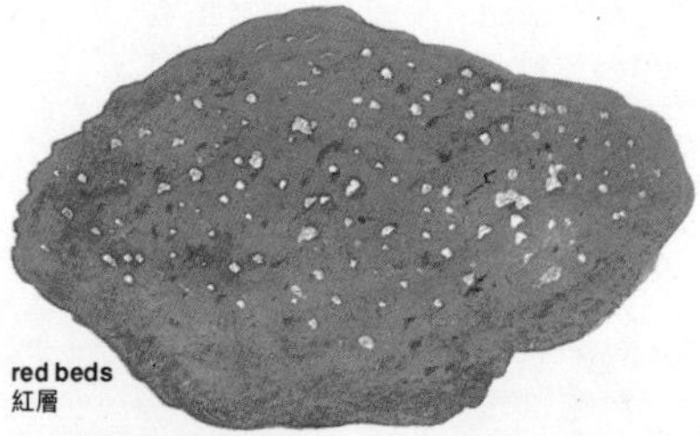
red beds
紅層

ironstone (*n*) a rock, usually sedimentary, that is made up largely of chemical compounds of iron.

clay-ironstone (*n*) an argillaceous (p.85) ironstone. The iron is usually present as siderite (p.51), iron carbonate ($FeCO_3$). Clay-ironstone occurs in beds (p.80) and as concretions (p.84).

black-band ironstone a clay-ironstone (↑) containing coal-like material. It can be burnt to produce iron.

富鐵岩石(名) 通常是沉積的並主要由鐵的化合物組成的一種岩石。

泥鐵礦(名) 一種泥質(第 85 頁)富鐵岩石。其中的鐵通常以菱鐵礦(第 51 頁)，即碳酸鐵($FeCO_3$)形式存在。泥鐵礦產於岩層(第 80 頁)中並以結核狀(第 84 頁)產出。

黑泥鐵礦 含煤狀物質的一種泥鐵礦(↑)，可以燃燒而產生鐵。

carbonaceous (*adj*) containing carbon. Carbonaceous rocks include coal (↓) and lignite (↓).

碳質的(形) 含碳的。碳質岩包含煤(↓)和褐煤(↓)。

peat (*n*) an accumulation of vegetable matter at the Earth's surface that has partly decomposed.

泥炭(名) 在地表上已部分分解的植物質堆積。

lignite 褐煤

lignite (*n*) a carbonaceous (↑) rock of a kind between peat (↑) and bituminous coal (↓); called 'brown coal'.

褐煤(名) 介於泥炭(↑)和烟煤(↓)之間的一種碳質(↑)岩，稱之"褐色煤"。

coal (*n*) plant remains that have been changed physically and chemically to form a hard, black substance that can be burned.

煤(名) 植物遺體經受物理和化學變化所形成的堅硬而黑色的可燃物質。

bituminous coal ordinary coal (↑) such as is used in the home for burning. It contains a relatively small proportion of carbon.

烟煤 普通煤(↑)，如家庭中常用的燃燒煤，含碳比例很低。

anthracite (*n*) a coal (↑) containing more than 90% of carbon.

無烟煤(名) 含碳高於 90% 的煤(↑)。

bitumen (*n*) a mineral composed of compounds of carbon and hydrogen (hydrocarbons) that looks like tar. Bitumen is found as a thick liquid or as a solid that breaks easily.

瀝青(名) 由碳和氫的化合物(碳氫化合物)所組成的一種礦物，外觀似焦油。瀝青呈濃稠液體或易碎固體產出。

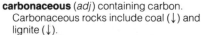

cannel coal 燭煤

asphalt (*n*) a brown or black almost solid hydrocarbon rather like bitumen (↑). Asphalt occurs naturally (e.g. in Trinidad) and is also formed in the production of petroleum (p.144).

地瀝青(名) 一種褐色或黑色近於固體的碳氫化合物，頗似**瀝青**(↑)。地瀝青可天然產出(例如在特立尼達)，也可在石油(第 144 頁)生產中產生。

cannel coal a fine-grained coal, dull in colour.

燭煤 細粒的煤，顏色暗淡。

jet (*n*) a type of cannel coal (↑) or lignite (↑), black in colour and hard. It can be polished and made into ornaments.

煤玉；黑玉(名) 一種煤(↑)或褐煤(↑)，色黑，質堅硬。可加工製成裝飾品。

sub-bituminous coal a coal (↑) of rank (↓) between bituminous coal (↑) and lignite (↑). Most coals of this type are of Mesozoic and Tertiary age (p.115).

次烟煤 煤級(↓)介於烟煤(↑)和褐煤(↑)之間的一種煤(↑)。這類煤大多數屬於中生代和第三紀(第 115 頁)。

coal seam a bed of coal.

煤層 煤的礦層。

coal measures strata (p.80) with coal seams (↑), or the coal seams themselves.

煤系 含煤層(↑)的地層(第 80 頁)，或煤層本身。

rank (of coal) the rank of a coal is a measure of the amount of carbon that it contains. Thus anthracite (↑), with a high proportion of carbon, is of high rank; bituminous coal, with a small proportion of carbon, is of low rank.

煤級 煤級是煤含碳量的一種量度。如無煙煤(↑)，含碳的比例高，屬高煤級；煙煤，含碳的比例低，屬低煤級。

hydrocarbon minerals (*n.pl.*) a general term for compounds of carbon and hydrogen that occur as minerals: coal, bitumens (↑), oil (p.144), nature gas (p.144), and asphalt (↑).

碳氫礦物(名、複) 碳和氫的化合物用的術語，用於呈礦物形式如：煤、瀝青(↑)、石油(第 144 頁)、天然氣(第 144 頁)和地瀝青(↑)產出的礦物。

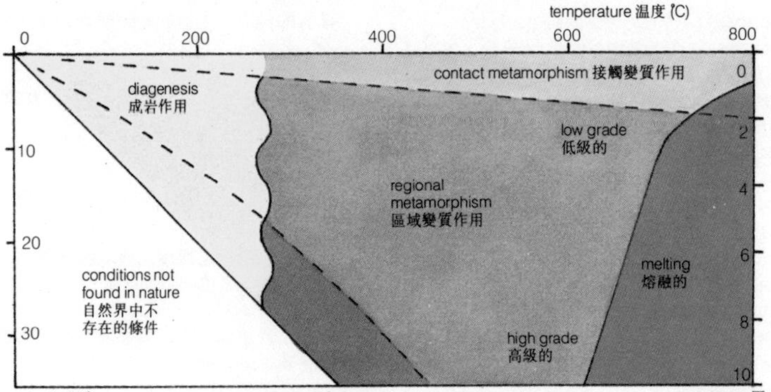

types of metamorphism in relation to temperature and pressure
變質作用類型與溫度和壓力的關係

metamorphism (*n*) the processes by which rocks are changed by the action of heat or pressure, or both. The rocks changed by metamorphism may be sedimentary (p.80) or igneous (p.62), or even metamorphic. The changes brought about by metamorphism are in the mineral composition, texture (p.72), or structure (p.122) of the rocks. Changes that take place at the Earth's surface (weathering or diagenesis (pp.20, 84)) are not included under metamorphism. **metamorphic** (*adj*), **metamorphose** (*v*).

變質作用(名) 岩石受熱力或壓力或兩者的作用而發生變化的過程。因變質作用而產生變化的岩石，可以是沉積(第 80 頁)岩或火成(第 62 頁)岩，或甚至是變質岩。變質作用引起岩石在礦物成分、結構(第 72 負)、或構造(第 122 頁)上發生變化。變質作用不包括在地球表面發生的那些變化(例如風化作用或成岩作用(第 20、84 頁))。(形容詞為 metamorphic，動詞為 metamorphose)

unmetamorphosed (*adj*), not metamorphosed (↑).

未變質的(形) 沒有變質的(↑)。

thermal metamorphism metamorphism (↑) produced by the action of heat on a rock, for example, by an igneous intrusion (p.64).

熱力變質作用 由熱力例如火成侵入體(第 64 頁)所產生的熱力作用於岩石所產生的變質作用(↑)。

contact metamorphism metamorphism (↑) produced by an igneous intrusion (p.64); largely thermal metamorphism (↑), although some deformation (p.122) may also take place.

接觸變質作用 火成侵入體(第 64 頁)所產生的變質作用(↑)：主要是熱力變質作用(↑)，但也可發生某些變形(第 122 頁)。

pyrometamorphism (*n*) an intense form of thermal metamorphism (↑) produced by contact with an igneous intrusion (p.64) at a very high temperature. **pyrometamorphic** (*adj*).

高熱變質作用(名) 強烈形式的熱力變質作用(↑)，是在極高溫度下與火成侵入體(第 64 頁)接觸所產生。(形容詞為 pyrometamorphic)

dynamic metamorphism metamorphism (↑) produced by mechanical stress (p.122), without heat.

動力變質作用 由機械應力(第 122 頁)而不是熱力所產生的變質作用(↑)。

regional metamorphism metamorphism (↑) produced by heat and pressure affecting the rocks of a large area (thousands of square kilometres in extent).

區域變質作用 由熱力和壓力所產生的變質作用(↑)，影響大面積岩石(範圍達數千平方公里)。

dynamothermal metamorphism = regional metamorphism (↑).

metasomatism (*n*) metamorphism (↑) in which material is added to a rock or taken away from it by liquids or gases passing through the rock. **metasomatic** (*adj*).

meta- (*prefix*) meta- at the beginning of a rock name shows that the rock concerned has been metamorphosed (↑), e.g. metaquartzite.

metamorphic grade the level that metamorphism (↑) has reached in a rock. It is thus possible to speak, for example, of 'high-grade metamorphism' or of 'metamorphism of medium grade'.

isograd (*n*) a line joining points at which the rocks are of the same metamorphic grade (↑).

prograde (*adj*) prograde metamorphism changes a rock from a lower to a higher metamorphic grade (↑); it is the normal form of metamorphism.

retrograde (*adj*) retrograde metamorphism changes a rock from a higher to a lower metamorphic grade (↑), for example, when a high-grade metamorphic rock is later heated for a long time at a temperature that is lower than the temperature at which the earlier metamorphism took place. *See also* **prograde** (↑).

動熱變質作用 同區域變質作用(↑)。

交代作用(名) 由滲入岩石的一些液體或氣體將物質加進岩石中或從岩石中帶走物質的變質作用(↑)。(形容詞為 metasomatic)

meta-(詞頭) 置於岩石名稱之前，表示該岩石已受到變質作用(↑)。例如變質石英岩。

變質等級 岩石中已達到的變質(↑)程度。譬如，可稱之屬"高級變質作用"或屬"中級變質作用"。

等變線(名) 連接岩石上屬於相同變質等級(↑)的各點的一條連線。

前進的(形) 指將岩石從較低變質等級改變成較高變質等級(↑)的變質作用；這是正常的變質作用形式。

退化的(形) 指將岩石從較高變質等級變成較低變質等級(↑)的變質作用。例如，高級變質岩石後來在低於早期發生變質時的溫度下長時間受熱的變質作用。參見**前進的**(↑)。

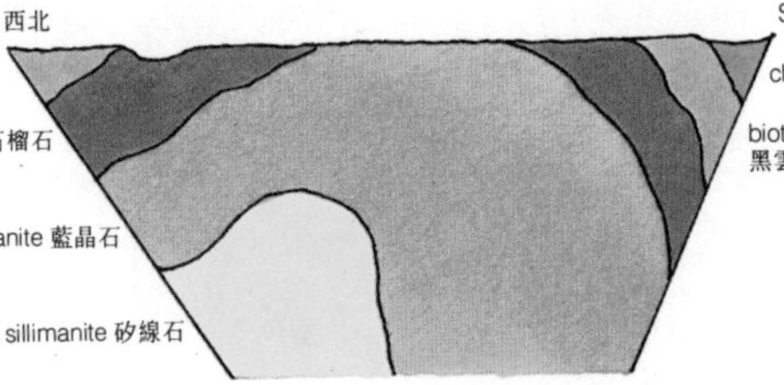

section across Scottish Highlands showing metamorphic zones
橫過蘇格蘭高地的剖面顯示各變質帶

metamorphic zones a metamorphic zone is an area of metamorphic rock that can be mapped in the field by some special character – e.g. the presence of an 'index' mineral (↓).

index mineral a mineral produced by metamorphism that is used to characterize a metamorphic zone (↑).

變質帶 變質帶是可以在野外根據某些特徵——例如一種"指示"礦物(↓)的出現，測繪到圖上的變質岩的一個地區。

指示礦物 由變質作用產生的礦物，此礦物可用來表示一個變質帶(↑)的特徵。

metamorphic facies a group of rocks of varying composition, all of which have been metamorphosed under similar conditions.

變質相 在同樣條件下遭受變質的一組成分各不相同的岩石。

greenschist facies a metamorphic facies (↑) developed at high pressures (from about 2 to 6 kbar) and relatively low temperatures (200 to 500°C). The minerals albite, biotite, muscovite, chlorite, epidote, tremolite, and actinolite (pp.55, 56, 57, 61) are typical of the facies. A characteristic green colour is given to the rocks by the chlorite and epidote.

綠色片岩相 在高壓(約 2 千巴至 6 千巴)和相對低溫(200 至 500℃)條件下發育的一種變質相(↑)。鈉長石、黑雲母、白雲母、綠泥石、綠簾石、透閃石和陽起石(第 55、56、57、61 頁)是此相的典型礦物。因綠泥石和綠簾石所具的特徵綠色而得名。

amphibolite facies a metamorphic facies (↑) developed at temperatures of 300 to 600°C and high pressure. Hornblende (p.57) and garnet (p.58) are characteristic minerals of the amphibolite facies.

角閃岩相 在 300 至 600℃ 溫度和高壓條件下發育的一種變質相(↑)。普通角閃石(第 57 頁)和石榴石(第 58 頁)是角閃岩相的特徵礦物。

granulite facies a metamorphic facies (↑) developed at temperatures of 700 to 800°C at high pressure and under dry conditions. Characteristic minerals are plagioclase, pyroxene, hornblende, and diopside (pp.55, 57).

麻粒岩相 在 700 至 800℃ 溫度和高壓及乾燥的條件下發育的一種變質相(↑)。其特徵礦物是斜長石、輝石類、普通角閃石和透輝石(第 55、57 頁)。

eclogite facies a metamorphic facies (↑) developed at high pressures (greater than 10 kbar) and high temperatures (600°C or more), characterized by the metamorphic rock eclogite (p.97).

榴輝岩相 在高壓(10 千巴以上)和高溫(600℃或更高溫度)條件下發育的一種變質相(↑)，變質岩榴輝岩(第 97 頁)為其特徵。

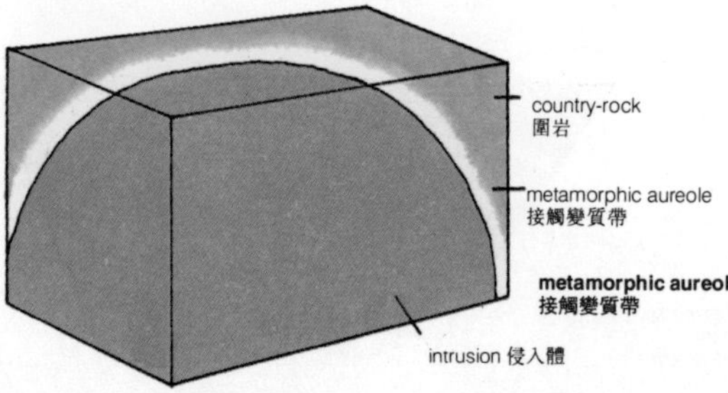

metamorphic aureole
接觸變質帶

metamorphic aureole an area round a plutonic igneous intrusion (p.64) in which the country-rock (p.65) has been metamorphosed.

接觸變質帶 深成火成侵入體(第 64 頁)周圍的一個地帶，帶內圍岩(第 65 頁)經已變質。

contact zone = metamorphic aureole (↑).

接觸帶 同接觸變質帶(↑)。

ghost stratigraphy signs of an earlier stratification (p.80) seen in highly metamorphosed rocks.

殘跡地層 在強烈變質岩石中所見的早先層理(第 80 頁)痕跡。

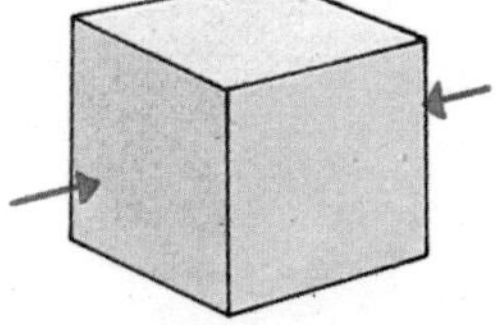

directed pressure 定向壓力

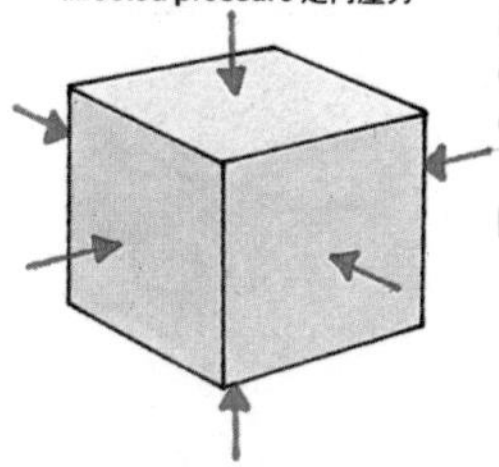

lithostatic pressure
地靜壓力

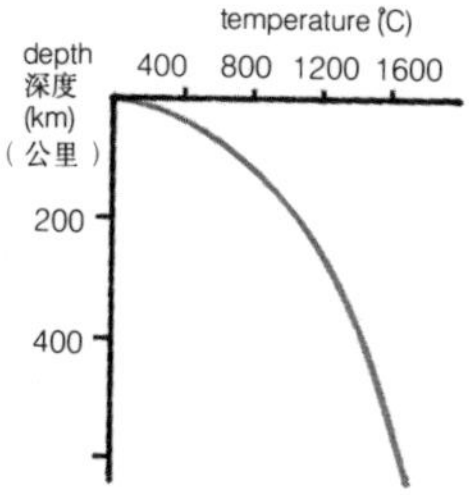

geothermal gradient
地熱增溫率

pressure (*n*) force per unit area.

壓力（名） 單位面積上受的力。

lithostatic pressure the pressure (force) at depth in the Earth that is due to the weight of rocks above (the *superincumbent load*). Lithostatic pressure increases rapidly with increasing depth.

地靜壓力 地球深處由於上覆岩石的重量（上覆負荷）而受到的壓力（力）。地靜壓力隨着深度的增加而迅速增大。

hydrostatic pressure the pressure (↑) at a point in a body of liquid that is due to the weight of liquid above it; *see* **lithostatic pressure** (↑).

靜水壓力 液體中某一點由於其上方液體重量而受到的壓力（↑）；參見**地靜壓力**（↑）。

load pressure = lithostatic pressure (↑).

負荷壓力 同地靜壓力（↑）。

confining pressure = lithostatic pressure (↑).

圍壓 同地靜壓力（↑）。

directed pressure a pressure (force) that is applied in such a way that it has a direction.

定向壓力 施加於一定方向的壓力（力）。

pore-fluid pressure when a rock is heated, minerals containing volatiles (p.18) such as water or carbon dioxide may break down, releasing these volatiles. The pressure that is produced by the volatiles is called the *pore-fluid pressure*. Like lithostatic pressure (↑) it is the same in all directions.

孔隙流體壓力 岩石受熱時，含水或二氧化碳之類揮發物（第 18 頁）的礦物分解釋放出這些揮發物所產生的壓力稱為"孔隙流體壓力"。如同地靜壓力（↑），此壓力的大小在一切方向都相同。

syntexis (*n*) a process in which rocks of more than one kind are melted down at great depth to form a new magma (p.62). Country-rock (p.65) may or may not be melted and taken into the mixture that forms the new magma. Compare **anatexis** (↓).

同熔作用（名） 不只一種岩石在極深之處被熔化形成新岩漿（第 62 頁）的過程。圍岩（第 65 頁）可以被熔化也可以不被熔化和帶入混合物中形成新岩漿。比較：**深熔作用**（↓）。

anatexis (*n*) the assimilation (p.62) and remelting of rocks by a magma (p.62). Compare **syntexis** (↑).

深熔作用（名） 岩石被岩漿（第 62 頁）同化（第 62 頁）和重熔。比較：**同熔作用**（↑）。

rheomorphism (*n*) the melting of part or the whole of a rock so that it is able to flow and be deformed (p.122).

柔流變質作用；軟流變質作用（名） 岩石部分或全部熔化，因而能流動和變形（第 122 頁）。

recrystallization (*n*) a process in which the original crystals in a rock are dissolved a little at a time and form a new set of crystals. The new crystals are usually larger than the original ones.

重結晶作用（名） 岩石中的原始晶體漸次溶解些少並形成新的一組晶體的過程。新晶體通常大於原始晶體。

geothermal heat flow the rate at which heat passes from the interior of the Earth to the atmosphere.

地熱流 熱從地球內部傳到大氣圈的速率。

geothermal gradient the rise in temperature that is found with increasing depth below the Earth's surface. The average gradient below the continents is about 30 °C km^{-1}.

地熱增溫率；地溫梯度 溫度隨着地面以下深度的增加而上升。大陸下面的平均增溫率大約每公里上升 30℃。

isotherm (*n*) a line joining points that are at the same temperature.

等溫線（名） 相同溫度的各點的連接線。

cataclasis (*n*) the mechanical breaking up of a rock by dynamic metamorphism (p.90). **cataclastic** (*adj*).

碎裂作用(名) 動力變質作用(第 90 頁)所造成的岩石機械破裂。(形容詞為 cataclastic)

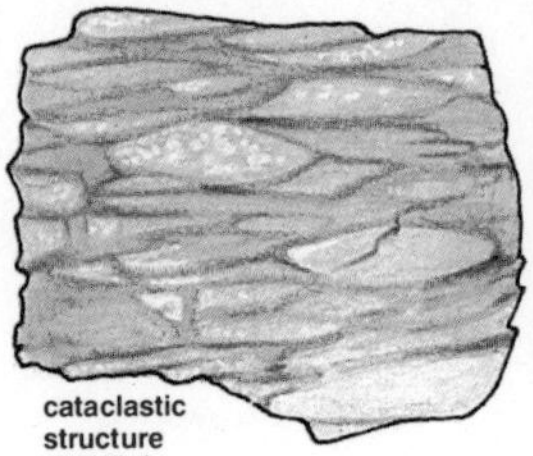

cataclastic structure
碎裂構造

idioblastic (*adj*) describes a metamorphic texture (p.72) in which the mineral grains show their full crystal form. **idioblast** (*n*).

自形變晶的(形) 描述礦物顆粒呈其完美晶形的一種變質結構(第 72 頁)。(名詞為 idioblast)

crystalloblastic (*adj*) describes a metamorphic texture produced by the recrystallization (p.93) of the minerals in a rock. **crystalloblast** (*n*).

變晶質的(形) 描述岩石內的礦物重結晶(第 93 頁)所產生的變質結構。(名詞為 crystalloblast)

porphyroblastic (*adj*) describes large euhedral (p.45) crystals in a metamorphic rock that are in a groundmass (p.73) of finer grain. **porphyroblast** (*n*).

斑狀變晶的(形) 描述變質岩內處於較細粒基質(第 73 頁)中的一些大的自形(第 45 頁)晶體。(名詞為 porphyroblast)

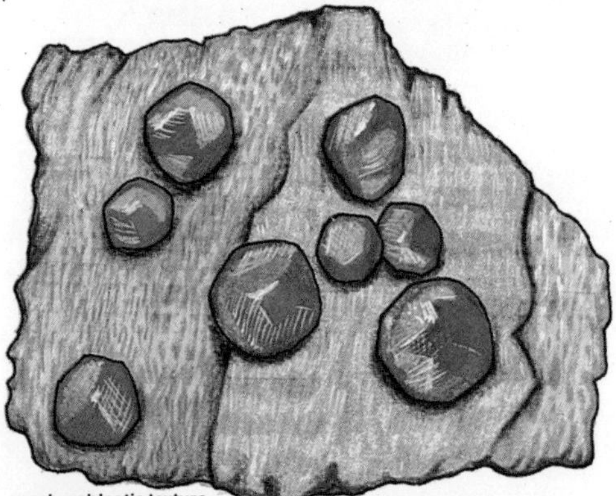

porphyroblastic texture
斑狀變晶結構

xenoblastic (*adj*) describes a metamorphic rock with mineral grains that do not show crystal faces (p.40). **xenoblast** (*n*).

他形變晶的(形) 描述所含礦物顆粒不顯示晶面(第 40 頁)的變質岩。(名詞為 xenoblast)

relict structure an original structure that can still be seen in a metamorphic rock after the original minerals have been replaced.

殘餘構造 原始礦物被交代後仍能在某種變質岩中見到的一種原始構造。

palimpsest structure a metamorphic structure showing something of the original texture of the rock. More or less the same as relict structure (↑).

變餘構造 顯示岩石原始結構的某些特點的變質構造。大體上同殘餘構造(↑)。

poikiloblastic (*adj*) describes a metamorphic texture in which new minerals have formed during recrystallization (p.93) around relics of the original minerals.

變嵌晶狀(形) 描述重結晶(第 93 頁)期間所形成的新礦物圍繞着殘餘的原始礦物的變質結構。

granoblastic (*adj*) describes an equigranular (p.73) metamorphic texture.

花崗變晶狀(形) 描述一種等粒狀(第 73 頁)變質結構。

foliation (*n*) the arrangement of mineral grains in layers in a rock. **foliated** (*adj*).

葉理(名) 礦物顆粒在岩石中成層狀的排列。(形容詞為 foliated)

lineation (*n*) a structure in or on a rock that forms lines.

綫理(名) 在岩石中或岩石表面上形成的線條狀構造。

schistosity (*n*) the arrangement of minerals in a coarse-grained metamorphic rock in layers that are parallel or almost parallel.

片理(名) 一些礦物在粗粒變質岩中成層狀相互平行或近於平行的排列。

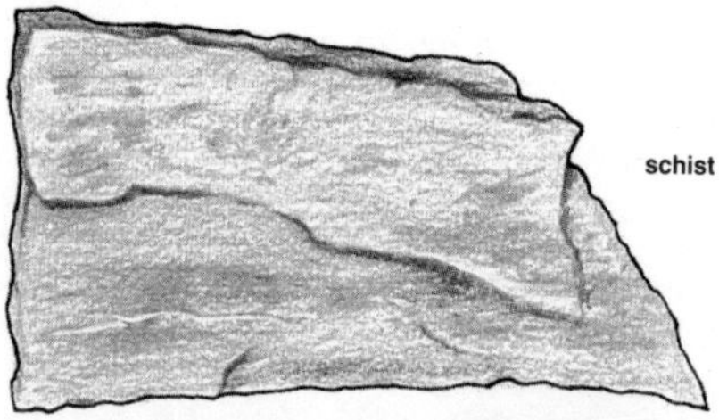

cleavage, rock cleavage if a rock will break or split along smooth planes that are more or less parallel it is said to show cleavage. If it is necessary to draw attention to the difference between this form of cleavage and cleavage as shown by crystals (p.40), the term 'rock cleavage' may be used.

劈理，岩石劈理 岩石沿一些近於平行的平滑面裂開或劈開，稱之顯示劈理。如必須將劈理和晶體(第 40 頁)解理區別，可使用"岩石劈理"這一術語。

slaty cleavage a form of cleavage (↑) that is developed in fine-grained rocks when they are put under great pressure. The cleavage planes are more or less parallel to the axial planes (p.123) of the folds and cut across the bedding planes (p.80).

板狀劈理 岩石處於巨大壓力下，發育於細粒岩石中的一種劈理(↑)。劈理面近似平行於褶皺的軸面(第 123 頁)，並橫割層面(第 80 頁)。

flow cleavage a form of cleavage (↑) in which new minerals grow and the bedding nearly disappears. Further recrystallization (p.93) leads to schistosity (↑).

流狀劈理 劈理(↑)的一種形式，新礦物在其中生長，層理幾乎消失。進一步重結晶(第 93 頁)則產生片理(↑)。

fracture cleavage fine-grained rocks that have been deformed (p.122) may tend to split along closely spaced parallel planes; this is fracture cleavage, also called *shear cleavage* or *false cleavage*.

破劈理 已變形(第 122 頁)的細粒岩石有沿密集平行面分裂的趨向；這是破劈理，也稱之"剪劈理"或"假劈理"。

shear cleavage = fracture cleavage (↑).

剪劈理 同破劈理(↑)。

false cleavage = fracture cleavage (↑).

假劈理 同破劈理(↑)。

ptygmatic (*adj*) a word used to describe free folding or flow folding, e.g. of quartzofeldspathic veins (p.145) in metamorphic or granitized rocks (p.63).

腸狀的(形) 描述自由褶皺或流狀褶皺用的一個詞。例如在變質岩或花崗岩化岩石(第 63 頁)中的一些長英質的岩脈(第 145 頁)。

quartzite (*n*) the metamorphic equivalent of a quartz sandstone; almost pure silica, SiO_2, recrystallized (p.93) into a mass of quartz crystals fitting closely together.

metaquartzite (*n*) = quartzite (↑).

marble (*n*) a metamorphosed limestone. The calcium carbonate ($CaCO_3$) of the limestone is recrystallized (p.93) as calcite (p.51). The word 'marble' is also used outside geology for various sedimentary and other rocks that can be polished.

hornfels (*n*) a hard, fine-grained rock produced by thermal metamorphism (p.90) of sediments. Hornfelses are found at the margins of igneous intrusions (p.64).

skarn (*n*) a metamorphic rock produced by the thermal metamorphism (p.90) and metasomatism (p.91) of an impure limestone or dolomite (p.51).

slate (*n*) a fine-grained argillaceous rock (p.85) that has a good cleavage (p.95), i.e. it splits easily into thin plates. Slates are produced by dynamic metamorphism (p.90) or regional metamorphism (p.90) of low grade (p.91).

phyllite (*n*) a metamorphic rock of higher grade (p.91) than a slate (↑) but of lower grade than a mica-schist (p.55, ↓). Phyllites are produced by regional metamorphism at low temperatures. They are of coarser grain than slates and show better cleavage than mica-schists. They show a characteristic lustre (p.47).

psammite (*n*) an arenaceous rock (p.85); usually a metamorphosed arenaceous rock. **psammitic** (*adj*).

石英岩(名) 與石英砂岩相當的變質岩：成分接近純二氧化矽(SiO_2)，係經重結晶(第 93 頁)的大量石英晶體緊密地鑲嵌在一起而成。

變質石英岩(名) 同石英岩(↑)。

大理岩(名) 變質的石灰岩。石灰岩中的碳酸鈣($CaCO_3$)重結晶(第 93 頁)成方解石(第 51 頁)。在地質學範疇之外，"大理岩"這個詞也用於指可磨光的各種沉積岩和其他岩石。

角頁岩；角岩(名) 沉積物經熱力變質作用(第 90 頁)而生成的堅硬、細粒岩石。角頁岩見於火成侵入體(第 64 頁)的邊緣。

夕卡岩(名) 由不純的石灰岩或白雲岩(第 51 頁)經熱力變質作用(第 90 頁)和交代作用(第 91 頁)生成的一種變質岩。

板岩(名) 具良好劈理(第 95 頁)的一種細粒泥質岩(第 85 頁)，這種岩石容易劈裂成一些薄板。板岩係由動力變質作用(第 90 頁)或低級(第 91 頁)區域變質作用(第 90 頁)做成的。

千枚岩；千層岩(名) 比板岩(↑)高級(第 91 頁)而比雲母(第 55 頁)片岩(↓)低級的變質岩。千枚岩係由低温的區域變質作用所產生。其顆粒比板岩粗，並顯示比雲母片岩更好的劈理。千枚岩顯示一種特有的光澤(第 47 頁)。

砂質岩(名) 一種砂質的岩石(第 85 頁)：通常為變質的砂質岩石。(形容詞為 psammitic)

marble
大理岩

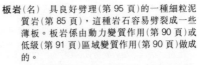

slate
板岩

gneiss
片麻岩

schist (*n*) a foliated (p.45) metamorphic rock of medium to coarse grain. Schists are produced by regional metamorphism (p.90). The name of a mineral may be added for a particular type of schist, e.g. mica-schist, hornblende-schist.

片岩(名) 中粒至粗粒的葉片狀(第 45 頁)變質岩。片岩係由區域變質作用(第 90 頁)生成。可冠上礦物名稱以表示某種特定的片岩類型。例如雲母片岩、普通角閃石片岩。

gneiss (*n*) a foliated (p.45) metamorphic rock of coarse grain (p.72) with a banded appearance. Gneisses are produced by regional metamorphism (p.90) of high grade (p.91). **gneissose** (*adj*).

片麻岩(名) 具條帶狀外表的葉片狀(第 45 頁)粗粒(第 72 頁)變質岩。片麻岩係由高級(第 91 頁)區域變質作用(第 90 頁)生成。(形容詞為 gneissose)

orthogneiss (*n*) a gneiss (↑) formed by the metamorphism of an igneous rock (p.62).

正片麻岩；火成片麻岩(名) 由火成岩(第 62 頁)變質形成的一種片麻岩(↑)。

paragneiss (*n*) a gneiss (↑) formed by the metamorphism of a sedimentary rock (p.80).

副片麻岩；水成片麻岩(名) 由沉積岩(第 80 頁)變質形成的一種片麻岩(↑)。

augen gneiss a gneiss (↑) in which the quartz-feldspar bands are in places thickened and of coarse grain (p.72), forming masses shaped like eyes.

眼球狀片麻岩 長石-石英條帶在適當部位增厚並具粗顆粒(第 72 頁)的一種片麻岩(↑)，形狀象一些眼狀團塊。

eclogite (*n*) a metamorphic rock composed mainly of red garnet (p.58) and green pyroxene (p.57); usually of coarse grain (p.72). The chemical composition of eclogite is similar to that of a basalt or gabbro (p.76). Eclogites are formed at high temperature and pressure. **eclogitic** (*adj*).

榴輝岩(名) 主要由紅石榴石(第 58 頁)和綠輝石類(第 57 頁)組成的一種變質岩，通常為粗粒(第 72 頁)。榴輝岩的化學成分類似玄武岩或輝長岩(第 76 頁)，它是在高温和高壓下形成的。(形容詞為 eclogitic)

migmatite (*n*) a mixed rock formed by the injection (p.64) of granitic material between sheets of an existing metamorphic rock. Migmatites are generally formed under regional metamorphism (p.90) of high grade (p.91). **migmatitic** (*adj*).

混合岩(名) 花崗質物質貫入(第 64 頁)現成變質岩層片之間所形成的一種混合的岩石。混合岩類一般是在高級(第 91 頁)區域變質作用(第 90 頁)下形成的。(形容詞為 migmatitic)

tectonite (*n*) a metamorphic rock in which the minerals tend to point in the same direction, i.e. have a *preferred orientation* or *deformational fabric*. The preferred orientation is the result of crystallization under stress (p.122).

構造岩(名) 所含礦物傾向於按同一方向排列，即具有一種"定向方位"或"變形組構"的變質岩。定向方位是在應力(第 122 頁)下結晶的結果。

mylonite (*n*) a metamorphic rock produced by pressure and rubbing, or *cataclasis* (p.94). The rocks are torn and rolled out to form a fine-grained rock that may show some foliation (p.95). **mylonitic** (*adj*).

糜稜岩(名) 受壓力和磨擦或"碎裂作用"(第 94 頁)產生的一種變質岩。岩石被撕碎和碾壓，形成一種可見其葉理(第 95 頁)的細粒岩石。(形容詞為 mylonitic)

granulite (*n*) a metamorphic rock with a granular texture (p.72), usually composed of quartz, feldspars, garnets, and pyroxenes (pp.55–8). Granulites are produced by regional metamorphism (p.90). *See also* **granulite facies** (p.92). **granulitic** (*adj*).

麻粒岩(名) 具粒狀結構(第 72 頁)的一種變質岩，通常由石英、長石、石榴石和輝石(第 55–58 頁)組成。麻粒岩係由區域變質作用(第 90 頁)產生的。參見**麻粒岩相**(第 92 頁)。(形容詞為 granulitic)

palaeontology (*n*) the study of the life of past geological times; the study of fossils (↓). **palaeontological** (*adj*).

paleontology (*n*) American spelling of palaeontology (↑). **paleontologic** (*adj*).

biosphere (*n*) that part of the world in which living things are present: the surface of the land, the soil, the seas, and the air.

organism (*n*) a living individual plant or animal.

fossil (*n*) the remains of an animal or plant preserved in a rock; a cast (↓) or impression or a trace (↓) of an animal or plant in a rock. **fossil, fossilized** (*adj*), e.g. a fossil fish, fossilized wood; **fossilize** (*v*).

fossiliferous (*adj*) containing fossils (↑).

fossil record all the remains of past animal and plant life found in the rocks.

mould (*n*) the impression left in a rock by a fossil (or other object). A mould may be *external* (an impression made by the outside of the fossil) or *internal* (a cast (↓) of the inside of a fossil).

古生物學(名)　研究遠古地質年代的生命的科學：研究化石(↓)的學科。(形容詞為 palaeonological)

古生物學(↑)美語拼寫為 **paleontology**。(形容詞為 paleontologic)

生物圈(名)　指地球上存在生命的部分，包括：陸地表面、土壤、海洋和空氣。

有機體；生物(名)　生活的植物或動物個體。

化石(名)　保存在岩石中的古代動、植物遺骸：保存在岩石中的古代動、植物的鑄型(↓)、印痕或痕跡(↓)。(形容詞為 fossil，fossilized)，例如：化石魚、木化石。(動詞為 fossilize)

含化石的(形)　指含有化石(↑)。

化石記錄　岩石中發現的全部古代動、植物的遺跡。

印模(名)　化石(或其他物體)在岩石中留下的印痕。印模有"外模"(化石外部構成的印痕)和"內模"(化石內部的鑄型(↓))之分。

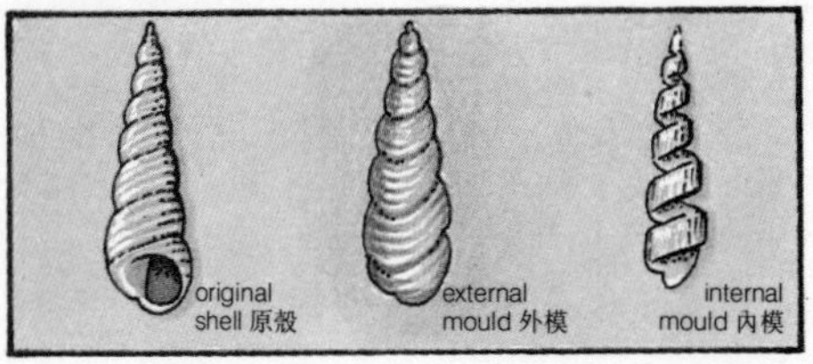

cast (*n*) a copy of a fossil (or other object) formed by the filling in of a mould (↑). A cast may be *internal* or *external*, i.e. of the inside or of the outside of the fossil.

trace fossil a sedimentary structure (p.83) formed by an animal moving across or moving in the sediment when it was being deposited, e.g. tracks, footprints, and burrows (holes made by animals).

fauna (*n*) the animals that live together in any one place or area at a particular time. A *fossil fauna* consists of all the animals that are found as fossils in a particular stratigraphical unit (p.113) – a bed, for example – in a particular area. **faunal** (*adj*). *See also* **flora** (p.110).

鑄型(名)　由印模(↑)填充物所形成的化石(或其他物體)的複製品。鑄型有"內鑄型"和"外鑄型"之分，即化石的內部或外部的鑄型。

遺跡化石　沉積物形成時由於動物走過沉積物或進入其中而形成的沉積岩構造(第 83 頁)。例如：蟲跡、足印、潛穴(動物所掘的孔洞)。

動物群(名)　指在任何一個地方或區域，某一特定時代共同生活的動物的總體。"化石動物群"指在一特定地區特定地層單位(第 113 頁)(例如在一層內)所發現的已成為化石的所有動物。(形容詞為 faunal)。參見**植物群**(第 110 頁)。

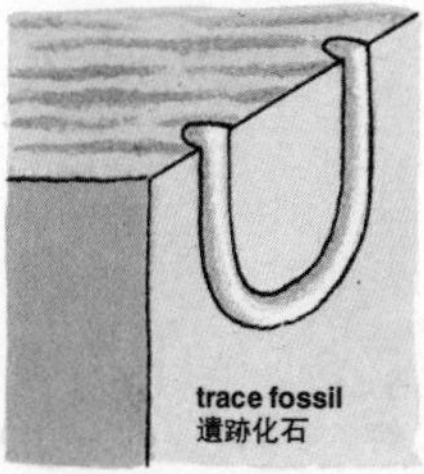

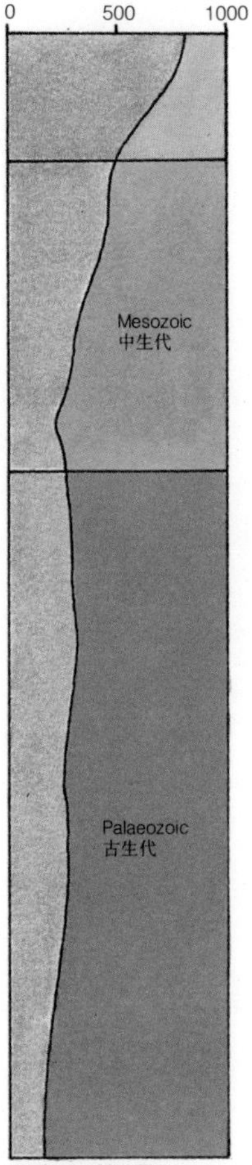

numbers of families of all groups during Phanerozoic time
顯生宙時各類生物科的數目

microfossil (*n*) a very small fossil (↑): one that can be seen only with a microscope. Some microfossils are important in stratigraphical palaeontology (p.111), e.g. the Foraminifera (p.104).
micropalaeonotology (*n*) the study of microfossils (↑). **micropalaeontological** (*adj*).
microfauna (*n*) a fauna (↑) of microfossils (↑).
taphonomy (*n*) the study of the ways in which fossils (↑) are formed.
taxonomy (*n*) the science of arranging animals and plants, whether living or fossilized (↑), in groups or classes according to their structure. **taxonomic** (*adj*).
taxon (*n*) (*taxa*) a taxonomic group or unit of classification, e.g. a genus (↓), a family (↓). **taxonomic** (*adj*).
phylum (*n*) (*phyla*) one of the main divisions of the animal or plant kingdom.
class (*n*) a division of a phylum (↑). e.g. Insecta (p.108), Mammalia (p.109).
order (*n*) a division of a class (↑).
superfamily (*n*) a division of an order (↑). Names of superfamilies end in -oidea.
family (*n*) a division of an order (↑). Names of animal families end in -idae.
subfamily (*n*) a division of a family (↑). Names of animal subfamilies end in -inae.
genus (*n*) (*genera*) a division of a family (↑), containing one or more species (↓). The name of a genus is a Latin word, written with a capital letter, e.g. *Lingula*.
species (*n*) a division of a genus (↑). The members of a species are all very much alike. In living forms they *interbreed* among themselves, i.e. they can become parents of young, who can in turn also become parents. Pairs from different species cannot, on the other hand, produce young. The name of a fossil species is a Latin word, written with a small letter, which comes after the name of the genus, e.g. *Didymograptus murchisoni*.
type (*n*) a fossil that represents the characters of a species (↑), a *holotype*; or a genus (↑), a *genotype*.

微體化石(名) 極微小的化石(↑)，即在顯微鏡下才能看見的化石。有些微體化石在地層古生物學(第 111 頁)上很重要，例如有孔蟲目(第 104 頁)。

微體古生物學(名) 研究微體化石(↑)的學科。(形容詞為 micropalaeontological)

微動物群(名) 微體化石(↑)中的一個動物群(↑)。

化石埋藏學(名) 研究化石(↑)形成途徑的學科。

分類學(名) 將動、植物，不論是現今生存的還是已成化石的(↑)，按其結構分門別類整理的科學。(形容詞為 taxonomic)

分類單位(名) 分類中的分類群或分類的單位。例如屬(↓)、科(↓)。(形容詞為 taxonomic，名詞複數為 *taxa*)

門(名) 動、植物界的主要劃分單位之一。(複數為 *phyla*)

綱(名) 門(↑)的劃分單位，例如昆蟲綱(第 108 頁)、哺乳動物綱(第 109 頁)。

目(名) 綱(↑)的劃分單位。

總科；超科(名) 目(↑)的劃分單位。總科的名稱詞尾為 -oidea。

科(名) 目(↑)的劃分單位。科的名稱詞尾為 -idae。

亞科(名) 科(↑)的劃分單位。亞科的名稱詞尾為 -inae。

屬(名) 科(↑)的劃分單位。一個屬包含一個或幾個種(↓)。屬名為拉丁詞，首字母大寫。如 *Lingular* 舌形貝屬。(複數為 *genera*)

種(名) 屬(↑)的劃分單位。一個種的所有成員間都非常相像。在現存各類型中，同屬一個種的成員之間才能配育，繁育有生殖能力的後代，而屬不同種的配對則不能繁育。化石的種名是一個拉丁詞，在屬名之後用小寫字母寫出。例如莫企遜對筆石 *Didymograptus murchisoni*。

模式(名) 能體現一個種(↑)或一個屬(↑)的特徵的代石，前者為"全型"，後者為"屬型"。

palaeoecology (*n*) the study of fossil animals and plants in relation to the conditions under which they lived – the environment (p.81).

古生態學(名) 研究動、植物化石與其生存條件(即環境(第 81 頁))關係的學科。

palaeobiogeography (*n*) the study of the way in which animals and plants were arranged in space on the surface of the Earth in the geological past.

古生物地理學(名) 研究遠古地質時代的動、植物在地球表面空間佈局方式的學科。

habitat (*n*) the environment (p.81) in which an animal or plant lives or lived.

生境(名) 指動物或植物現時或往昔所生活的環境(第 81 頁)。

assemblage (*n*) (1) all the fossils that are present in a particular bed or stratum (p.80); (2) the fossils of a species (p.99), or some other small group, from a particular horizon (p.112) or place; (3) a group of fossils found by themselves that are thought to belong to one animal.

組合；集合(名) (1)某特定層或地層(第 80 頁)中存在的全部化石；(2)從一個特定層位(第 112 頁)或地點取得的同一個種(第 99 頁)或若干其他小種群的一組化石；(3)依化石本身查明而認為同屬一隻動物的一組化石。

fossil community a group of fossils found in the same place where they lived together.

化石群落 在原共同生活處發現的一組化石。

a shell-bed community with lamellibranchs, ammonites, echinoids, etc.
包含瓣鰓類，菊石，海膽等的貝殼層群落

plankton (*n*) all the organisms that float in the sea or in a lake and are carried about by the movement of the water; e.g. Foraminifera (p.104) and Radiolaria (p.104). **planktonic** (*adj*).

浮游生物(名) 指漂游於海洋或湖泊中並被水體運動所攜帶的一切生物。例如有孔蟲(第 104 頁)和放射蟲(第 104 頁)。(形容詞為 planktonic)

microplankton (*n*) the smallest members of the plankton (↑); those that cannot easily be seen with the unaided eye.

小型浮游生物(名) 浮游生物(↑)中最小型的成員；肉眼難以看見的浮游生物。

phytoplankton (*n*) all the plant forms that float in the sea or in lakes. *See also* **plankton** (↑).

浮游植物(名) 指漂游於海洋或湖泊中的一切植物形式，參見**浮游生物**(↑)。

nekton (*n*) all the animals that swim in the sea or in lakes.

自泳生物(名) 於海洋或湖泊中營游泳生活的一切動物。

benthos (*n*) all the animals and plants that live on the sea floor. **benthonic** (*adj*).

底棲生物(名) 指所有生活於海底的動物和植物。(形容詞為 benthonic)

pelagic (*adj*) animals that live in the open sea but not on the sea floor are called pelagic. They include the nekton (↑) and the plankton (↑).

遠洋的(形) 指生活於開濶海洋而非底棲的動物，稱為遠洋的動物，包括自泳生物(↑)和浮游生物(↑)。

a sessile fossil: a crinoid
一種固着的化石：海百合

aerobic (*adj*) needing free oxygen in order to live or be active. *See also* **anaerobic** (↓).

anaerobic (*adj*) not needing oxygen in order to live or be active. *See also* **aerobic** (↑).

epifauna (*n*) a fauna (p.98) that in life is fixed to another and larger organism (p.98).

epizoon (*n*) an organism of an epifauna (↑). **epizoan** (*adj*).

sessile (*adj*) describes an organism (p.98) that is closely attached to a surface, such as the sea floor or another organism. Applied to benthos (↑), 'sessile' means attached to the sea floor. *See also* **sedentary** (↓).

sedentary (*adj*) describes an organism (p.98) that is attached, as, for example, an oyster. *See also* **sessile** (↑).

biocoenose (*n*) an assemblage (↑) of organisms (p.98) that live together as one community (↑).

biolith (*n*) a deposit of organic (p.17) material or material formed by the activities of organisms.

喜氧的(形) 生存或活動需要有游離氧的。參見**厭氧的**(↓)。

厭氧的(形) 生存或生活不需要氧的。參見**喜氧的**(↑)。

體外寄生動物群(名) 固着在其他較大生物體(第 98 頁)上為生的動物群(第 98 頁)。

體外寄生物(名) 體外寄生動物群(↑)中的一個生物體。(形容詞為 epizoan)

固着的(形) 形容一類生物(第 98 頁)緊附着於一個表面，例如緊附於海底或其他生物體。這個詞用於底棲生物(↑)時，是指附着於海底。參見**定居的**(↓)。

定居的(形) 形容一種被固定的生物體(第 98 頁)，如牡蠣類。參見**固着的**(↑)。

生物群落(名) 作為一個群落(↑)共同生活之生物(第 98 頁)組合(↑)。

生物岩(名) 有機(第 17 頁)物質或生物活動所形成物質的沉積物。

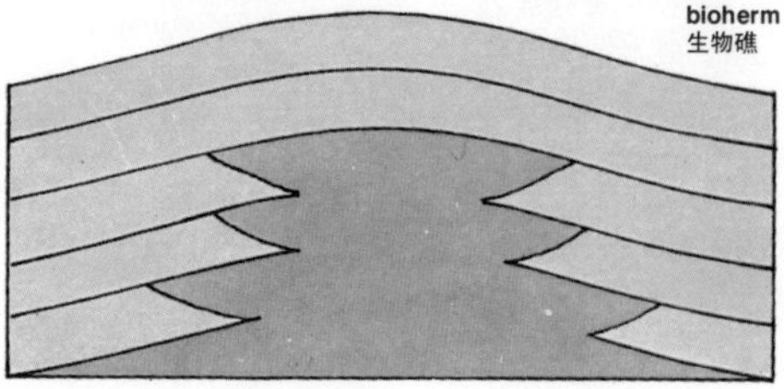

bioherm (*n*) an organic deposit (p.80) built largely or entirely of the remains of fixed organisms (p.98); a fossil reef (p.38). A bioherm is usually shaped like a small hill. It is a special type of biolith (↑).

biostrome (*n*) a mass of organic material in the form of a sheet or bed (p.80) built by sedentary organisms (↑, p.98) that have been preserved in place. *See also* **bioherm** (↑).

biogenic (*adj*) produced by organisms (p.98).

stromatolites (*n*) rounded sedimentary structures (p.83) formed by the plants called algae (p.110), which live in water. The oldest stromatolites are of Pre-Cambrian age (p.114) and are among the oldest fossils known.

生物礁(名) 大部或全部由固着生物體(第 98 頁)的遺骸構成的有機沉積物(第 80 頁)；一種化石礁(第 38 頁)。生物礁之形態通常像一座小丘，它是生物岩(↑)中的一種特殊類型。

生物層(名) 由原地保存的定居生物體(↑、第 98 頁)所構成的成層(第 80 頁)或成片的有機物質。參見**生物礁**(↑)。

生物成因的(形) 由生物體(第 98 頁)形成的。

疊層石(名) 由生活於水中的藻類(第 110 頁)植物所形成的圓形沉積構造(第 83 頁)，最古老的疊層石時代是前寒武紀(第 114 頁)，它亦是已知的最古老的化石。

evolution (*n*) the process by which new forms of living things can develop from earlier forms by passing on small changes from one generation to the next. **evolutionary** (*adj*), **evolve** (*v*).

adaptation (*n*) a character of an organism (p.98) that fits it for a particular environment (p.81); (2) the process by which an organism (p.98) is changed to become more fit for its environment (p.81). **adaptive** (*adj*), **adapt** (*v*).

natural selection the process that according to Darwin's theory controls which members of a population of animals or plants will live to produce young and pass on their genes (p.156) to the next generation.

adaptive radiation the development of new species (p.99) that takes place when the descendants of a taxon (p.99) evolve (↑) by natural selection (↑) in fitting themselves to various environments (p.81).

mutation (*n*) a discontinuous change in a gene (p.156) or an organism (p.98) that can be inherited, i.e. passed on to its descendants. **mutate** (*v*).

ancestral (*adj*) referring to organisms (p.98) from which later organisms are descended.

ontogeny (*n*) the course followed by the life history of an individual animal or plant.

phylogeny (*n*) the course followed by the evolution (↑) of a species (p.99) or other taxonomic (p.99) group. **phylogenetic** (*adj*).

diversity (species) the range of variation that is shown by a species (p.99).

diversification (*n*) the process of becoming more diverse, that is, more different. **diversify** (*v*).

lineage (*n*) a line of descent from earlier members of the same or a similar group of animals or plants; a series of fossils that show a course of evolution (↑).

divergence (*n*) the development of a new population of organisms from an earlier one. **diverge** (*v*), **divergent** (*adj*).

radiation (*n*) the evolutionary divergence (↑) of a group of species. **radiate** (*v*), **radiating** (*adj*).

extinction (*n*) the dying out of a (whole) group of animals or plants. **extinct** (*adj*).

演化；進化(名)　生物體將細小變化從一代傳給下一代而從早期類型發展成新類型的過程。(形容詞為 evolutionary，動詞為 evolve)

適應(名)　生物體(第 98 頁)使本身適合某種特定環境(第 81 頁)的一種特性；(2)使生物體(第 98 頁)改變成更適應其生活環境(第 81 頁)的過程。(形容詞為 adaptive，動詞為 adapt)

自然選擇；天擇　根據達爾文學說，支配一個動物或植物種群的那些成員將生存到產生後裔，並將其基因(第 156 頁)傳遞給下一代的過程。

適應輻射　一個分類單位(第 99 頁)的生物後裔，在適應各種環境(第 81 頁)的過程中藉自然選擇(↑)而演化(↑)時，出現新種(第 99 頁)的發展。

突變(名)　指某一基因(第 156 頁)或某一生物體(第 98 頁)內可以遺傳，即可傳遞給其後代的不連續變化。(動詞為 mutate)

祖先的(形)　指傳宗接代的生物(第 98 頁)。

個體發育(名)　由單個動物或植物個體的生活史所繼承的過程。

系統發育(名)　一個物種(第 99 頁)或其他分類級(第 99 頁)群體的演化(↑)所繼承的過程。(形容詞為 phylogenetic)

種的變異度　一個物種(第 99 頁)所顯示的變異範圍。

異化(名)　導致產生更大變異，即更大差別的過程。(動詞為 diversify)

譜系；世系(名)　從同類或相似的一群動物或植物的早期成員延續的一系列後裔；顯示演化(↑)進程的一系列化石。

趨異(名)　從一個較早的生物群體演變發展成一個新群體的過程。(動詞為 diverge，形容詞為 divergent)

輻射(名)　一個種群的演化趨異(↑)。(動詞為 radiate，形容詞為 radiating)

滅絕(名)　一(整)群動物或植物的死亡消失。(形容詞為 extinct)

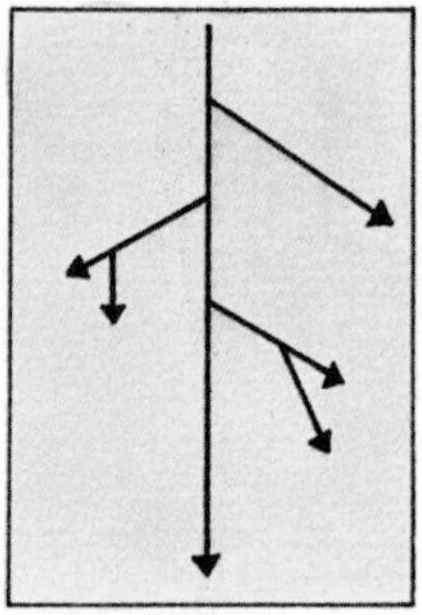

phylogenetic tree
系統發育樹

extinctions and appearances of new forms during
滅絕與新類型的出現

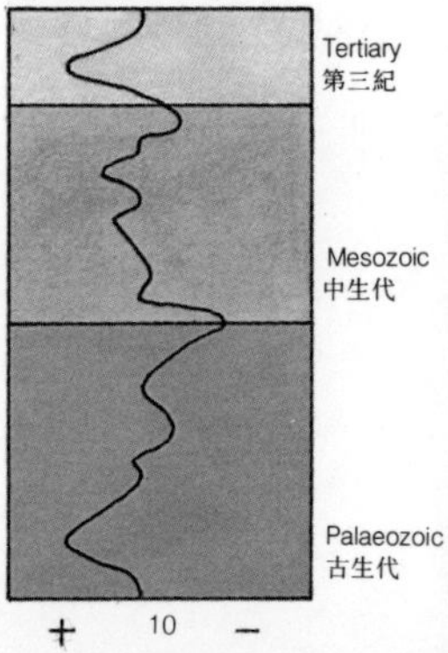

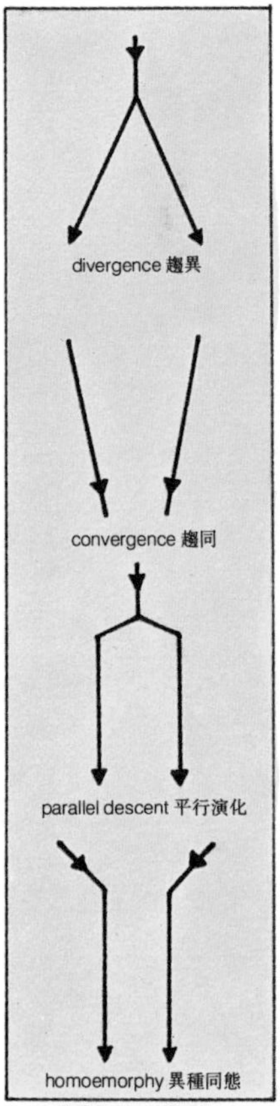

evolutionary patterns
演化型式

convergence (*n*) the development of similar forms in different groups of plants or animals at different places or at different times because of the effects produced by similar environments on their separate evolutions (↑). **convergent** (*adj*).

polyphyletic (*adj*) a group of organisms (p.98) is polyphyletic if its members have evolved (↑) from different series of earlier forms by convergent evolution (↑).

trend (*n*) the evolution (↑) of a particular structural feature within a group of organisms (p.98).

transient (*n*) a stage in the phylogeny (↑) of a species; a stage in any closely spaced evolutionary (↑) series (p.159).

homoeomorphy (*n*) the occurrence of similar forms (shapes) in members of the same phylum (p.99).

explosive evolution a diversification (↑) that for a time takes place much more rapidly than at other times; e.g. the very rapid evolution of the fishes in the late Silurian and early Devonian (p.114).

evolutionary burst = explosive evolution (↑).

quantum evolution the sudden appearance within a short space of geological time of large taxonomic (p.99) units, e.g. orders (p.99). *See also* **explosive evolution** (↑).

speciation (*n*) the production of new species (p.99) by the splitting or division of earlier species in the course of evolution (↑).

bioseries (*n*) an evolutionary (↑) series of fossils. It may be a series (p.159) of whole individuals or a series of specimens that show trends in particular features.

cladogenesis (*n*) the development of species (p.99) by division of the line of descent. *See diagram.*

cladistics (*n*) a cladistic classification is one that is based on the branching pattern (the *cladogram*) of the evolution of a group of animals. Groups that have separated from each other more recently are put closer together than those that have separated at earlier times. This type of classification is unlike those that have been used in the past.

趨同(名) 在不同地點或不同時期的不同動物或植物群體，由於相似環境對其各別演化(↑)的影響而產生相似形態的發展過程。(形容詞為 convergent)

多系列演化的；多源演化的(形) 若一個生物種群的成員係由不同系列的早期類型通過趨同演化(↑)而進化(↑)的，則稱此生物種群(第 98 頁)是多系列演化的。

趨勢(名) 一個生物(第 98 頁)種群之內某一特定結構特徵的演化(↑)。

漸變階段(名) 一個生物種的系統發育(↑)中的一個階段；任何緊密演化(↑)系列(第 159 頁)中的一個階段。

異種同態(名) 同一門(第 99 頁)的成員中相似類型(形態)的出現。

突發式演化 指異化(↑)在某一段時間內發生較其他時間發生迅速得多。例如魚類在晚志留世和早泥盆世(第 114 頁)的演化極快。

演化突變 同突發式演化(↑)。

量子式演化 大分類(第 99 頁)單位如目(第 99 頁)在很短的地質年代間隔內突然出現。參見**突發式演化**(↑)。

物種形成(名) 演化(↑)過程中由於早期種的分裂或分化導致產生新種(第 99 頁)。

生物系列(名) 化石的演化(↑)系列。可以是一個系列(第 159 頁)的全部個體，也可以是顯示個別特徵趨勢的一系列標本。

分枝演化(名) 物種(第 99 頁)由於世系分化而發生的進化(見圖)。

親緣分枝法(名) 親緣分枝法的分類是根據一個動物種群的演化分枝模式(進化分枝圖)作的分類。相互分離時間較近的種群擺放在比那些更早時間已相互分離的種群更為接近的位置。這種分類與已往所使用的分類不同。

Invertebrata (*n.pl.*) the invertebrates: all animals without backbones (the long row of bones in the middle of the back). **invertebrate** (*adj*).

Protozoa (*n.pl.*) the phylum (p.99) of the simplest animals, consisting of only one cell. Most of them are very small. The Protozoa include the Foraminifera (↓) and Radiolaria (↓). **protozoan** (*adj*).

Foraminifera (*n.pl.*) Protozoa (↑) with shells, which are commonly divided into parts (*chambers*). The shell may be made of calcite (p.51) or arenaceous material (p.85). Most Foraminifera are marine. The Foraminifera are valuable in stratigraphical correlation (p.117), especially in the Tertiary (p.115).

forams (*n.pl.*) short for Foraminifera (↑).

無脊椎動物（名、複） 無脊椎動物是指一切沒有脊柱（背部中央的長排骨骼）的動物。（形容詞為 invertebrate）

原生動物門（名、複） 只由一個細胞組成的最簡單動物的一門（第 99 頁）。其中大部分都十分細小，原生動物門包括有孔蟲目（↓）和放射蟲目（↓）。（形容詞為 protozoan）

有孔蟲目（名、複） 具殼的原生動物（↑），其殼一般分隔成腔室。殼體由方解石（第 51 頁）或砂質物質（第 85 頁）構成。有孔蟲類大部分是海洋動物，有孔蟲在地層對比（第 117 頁）上，特別是在第三系（第 115 頁）地層對比上很有價值。

有孔蟲（名、複） 有孔蟲目（↑）的簡稱。

Foraminifera 有孔蟲

Radiolaria (*n.pl.*) marine planktonic (p.100) Protozoa (↑). They have skeletons of silica (p.16). Radiolaria are not common as fossils but are known from the Pre-Cambrian (p.114) onwards.

Metazoa (*n.pl.*) animals with many cells (p.153). **metazoan** (*adj*).

Porifera (*n.pl.*) the phylum (p.99) of the *sponges*, the simplest of the Metazoa (↑). They live in water, most of them in the sea. Small pieces of the internal skeletons of sponges occur as fossils (sponge spicules).

Archaeocyatha (*n.pl.*) marine animals of a simple kind with a calcareous (p.86) skeleton, usually shaped like a cone. They lived on the bottom of the sea but are now extinct.

放射蟲目（名、複） 海洋浮游的（第 100 頁）原生動物（↑）。這些動物具有二氧化矽質的骨骼（第 16 頁）。放射蟲通常不形成化石，它自前寒武紀起（第 114 頁）就已存在。

後生動物（名、複） 具多細胞（第 153 頁）的動物。（形容詞為 metazoan）

多孔動物門（名、複） 為海綿類動物門（第 99 頁）。是最簡單的後生動物（↑）。係水生，大部分生活於海洋中，海綿的內骨骼小碎片可成為化石（海綿骨針）。

古杯動物門（名、複） 具鈣質（第 86 頁）骨骼，通常狀似錐體的簡單的海洋動物。這類動物生活於海底，但現已滅絕。

Coelenterata *(n.pl.)* a phylum of animals with radial symmetry (p.42): symmetry like that of a wheel. They have a mouth and a space inside the body (the *coelenteron*) in which they digest their food. The Coelentera include the classes Scyphozoa (jellyfish), Hydrozoa (hydra and obelia), and Anthozoa (corals and sea-anemones). The corals (↓) are the most important fossil coelenterates.

腔腸動物門(名、複) 身體呈車輪狀輻射對稱(第 42 頁)的動物門，這些動物有口腔，體內尚有一個用於消化食物的空腔(腸腔)。腔腸動物門包括鉢水母綱(水母)、水螅綱(水螅和藪枝蟲)和珊瑚蟲綱(珊瑚和海葵)。珊瑚(↓)是最重要的腔腸動物化石。

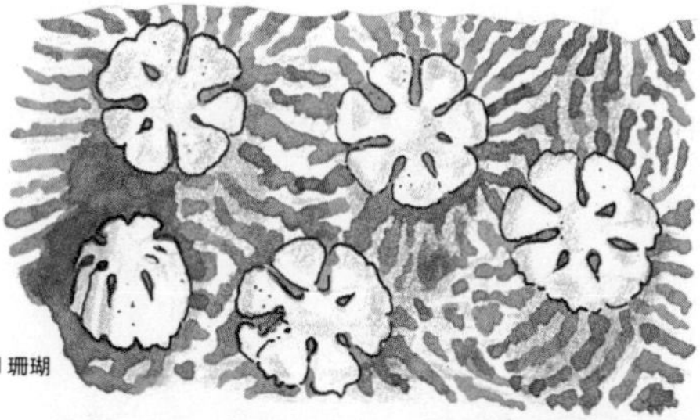

coral 珊瑚

corals (*n.pl.*) marine animals living on the sea floor with skeletons made of calcium carbonate, $CaCO_3$. They belong to the class Anthozoa: see Coelentera (↑). As fossils the corals are important in two ways: as builders of coral reefs (rising to just below or just above the surface of the water) and as zone fossils (p.111).

珊瑚(名、複) 生活於海底、其骨骼由碳酸鈣($CaCO_3$)構成的海洋動物。珊瑚屬珊瑚蟲綱：見腔腸動物門(↑)。作為化石，珊瑚有兩個重要方面：即作為珊瑚礁(剛好上升到水面上下)的營造者和作為分帶化石(第 111 頁)。

Vermes (*n.pl.*) worms. Animals with soft bodies, found only rarely as fossils.

蠕形動物；蠕蟲總門(名、複) 蠕蟲類，具軟體的動物，其化石罕見。

worms *see* **Vermes** (↑).

蠕蟲類 見**蠕形動物**(↑)。

Annelida (*n.pl.*) the annelids segmented worms (↑), i.e. worms whose bodies are made up of a number of parts or *segments*. Most of them live in the sea. They occur as fossils.

環節動物門(名、複) 環節動物為節狀蠕蟲(↑)，即身體由一些節狀部分組成的蠕蟲，主要生活在海洋中，有化石產出。

Brachiopoda (*n.pl.*) the brachiopods, a phylum (p.99) of marine animals with a shell in two parts, called *valves*, one upper and one lower. A typical brachiopod has a muscle, called the *pedicle*, which passes through a hole at one end of the shell and is fixed to the sea bed. There are two classes (p.99) of Brachiopoda: the Inarticulata and the Articulata. In the Articulata the two halves of the shell are joined by a hinge; the Inarticulata have no hinge. Brachiopods are important as fossils.

腕足動物門(名、複) 腕足類，是具殼體的海洋動物的一門(第 99 頁)，殼體分成上、下兩部分，稱為"殼瓣"。典型的腕足類有一個叫肉莖的肌肉，肉莖穿過殼體一端的孔並固着在海底。腕足動物門分無鉸綱和有鉸綱兩綱(第 99 頁)。有鉸綱中的貝殼，其兩半殼由鉸合部聯結起來，無鉸綱沒有鉸合部。作為化石，腕足類具重要的意義。

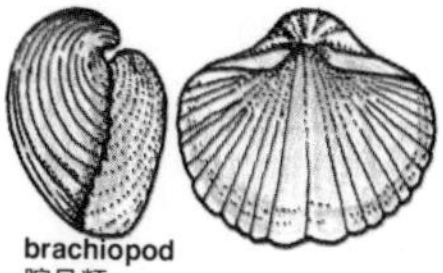

brachiopod
腕足類

Bryozoa (*n.pl.*) very small marine animals that form groups or colonies. They have skeletons made of calcium carbonate, $CaCO_3$.

苔蘚動物門(名) 形成群體的極細小海洋動物，其骨架由碳酸鈣($CaCO_3$)構成。

Polyzoa (*n.pl.*) = Bryozoa (↑).

苔蘚蟲(名、複) 同苔蘚動物門(↑)。

Echinodermata (*n.pl.*) the echinoderms, a phylum (p.99) of marine animals with skeletons made of calcium carbonate, $CaCO_3$. The skeletons are made up of plates and rods. All echinoderms show radial symmetry (p.42), like that of a wheel, usually fivefold. Two phyla, the Crinoidea and the Echinoidea (↓), are of geological importance.

棘皮動物門(名、複) 棘皮類動物，屬這一門(第 99 頁)的海洋動物，其骨骼由碳酸鈣($CaCO_3$)構成。它的骨骼由板狀和棒狀體組成，所有棘皮動物都呈現車輪狀軸射對稱(第 42 頁)，通常為五輻對稱。海百合綱和海膽綱(↓)這兩綱具地質意義。

Crinoidea (*n.pl.*) the crinoids, a class (p.99) of the Echinodermata (↑). The Crinoidea include the modern sea-lilies. Some crinoids are sessile (p.101); others swim freely. The fixed crinoids have a stem like a flower with a cup (called the *calyx*) at the top. Crinoids are important as index fossils (p.111).

海百合綱(名、複)海百合屬棘皮動物門(↑)的一綱(第 99 頁)。海百合綱包括現代海百合。有些海百合是固着的(第 101 頁)，有些則是自游的。固着的海百合有一個柄，像一支花，頂端有一杯狀物，稱為萼。海百合是重要的標準化石(第 111 頁)。

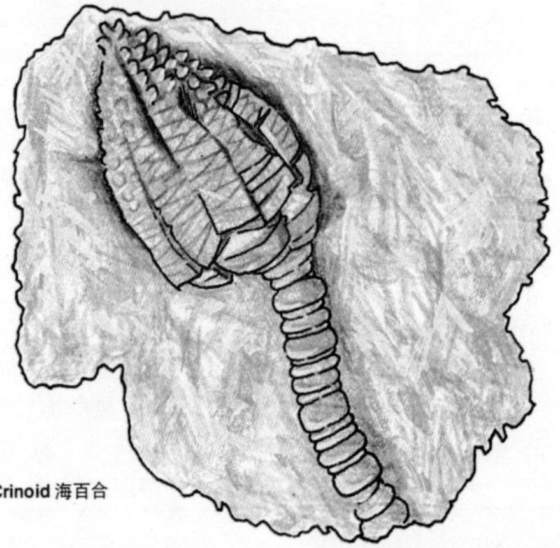

Crinoid 海百合

Echinoidea (*n.pl.*) the echinoids or sea-urchins. A class of the Echinodermata (↑). The skeleton or *test* is made up of plates of calcium carbonate, $CaCO_3$, and is either round or heart-shaped. It carries rows of sharp *spines*. Echinoids are valuable as zone fossils (p.111) in the Mesozoic and Tertiary (p.115).

海膽綱(名、複) 海膽屬棘皮動物門(↑)的一綱。其骨骼或外殼呈圓形或心形。由碳酸鈣($CaCO_3$)骨板組成。它具有排列成行的骨針。海膽是很寶貴的中生代和第三紀(第 115 頁)分帶化石(第 111 頁)。

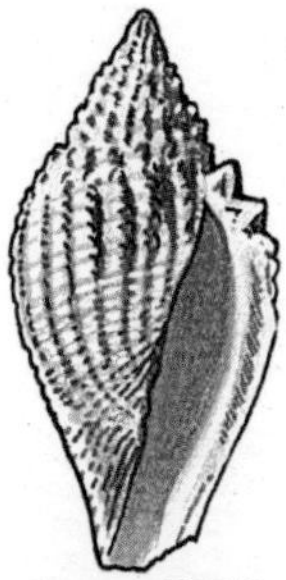
Gastropod 腹足類

Lamellibranch 瓣鰓類

Ammonite 菊石

Mollusca (*n.pl.*) the molluscs, a phylum of invertebrates most of which have shells. Some live on land, some in fresh water, and some in the sea. The Mollusca are divided into five classes, of which the Gastropoda (↓), the Lamellibranchiata (↓), and the Cephalopoda (↓) are of geological importance.

軟體動物門（名、複） 軟體動物屬無脊椎動物的一門，大部分具有貝殼。有生活在陸上的，有生活在淡水中的，也有生活在海水中的。軟體動物劃分為五綱，其中腹足綱（↓）。瓣鰓綱（↓）和頭足綱（↓）具有重要的地質意義。

Gastropoda (*n.pl.*) the gastropods, a class of the Mollusca (↑). They have shells in one piece which are usually twisted like a screw.

腹足綱（名、複） 腹足動物屬軟體動物門（↑）的一綱，此類動物只有一個通常扭曲成螺旋狀的殼。

Lamellibranchiata (*n.pl.*) the lamellibranchs, a class of the Mollusca (↑). Lamellibranchs have a shell of calcium carbonate, $CaCO_3$, which is in two parts or *valves*, one on each side of the animal. The valves are joined by a hinge and both valves have 'teeth' at the hinge. Lamellibranchs are important as fossils.

瓣鰓綱（名、複） 瓣鰓類屬軟體動物門（↑）的一綱。它具有碳酸鈣質（$CaCO_3$）貝殼，貝殼分二部分或兩瓣，每側各一瓣，兩瓣由鉸合部連接在一起，每一瓣之鉸合部均有"齒"。瓣鰓類化石是重要的化石。

Pelecypoda, pelecypods (*n.pl.*) = Lamellibranchiata (↑).

斧足綱，斧足類（名、複） 同瓣鰓綱（↑）。

Cephalopoda (*n.pl.*) the cephalopods, a class of the Mollusca (↑). The Cephalopoda have a shell that is divided into separate parts (called *chambers*) by walls (called *septa*; sing. *septum*). All cephalopods are marine.

頭足綱（名、複） 頭足類屬軟體動物門（↑）的一綱。頭足綱動物的殼被內壁（隔壁）分隔成許多不相連部分（房室）。頭足類動物全都是海洋動物。

Nautiloidea (*n.pl.*) the nautiloids, an order (p.99) of the Cephalopoda (↑). They have shells that are shaped like a long cone, which may be straight or curved. The septa (↑ Cephalopoda) that divide up the shell are gently curved where they meet the shell and there is a tube (called the *siphuncle*) that runs through the shell, passing through the centres of the septa.

鸚鵡螺亞綱（名、複） 鸚鵡螺屬頭足綱（↑）的一目（第 99 頁），此類動物具有直的或彎曲的長錐形殼體，分隔殼體的隔壁（↑ 頭足綱）與外殼之接觸處微微彎曲，同時具有一條穿過隔壁中央貫穿整個殼體的管狀物（體管）。

Ammonoidea (*n.pl.*) the ammonites, an extinct order (p.99) of the Cephalopoda (↑). The septa (↑ Cephalopoda) that divide up the shell are less simple in shape than those of the Nautiloidea (↑) and the siphuncle (↑) is at the outer edge of the shell, unlike that of a nautiloid. The ammonites are important as fossils.

菊石亞綱（名、複） 菊石屬頭足綱（↑）的一目（第 99 頁），現已滅絕。分隔殼體的隔壁（↑ 頭足綱）比鸚鵡螺（↑）複雜，與鸚鵡螺不同，其體管（↑）位於殼體之外緣。菊石是重要的化石。

Belemnoidea (*n.pl.*) an extinct order (p.99) of the Cephalopoda (↑). The most important members geologically are the belemnites. Their fossil shells are in the shape of a long cone with a conical hole at the broad end.

箭石目（名、複） 屬頭足綱（↑）中的一個目（第 99 頁），現已滅絕。箭石是地質上最重要的成員，其化石殼呈長錐形，在其寬端有一圓錐狀的穴。

belemnites (*n.pl.*) *see* **Belemnoidea** (↑).

箭石（名、複） 見箭石目（↑）。

Arthropoda (*n.pl.*) the arthropods, a large phylum (p.99) of invertebrates (p.104): animals with bodies divided into a number of parts called *segments* and with an external skeleton or *carapace* made of chitin (a hard compound of carbon, hydrogen, nitrogen, and oxygen) and with jointed limbs. The classes (p.99) that are of geological importance are the Trilobita (↓), the Eurypterida (↓), the Ostracoda (↓), and the Crustacea (↓). The arthropods also include the Insecta (insects), the Myriapoda (centipedes and millipedes), and the Arachnida (spiders, scorpions, and mites).

節肢動物門(名、複) 節肢動物是無脊椎動物(第 104 頁)的一大門(第 99 頁)，動物身體分成幾個節，覆以幾丁質(碳、氫、氮和氧組成的硬質化合物)的外骨骼或背甲並有節肢。地質上具重要意義的綱(第 99 頁)是三葉蟲綱(↓)、板足鱟亞綱(↓)、介形亞綱(↓)和甲殼綱(↓)。節肢類動物也包括昆蟲綱(昆蟲類)、多足綱(蜈蚣和千足蟲)以及蜘蛛綱(蜘蛛、蝎和蟎)。

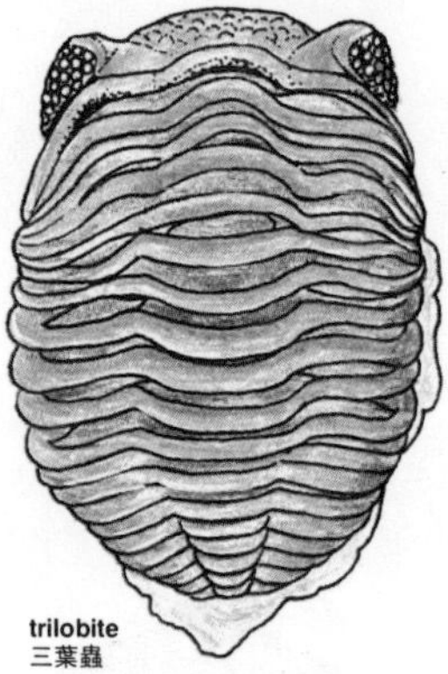
trilobite 三葉蟲

Trilobita (*n.pl.*) the trilobites, an extinct (p.102) class (p.99) of the Arthropoda (↑). The part of the external skeleton covering the back of the trilobite is divided into three parts. The body of the animal is also divided into a head (called the *cephalon*), a thorax and a tail (the *pygidium*). Trilobites are important in the Palaeozoic (p.114).

三葉蟲綱(名、複) 三葉蟲屬節肢動物門(↑)的一目(第 99 頁)，現已滅絕(第 102 頁)。三葉蟲背上覆蓋的外骨骼分為三部分。動物身體也分為頭(頭部)、胸和尾(尾部)。三葉蟲是重要的古生代(第 114 頁)動物。

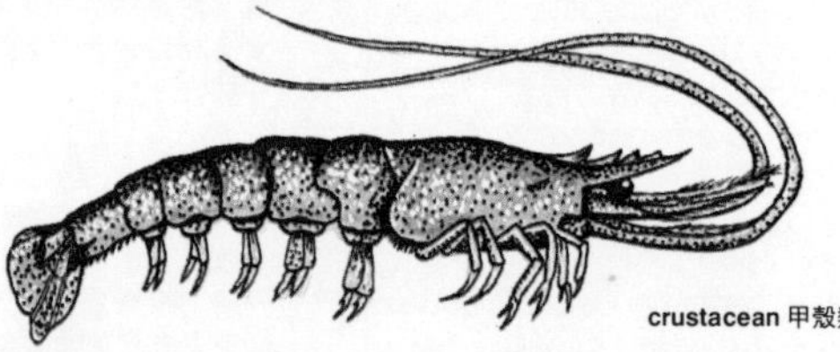
crustacean 甲殼類

Eurypterida (*n.pl.*) the eurypterids, an extinct class (p.99) of the Arthropoda (↑). They lived in fresh water and some were up to 2 m long.

Ostracoda (*n.pl.*) the ostracods, small arthropods (↑) living in fresh and salt water. The animal lives inside two *valves* (halves) of the *carapace* (shell). Ostracods are used as zone fossils (p.111), for example in the Jurassic (p.115).

Crustacea (*n.pl.*) the crustaceans, a class of the Arthropoda (↑). Crabs and lobsters are modern examples. Most crustaceans are marine.

Arachnida (*n.pl.*) the arachnids, a division of the Arthropoda (↑) that includes the modern spiders, scorpions, and mites.

Insecta (*n.pl.*) the insects, a class of the Arthropoda (↑). Fossils are uncommon.

板足鱟亞綱(名、複) 板足鱟屬節肢動物門(↑)的一目(第 99 頁)，現已滅絕。它們生活在淡水中，有些可長達 2 米。

介形亞綱(名、複) 介形類是生活在淡水或鹹水中的小型節肢動物(↑)。它們居住於兩瓣(對稱的兩個)背甲(貝殼)之中。介形類用作分帶化石(第 111 頁)，例如作為侏羅紀(第 115 頁)的分帶化石。

甲殼綱(名、複) 甲殼類屬節肢動物門(↑)的一綱。螃蟹和大螯蝦是現成的例子。大部分甲殼類是海生動物。

蜘蛛綱(名、複) 蜘蛛類屬節肢動物門(↑)的一個類別，包括現代生存的蜘蛛類、蝎類和蟎類。

昆蟲綱(名、複) 昆蟲屬節肢動物門(↑)的一綱，其化石不常見。

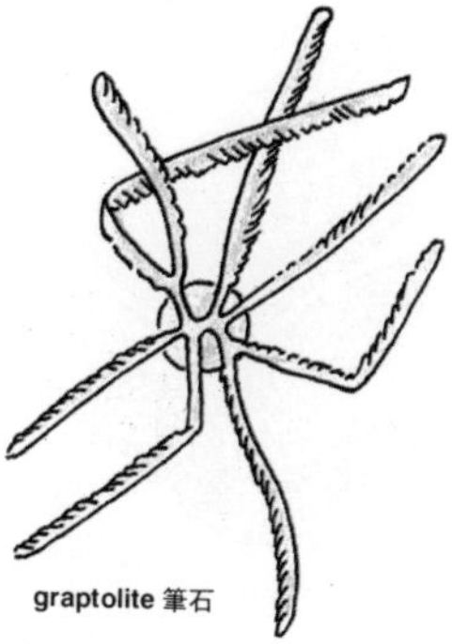
graptolite 筆石

Chordata (*n.pl.*) the chordates, a phylum (p.99) of animals that have at some time in their life-history a hollow tube that can be bent (the *notochord*) or a series of jointed pieces of bone or other material (*vertebrae*) inside the body. Two groups are of geological importance: the graptolites (↓) and the vertebrates (↓).

脊索動物門（名、複） 脊索動物是動物界的一門（第 99 頁），其生活史的某個階段體內有一條可彎曲的空心管（脊索）或一系列有節的骨片或其他物質（脊椎）。筆石（↓）和脊椎動物（↓）這兩個種群在地質上具有重要意義。

Graptolithina (*n.pl.*) the graptolites. Extinct branching organisms that lived closely together. They are found in rocks of Lower Ordovician to Lower Devonian ages (p.114).

筆石綱（名、複） 筆石是緊密聚生在一起的分枝狀生物，現已滅絕。它們被發現於早奧陶世至早泥盆世（第 114 頁）時期的岩石。

Vertebrata (*n.pl.*) vertebrates, the subphylum (p.99) of animals with a skeleton of *vertebrae* inside the body. This skeleton may be made of cartilage (p.152) or of bone. The Vertebrata can be divided into two superclasses, the Agnatha (↓) and the Gnathostoma (↓).

脊椎動物亞門（名、複） 脊椎動物是體內帶脊椎骨骼的動物界的一亞門（第 99 頁）。其骨骼可由軟骨（第 152 頁），也可由硬骨構成。脊椎動物亞門又劃分為兩個超綱：無頜超綱（↓）和有頜超綱（↓）。

Pisces (*n.pl.*) the fishes. They are the earliest and simplest vertebrates (↑). The earliest fossil fish are found in Ordovician rocks (p.114).

魚綱（名、複） 魚類是最早、最簡單的脊椎動物（↑）。最早的化石魚發現於奧陶系岩石（第 114 頁）。

Agnatha (*n.pl.*) a superclass of the Vertebrata (↑): jawless fish. Fossil forms are found in rocks of Devonian age (p.114).

無頜超綱（名、複） 脊椎動物亞門（↑）的一個超綱：無顎的魚。其化石發現於泥盆紀（第 114 頁）岩石。

Gnathostoma (*n.pl.*) a superclass of the Vertebrata (↑): vertebrates with jaws. Examples are known from the Devonian (p.114).

有頜超綱（名、複） 脊椎動物亞門（↑）的一個超綱：有顎的的脊椎動物。例證發現於泥盆紀（第 114 頁）岩石。

Amphibia (*n.pl.*) the amphibians, a class of vertebrates (↑) living on land and in water. Fossil amphibia are found in rocks of Upper Devonian to Recent age (p.115). **amphibian** (*adj*).

兩棲綱（名、複） 兩棲類是脊椎動物（↑）的一綱，既在陸上生活又在水中生活。兩棲類的化石發現於晚泥盆世至近代（第 115 頁）岩石。（形容詞為 amphibian）

Reptilia (*n.pl.*) the reptiles, a class of vertebrates (↑) which evolved (p.102) from the Amphibia (↑) in Upper Carboniferous times (p.114). **reptilian** (*adj*).

爬蟲綱（名、複） 爬蟲類屬脊椎動物（↑）的一綱，自晚石炭世（第 114 頁）兩棲綱（↑）演化（第 102 頁）而來。（形容詞為 reptilian）

dinosaurs (*n.pl.*) a group of Mesozoic reptiles (↑). Some were very large: up to 35 m long.

恐龍類（名、複） 中生代爬蟲動物（↑）的一個種群，有些身軀巨大，長達 35 米。

Aves (*n.pl.*) the birds: a class of the Vertebrata (↑). Very few fossils are known. The earliest are of Upper Jurassic age (p.115). **avian** (*adj*).

鳥綱（名、複） 鳥類屬脊椎動物亞門（↑）的一綱。已知化石很稀少，發現的最早化石屬晚侏羅世（第 115 頁）。（形容詞為 avian）

Mammalia (*n.pl.*) the mammals, a class of warm-blooded vertebrates with hair and teeth of various shapes. The mother feeds the young with her milk. Fossil mammals range from Jurassic to Recent (p.115) and are most important in the Tertiary (p.115). **mammalian** (*adj*).

哺乳綱（名、複） 哺乳類動物屬有毛髮和各類形態牙齒的溫血脊椎動物的一綱。母親以其乳汁喂養幼體。哺乳類化石自侏羅紀延伸至全新世（第 115 頁），而在第三紀（第 115 頁）尤其重要。（形容詞為 mammalian）

palaeobotany (*n*) the study of fossil plants. **palaeobotanical** (*adj*).

flora (*n.pl.*) the plants of a particular place or time. In palaeobotany (↑), the plants in a stratigraphical (p.112) unit or a geological area.

algae (*n.pl.*) the sea weeds and related plants. Fossil algae are known from the Pre-Cambrian (p.114).

Diatomaceae (*n.pl.*) the diatoms; single-celled algae (↑) whose cell walls are full of silica, SiO_2.

vascular plants plants having a *vascular system*: a system of cells and tissues (p.160) for carrying water, mineral salts, and food through the plant. The vascular system also gives strength and support to the plant. The vascular plants are divided into two groups: the Pteridophyta, or spore-bearing plants, and the Spermatophyta, or seed-bearing plants (↓).

Pteridophyta (*n.pl.*) the spore-bearing plants, one of the main divisions of the plant kingdom. It includes the modern ferns (Filicales), horsetails (Equisetales), and club-mosses (Lycopodiales).

Pteridospermae (*n.pl.*) the seed-ferns: an extinct group of plants that were shaped like ferns, a group of flowerless plants but that bore seeds. They were important in the Carboniferous and Permian periods (p.114).

Spermatophyta (*n.pl.*) the spermatophytes or seed-bearing plants. They can be divided into the Gymnospermae (↓) or gymnosperms – the conifers etc. – and the Angiospermae (↓) or angiosperms – the flowering plants.

Gymnospermae (*n.pl.*) the gymnosperms, one of the main divisions of the plant kingdom. It includes the conifers (Coniferales), ginkgos (Ginkgoales), and the cycads (Cycadales).

Angiospermae (*n.pl.*) the angiosperms or flowering plants. One of the main divisions of the plant kingdom and the most important group since the Cretaceous (p.115).

palynology (*n*) the study of fossil spores and pollen (the cells and powder by which plants reproduce themselves). Study of fossil spores and pollen is useful in stratigraphical correlation (p.117).

古植物學(名)　研究植物化石的學科。(形容詞為 palaeobotanical)

植物群；植物區系(名、)　一個特定地區或特定時期內的植物。古植物學(↑)上是指某一地層(第 112 頁)單位或某一地質區內的植物。

藻類(名、複)　海草以及和海草有關的植物，前寒武紀(第 114 頁)地層中就有藻類化石。

矽藻綱(名、複)　矽藻屬單細胞藻類(↑)，其細胞壁充滿二氧化矽(SiO_2)。

維管植物　一種具有維管系統的植物。維管系統是一種輸送水、礦物鹽和食物到植物各部分的細胞和組織系統(第 160 頁)。它具有一定強度，可支撐植物。維管植物劃分為兩類：真蕨植物門(即具孢子的植物)和種子植物門(即具種子的植物(↓))。

真蕨植物門(名、複)　具孢子的植物，是植物界的主要劃分之一，包括現代羊齒(真蕨目)，木賊(木賊目)和石松(石松目)。

種子蕨綱(名、複)　種子蕨類植物：是一個形態像蕨類、現已滅絕的植物種群，也是一個無花却有種子的植物種群。是石炭紀和二疊紀(第 114 頁)的重要植物。

種子植物門(名、複)　種子植物類，即具種子的植物。可劃分為裸子植物亞門(↓)或裸子植物類(針葉樹等)和被子植物亞門(↓)或被子植物類(開花的植物)。

裸子植物亞門(名、複)　裸子植物類是植物界的主要劃分之一，包括針葉樹(松柏目)、銀杏(銀杏目)、蘇鐵(蘇鐵目)。

被子植物亞門(名、複)　被子植物類，即開花的植物，是植物界主要劃分之一，係自白堊紀(第 115 頁)以來最重要的種群。

孢粉學(名)　研究化石孢子和花粉的學科。孢子和花粉是植物進行繁殖的一些細胞和粉末，孢子和花粉化石的研究對於地層對比(第 117 頁)很有用處。

fossil plants 植物化石

stratigraphical palaeontology the study of fossils in order to understand the geographical distribution of animals and plants in the geological past and the history of life through geological time.

地層古生物學 研究化石以了解動、植物在地質史上的地理分佈及其在整個地質時代的生活史的學科。

zone fossil a fossil species (p.99) that is chosen as characteristic of those that are present in a particular stratum (p.80). The name of the zone fossil is given to the zone (p.117). A zone fossil should be found only in the stratum that is named after it.

分帶化石 在特定地層(第 80 頁)所存在的各種化石中被選作特徵化石的一個化石種(第 99 頁)。帶(第 117 頁)是以分帶化石的名稱命名。分帶化石必須是只在以之命名的地層中找到者。

index fossil = zone fossil (↑).

標準化石 同分帶化石(↑)。

range (*n*) the range of a fossil is the distance in time covered by its occurrence in the rocks, from its first appearance to its last.

延續時限；延限(名) 一種化石的延續時限，是指它在岩石中的產出即從開始出現到最後消失所經歷的時間間隔。

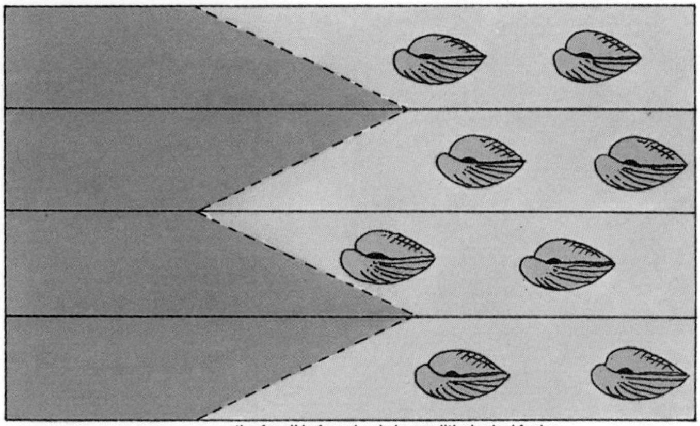

the fossil is found only in one lithological facies
此類化石僅發現於一種岩相中

facies fossil, facies fauna a facies fossil or a facies fauna is one that occurs only in rocks of a particular rock type or facies (p.117). Such fossils are of little or no use for stratigraphical correlation (p.117) and it is important to know that they are facies fossils.

指相化石，指相動物群 指相化石或指相動物群是只產出於某一特定岩石類型或岩相(第 117 頁)中的化石。這種化石對地層對比(第 117 頁)沒多大用處，重要的是要明白這些化石是指相化石。

derived fossil a fossil that has been eroded (p.20) from rocks that were deposited earlier and has been deposited again in a younger bed (p.80). A derived fossil is therefore not of the same age as the rocks in which it is found.

移積化石；轉生化石 指一個化石從早先沉積的岩石中剝蝕(第 20 頁)出來後又沉積在較新的地層(第 80 頁)中。所以移積化石與包含它的岩層並不屬同一時代。

stratigraphy (*n*) the study of stratified rocks (p.80), their nature, their occurrence, their relationships to each other and their classification. **stratigraphical, stratigraphic** (*USA*) (*adj*).

historical geology the study of the history of the Earth. It includes stratigraphy (↑).

Uniformitarianism (*n*) the view that geological processes were of the same kind in the past as they are today and produced similar results. *See also* **Catastrophism** (↓).

Catastrophism (*n*) the view, no longer held in geology, that the history of the Earth has to be explained by a series of violent events or catastrophes. *See also* **Uniformitarianism** (↑).

succession (*n*) the order in which rock-groups appear. When a succession is set out in the form of a table the beds (p.80) are shown with the oldest at the bottom and the youngest at the top.

superposition (*n*) the order in which rocks are placed one above the other. The *principle* or *law of superposition* is that in a layered succession (↑) of rocks the lower beds (p.80) will be the older and the upper beds will be the younger (unless the rocks have been turned upside down).

time plane a surface within a series of sedimentary rocks that marks a particular moment in geological time.

horizon (*n*) (1) a plane of stratification (p.80) that was once horizontal and continuous; (2) a time plane (↑) within a sedimentary series (↓) or a bed (p.80) (usually a thin bed) that contains characteristic fossils or has a characteristic lithology (p.85).

sequence (*n*) a succession (↑) of bedded rocks; the stratigraphical (↑) order in which beds appear.

cyclic sequence a sequence (↑) of sediments (p.80) repeated in a particular order, e.g. sandstone – shale – limestone. A cyclic sequence is commonly the result of marine transgression (p.119) and regression (p.119).

cyclothem (*n*) a unit of a cyclic sequence (↑).

rhythmic sequence a cyclic sequence (↑) on a small scale.

地層學(名)　研究層狀岩石(第 80 頁)及其性質、賦存狀況、相互關係和分類的學科。(形容詞為 stratigraphical，stratigraphic (美語))

歷史地質學　研究地球歷史，包括地層學(↑)的學科。

均變說；天律不變說(名)　認為地質變化過程在過去和現在一樣具有相同的形式並產生類似的結果的一種論點。參見**災變說**(↓)。

災變說(名)　在地質學上已不受支持的一種論點，它認為必須用一系列激烈事件或災變來闡明地球的歷史。參見**均變說**(↑)。

序列(名)　岩組出現的順序。用圖表形式表示一個序列時，地層(第 80 頁)的表達是最老的在下層，最新的在上層。

疊覆(名)　岩石一層疊一層的順序。"疊覆原理"或"疊覆律"是指在層狀岩石的疊覆序列(↑)中，下部層(第 80 頁)一定是較老的層，上部層則是較新的層(除非岩層發生倒轉)。

同時面；時間面　沉積岩系列中的某一個面，它標誌地質時期內的某一特定瞬間。

層位(名)　(1)已往是水平而又連續的一個層理(第 80 頁)面；(2)沉積系列(↓)中的某一個同時面(↑)，或含有特徵化石或具特徵岩性(第 85 頁)一個層(第 80 頁)(通常是一個薄層)。

層序(名)　層狀岩石的序列(↑)；地層(↑)呈現的順序。

旋迴層序　按一種特定次序，例如砂岩 - 頁岩 - 石灰岩重複的沉積物(第 80 頁)層序(↑)，旋迴層序一般是海侵(第 119 頁)和海退(第 119 頁)造成的。

韻律層(名)　旋迴層序(↑)的一個單位。

韻律層序　一種小規模的旋迴層序(↑)。

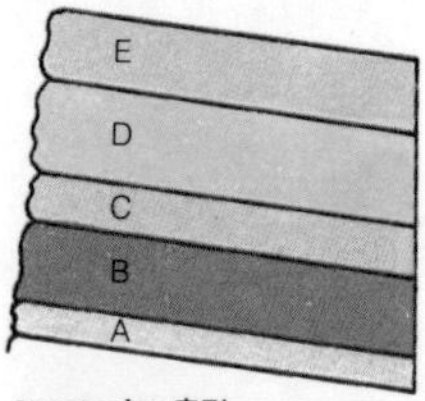

succession 序列

bed A is the oldest
bed E the youngest
(if not inverted)
A 層最老
E 層最新
（假定沒有倒轉）

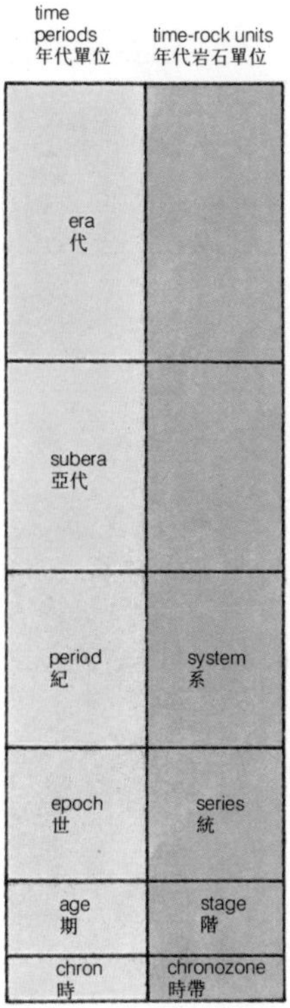

aeon, eon (*n*) the largest division of geological time. An aeon is made up of several eras (↓).

era (*n*) a division of geological time; made up of several periods (↓) or sub-eras (↓).

sub-era (*n*) a division of an era (↑).

period (*n*) a large division of geological time; it corresponds to a system (↓).

epoch (*n*) a division of geological time; part of a period (↑); it corresponds to a series (↓).

age (*n*) a division of geological time; part of an epoch (↑); it corresponds to a stage (↓).

chron (*n*) the smallest division of geological time; part of an age.

chronostratigraphical (*adj*) a chronostratigraphical unit is a division of the geological column (↓) that is based on geological time. *See also* **lithostratigraphical** (↓), **biostratigraphical** (p.117).

system (*n*) one of the major stratigraphical (↑) divisions of the geological column (↑); it corresponds to a geological period (↑).

series (*n*) a stratigraphical division within a system (↑); it corresponds to an epoch (↑).

stage (*n*) a stratigraphical division within a series (↑); it corresponds to an age (↑).

lithostratigraphical (*adj*) a lithostratigraphical unit is one that is based on lithological (p.85) characters rather than on geological time or fossils. *See also* **chronostratigraphical** (↑), **biostratigraphical** (p.117).

rock-stratigraphical (*adj*) = lithostratigraphical (↑).

group (*n*) a lithostratigraphical term (↑) for a rock unit consisting of two or more formations (↓) that are next to each other in a succession (↑) and are related to each other.

formation (*n*) a term for the basic lithostratigraphical (↑) division.

member (*n*) a lithostratigraphical (↑) term for a part of a formation (↑).

bed (*n*) the smallest lithostratigraphical (↑) division. *See also p.80.*

geological column a diagram that shows the divisions of geological time and the succession (↑) for a given area.

宙(名)　地質年代的最大劃分單位。一個宙由幾個代(↓)組成。

代(名)　地質年代的一個劃分單位，由幾個紀(↓)或亞代(↓)組成。

亞代(名)　代(↑)的劃分單位。

紀(名)　地質年代的一個大劃分單位；紀與系(↓)相對應。

世(名)　地質年代的劃分單位；紀(↑)的一個部分；世與統(↓)相對應。

期(名)　地質年代的一個劃分單位；世(↑)的一部分，期與階(↓)相對應。

時(名)　地質年代的最小劃分單位；期的一部分。

年代地層的(形)　年代地層單位是以地質年代為依據的地質柱狀圖(↓)的一個劃分單位。參見**岩性地層的**(↓)，**生物地層的**(第 117 頁)。

系(名)　地質柱狀圖(↑)中大的地層(↑)劃分單位之一；系與紀(↑)相對應。

統(名)　系(↑)內的地層劃分單位；統與世(↑)相對應。

階(名)　統(↑)內的地層劃分單位；階與期(↑)相對應。

岩性地層的(形)　指岩性地層單位以岩性(第 85 頁)特徵為依據，而不是以地質年代或化石為依據的。參見**年代地層的**(↑)，**生物地層的**(第 117 頁)。

岩石地層的(形)　同岩性地層的(↑)。

群(名)　岩性地層學的術語(↑)，指一個岩石單位，它包括兩個或兩個以上的組(↓)。這些組在序列(↑)上是互相緊連，並且相互有關。

組(名)　是岩性地層(↑)基本劃分單位的術語。

段(名)　組(↑)的一個部分所用的岩性地層學(↑)術語。

層(名)　岩性地層(↑)的最小劃分單位。參見第 80 頁)。

地質柱狀圖　表示指定地區的地質年代劃分及地層序列(↑)的一種圖表。

Precambrian, Pre-Cambrian (*n, adj*) the period of time before the Cambrian (↓), i.e. from the formation of the Earth until about 570 million years ago: about 4000 million years.

Proterozoic (*n, adj*) (1) one of two aeons (p.113) into which the Precambrian (↑) is divided, ranging from 2500 to 570 million years ago. (2) the whole of the Precambrian.

Archaean (*n, adj*) (1) the earlier of two aeons (p. 113) into which the Precambrian (↑) is divided, covering the period from the formation of the Earth until 2500 million years ago; (2) the whole of the Precambrian.

basement complex, basement a general term for igneous (p. 62) or metamorphic (p. 90) rocks, usually Precambrian (↑), which cover a wide area and on which rest unmetamorphosed (p.90) sediments of later age.

Phanerozoic (*n, adj*) the stratigraphical systems from the Cambrian (↓) to the Recent (↓).

Palaeozoic (*n, adj*) the era of geological time that ranges from 570 to 230 million years ago. It is divided into the Lower Palaeozoic, consisting of the Cambrian (↓), Ordovician (↓), and Silurian (↓) periods, and the Upper Palaeozoic, consisting of the Devonian (↓), Carboniferous (↓), and Permian (↓) periods.

Cambrian (*n, adj*) the earliest period of the Palaeozoic era, dating from about 570 million years ago to 500 million years ago.

Ordovician (*n, adj*) a period of the Palaeozoic Era (↑), dating from 500 to 435 million years ago.

Silurian (*n, adj*) a period of the Palaeozoic Era (↑), dating from 435 to 400 million years ago.

Devonian (*n, adj*) a period of the Palaeozoic Era (↑), dating from 400 to 345 million years ago.

Carboniferous (*n, adj*) a period of the Palaeozoic Era (↑), dating from 345 to 280 million years ago. In the USA it is divided into the Mississippian (below) and the Pennsylvanian (above).

Mississippian (*n, adj*) *see* **Carboniferous** (↑).

Pennsylvanian (*n, adj*) *see* **Carboniferous** (↑).

Permian (*n, adj*) the latest period of the Palaeozoic Era (↑), dating from 280 to 230 million years ago.

前寒武紀(的)(名、形)　寒武紀(↓)之前的地質時期，即自地球形成至大約 5.7 億年以前，共約 40 億年。

元古宙(的)(名、形)　(1)劃分前寒武紀(↑)的二個宙(第 113 頁)之一，距今 25 億年至 5.7 億年；(2)指整個前寒武紀。

太古宙(的)(名、形)　(1)劃分前寒武紀(↑)的兩個宙(第 113 頁)中的較早的一個，包括從地球形成至距今 25 億年的這一段時期；(2)指整個前寒武紀。

基底雜岩，基底　火成岩(第 62 頁)或變質岩(第 90 頁)用的一個通用術語。一般是指前寒武紀(↑)，分佈廣，其上覆蓋有時代較晚的不變質(第 90 頁)沉積岩。

顯生宇(的)(名、形)　自寒武紀(↓)到全新世(↓)的地層系統。

古生代(的)(名、形)　地質年代的代，距今 5.7 億至 2.3 億年。劃分為下古生代和上古生代，下古生代包括寒武紀(↓)、奧陶紀(↓)及志留紀(↓)，上古生代包括泥盆紀(↓)、石炭紀(↓)和二疊紀(↓)。

寒武紀(的)(名、形)　古生代最早的紀，年代距今約 5.7 億至 5 億年。

奧陶紀(的)(名、形)　古生代(↑)的一個紀，年代距今 5 億至 4.35 億年。

志留紀(的)(名、形)　古生代(↑)的一個紀，年代距今 4.35 億至 4 億年。

泥盆紀(的)(名、形)　古生代(↑)的一個紀，年代距今 4 億至 3.45 億年。

石炭紀(的)(名、形)　古生代(↑)的一個紀，年代距今 3.45 億至 2.8 億年。在美國劃分為密西西比紀(下部)和賓西法尼亞紀(上部)。

密西西比紀(的)(名、形)　見**石炭紀**(↑)。

賓西法尼亞紀(的)(名、形)　見**石炭紀**(↑)。

二疊紀(的)(名、形)　古生代(↑)最晚的一個紀，年代距今 2.8 億至 2.3 億年。

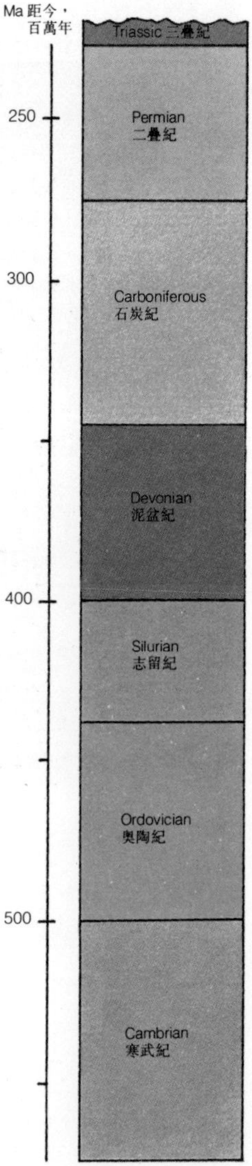

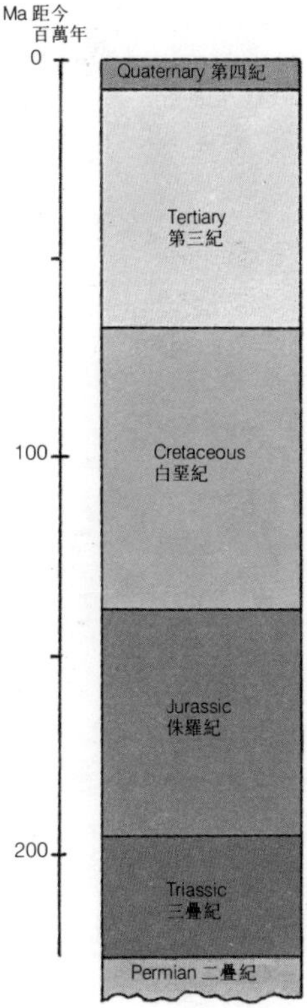

Mesozoic (*n, adj*) the era between the Palaeozoic (↑) and the Cainozoic (↓), ranging from 230 to 65 million years ago. It is made up of the Triassic, Jurassic, and Cretaceous periods.

Triassic (*n, adj*) the earliest period of the Mesozoic Era (↑), ranging from 230 to 195 million years ago.

Trias = Triassic (↑).

Jurassic (*n, adj*) one of the three periods of the Mesozoic Era (↑), ranging from 195 to 140 million years ago.

Cretaceous (*n, adj*) the youngest of the three periods of the Mesozoic Era (↑), ranging from 140 to 65 million years ago.

Cainozoic, Cenozoic (*n, adj*) the era of geological time that follows the Mesozoic (↑), ranging from 65 million years ago to the present. It is made up of the Tertiary and Quaternary sub-eras (↓).

Tertiary (*n, adj*) the sub-era between the Mesozoic era (↑) and the Quaternary sub-era (↓), ranging from 65 million years ago to 2 million years ago. It is divided into two periods, the Palaeogene and the Neogene, and five epochs: the Palaeocene, Eocene, Oligocene, Miocene, and Pliocene.

Palaeocene (*n, adj*) *see* **Tertiary** (↑).

Eocene (*n, adj*) *see* **Tertiary** (↑).

Oligocene (*n, adj*) *see* **Tertiary** (↑).

Miocene (*n, adj*) *see* **Tertiary** (↑).

Pliocene (*n, adj*) *see* **Tertiary** (↑).

Palaeogene (*n, adj*) the earlier of the two periods of the Tertiary sub-era (↑). It consists of the Palaeocene, Eocene, and Oligocene epochs (↑).

Neogene (*n, adj*) the later of the two periods of the Tertiary sub-era (↑). It consists of the Miocene and Pliocene epochs (↑).

Quaternary (*n, adj*) the period from 2 million years ago to the present; a subdivision (sub-era) of the Cainozoic Era (↑). It is divided into two epochs: the Pleistocene (↓) and Holocene (↓).

Pleistocene (*n, adj*) an epoch of the Quaternary sub-era (↑); the time of the last ice age.

Holocene (*n, adj*) the latest epoch of the Quaternary sub-era (↑); it includes the present time.

Recent = Holocene (↑).

中生代(的)(名、形)　介於古生代(↑)和新生代(↓)之間的代，自距今 2.3 億年延續至 6500 萬年前。中生代由三疊紀、侏羅紀和白堊紀構成。

三疊紀(的)(名、形)　中生代(↑)最早的紀，自距今 2.3 億年延續至 1.95 億年前。

三疊紀(↑)的另一英文拼寫為 Trias。

侏羅紀(的)(名、形)　中生代(↑)的三個紀之一，自距今 1.95 億年延續至 1.4 億年前。

白堊紀(的)(名、形)　中生代(↑)三個紀中最年青的紀，自距今 1.4 億年延續至 6500 萬年前。

新生代(的)(名、形)　接於中生代(↑)之後的地質年代中的代，自距今 6500 萬年延續至現代。由第三亞代和第四亞代(↓)組成。

第三亞代(的)(名、形)　中生代(↑)和第四亞代(↓)之間的亞代，自距今 6500 萬年延續到 200 萬年前，劃分為早第三紀和晚第三紀兩紀，包括古新世、始新世、漸新世、中新世、上新世五個世。

古新世(的)(名、形)　見第三紀(↑)。

始新世(的)(名、形)　見第三紀(↑)。

漸新世(的)(名、形)　見第三紀(↑)。

中新世(的)(名、形)　見第三紀(↑)。

上新世(的)(名、形)　見第三紀(↑)。

早第三紀(的)(名、形)　第三亞代(↑)兩個紀中較早的一個紀，由古新世、始新世和漸新世(↑)組成。

晚第三紀(的)(名、形)　第三亞代(↑)兩個紀中較晚的一個紀，由中新世和上新世(↑)組成。

第四亞代(的)(名、形)　自距今二百萬年至現代的紀，是新生代(↑)的再分單位(亞代)，劃分為兩個世：更新世(↓)和全新世(↓)。

更新世(的)(名、形)　第四亞代(↑)的一個世：最後冰期時代。

全新世(的)(名、形)　第四亞代(↑)最晚的世：它包括現代時期。

近代　同全新世(↑)。

orogenic period a period of mountain-building. *See also p.132.*
Caledonian (*adj*) relating to a period of mountain-building in Ordovician and Devonian times (p.114). The general trend of the Caledonian structures is north-east – south-west.
Caledonides (*n*) the former range of mountains that was formed during the Caledonian orogeny (↑), reaching from Norway to Scotland and Ireland.
Hercynian (*adj*) relating to the period of mountain-building that took place in late Palaeozoic times (p.114) in Europe.
Variscan (*adj*) (1) = Hercynian (↑); (2) relating to a period of mountain-building from the Carboniferous (p.114) to the Triassic (p.115).
Kimmerian (*adj*) relating to a period of mountain-building that took place in Jurassic times (p.115) in Europe.
Alpine (*adj*) relating to the period of mountain-building in the Tertiary period (p.115) that formed the Alps in Europe.
Taconic (*adj*) relating to a period of mountain-building that took place in late Ordovician times (p.114) in North America.
Acadian (*adj*) relating to a period of mountain-building that took place in ? late Devonian to end Permian times (p.114) in North America.
Appalachian (*adj*) relating to a period of mountain-building that took place in late Palaeozoic times in North America.
Laramide (*adj*) relating to a period of mountain-building that took place in late Cretaceous (? Jurassic) to early Eocene times (p.115) in North America.
synorogenic (*adj*) taking place at the same time as a period of mountain-building.
post-orogenic (*adj*) taking place after a period of mountain-building.
syntectonic (*adj*) taking place at the same time as a period of deformation (p.122).
synkinematic (*adj*) = syntectonic (↑).
post-tectonic (*adj*) taking place after a period of deformation (p.122).
postkinematic (*adj*) = post-tectonic (↑).

造山期　山脈形成的時期。參見第 132 頁。

加里東期的(形)　與奧陶紀和泥盆紀年代內(第 114 頁)的造山運動期有關的。加里東構造總的走向是北東 - 南西向。

加里東造山帶(名)　以前在加里東造山運動期(↑)形成的山脈，自挪威延伸到蘇格蘭和愛爾蘭。

海西期的(形)　與歐洲晚古生代(第 114 頁)發生的造山運動期有關的。

華里西期的(形)　(1)同海西期的(↑)；(2)與從石炭紀(第 114 頁)到三疊紀(第 115 頁)的造山運動期有關的。

基米里期的(形)　與歐洲侏羅紀年代(第 115 頁)發生的造山運動期有關的。

阿爾卑斯期的(形)　與第三紀時期(第 115 頁)在歐洲形成阿爾卑斯山脈的造山運動期有關的。

塔康期的(形)　與北美洲奧陶紀晚期(第 114 頁)發生的造山運動有關的。

阿卡德期的(形)　與在北美洲發生於晚泥盆紀(？)到二疊紀(第 114 頁)末期內的造山運動期有關的。

阿帕拉契期的(形)　與北美洲晚古生代發生的造山運動期有關的。

拉臘米期的(形)　與北美洲晚白堊世(？侏羅紀)到早始新世(第 115 頁)時期發生的造山運動期有關的。

同造山期的(形)　與造山運動期同時發生的。

造山期後的(形)　造山運動期之後發生的。

同構造的(形)　與形變期(第 122 頁)同時發生的。

同造山運動的(形)　同構造的(↑)。

構造後的(形)　構造形變(第 122 頁)期後發生的。

造山運動後的(形)　同構造後的(↑)。

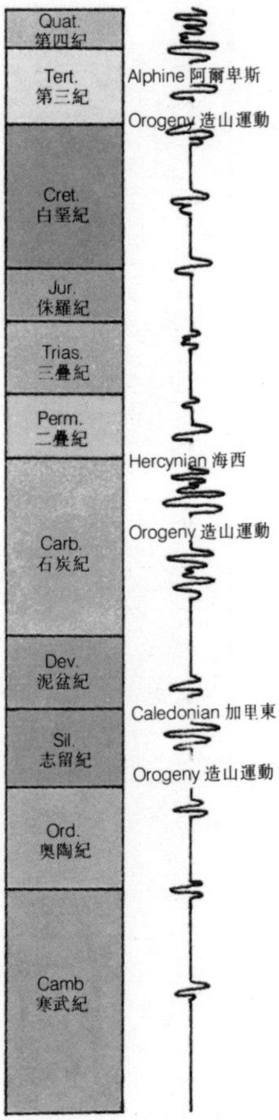

orogenies
造山運動

biostratigraphical (*adj*) a biostratigraphical unit is one that is based on fossils rather than on lithological (p.85) characters or on geological time. *See also* **lithostratigraphical** (p.113), **chronostratigraphical** (p.113).

zone (*n*) a biostratigraphical (↑) division: a stratigraphical division (p.111) with characteristic fossils. One of the fossils present – the zone fossil – gives the name to the zone. **zonal** (*adj*).

hemera (*n*) a small unit of geological time as marked by the rise and fall of a particular species fossil. The word is not now in common use. **hemeras, -ai, -ae** (*pl.*).

epibole (*n*) a stratigraphical term for the rocks deposited during a hemera (↑); i.e., the time-rock unit corresponding to a hemera. The word is not now in common use.

correlation (*n*) in stratigraphy, the matching of rocks of a particular age that are found in one place with other rocks found elsewhere. Fossils (p. 98) are generally used for stratigraphical (p. 112) correlation. **correlate** (*v*).

provenance (*n*) the source area of the materials that form a sedimentary rock; the nature of the rocks from which it has been formed.

facies (*n*) (*facies*) the general characters of a sedimentary rock, especially those that indicate the environment (p.81) in which it was deposited.

lithofacies (*n*) a facies (↑) that is characterized by a particular rock type.

biofacies (*n*) a facies (↑) that is characterized by a particular assemblage (p.100) of fossils.

diachronous (*adj*) 'across time'. A word used to describe a bed (p.80) or a stratigraphical unit that is of different ages in different places and cuts across the time planes.

生物地層的(形) 指生物地層單位是根據化石而不是根據岩性(第 85 頁)特徵或地質年代確定的。參見**岩性地層的**(第 113 頁)、**年代地層的**(第 113 頁)。

帶(名) 生物地層(↑)的劃分，即具特徵化石的一個地層劃分(第 111 頁)。將所出現的化石之一(分帶化石)作為該帶的名稱。(形容詞為 zonal)

極盛時期(名) 地質年代的一個小單位，以一個特定種的化石的盛衰作標誌。這詞現不常用。(複數為 hemeras，hemerai，hemerae)

極盛帶(名) 用於極盛時期(↑)沉積的岩石的一個地層學術語：即與極盛時期相對應的年代岩石地層單位。這詞現不常用。

對比(名) 地層學上指在某地找到的特定時代的岩石與在別處找到的岩石的對照。常用化石(第 98 頁)作地層(第 112 頁)對比。(動詞為 correlate)

源岩區；源岩(名) 形成某種沉積岩物質的來源區；構成該沉積岩的物質來源處的岩性。

相(名) 沉積岩的總體特徵，特別是那些指示沉積環境(第 81 頁)的特徵。

岩相(名) 具特定岩石類型特徵的相(↑)。

生物相(名) 具特定的化石組合(第 100 頁)特徵的相(↑)。

穿時的；歷時性的(形) "穿過時代"的。這個詞用來描述某一地層(第 80 頁)或某一地層單位在不同的地方屬不同的時代並且超越幾個年代面。

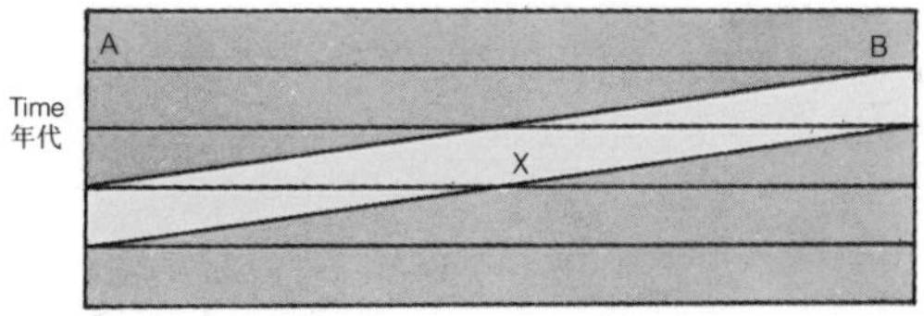

a diachronous formation
X is older at A than at B
一個穿時的組
X 在 A 處比 B 處老

type-area, type-locality a place that is chosen as the example for a stratigraphical unit (p.112).

outlier (*n*) an outcrop (p.122) of younger rocks with older rocks all round them, the younger rocks being separated from their main outcrop.

inlier (*n*) an outcrop (p.122) of older rocks with younger rocks all round them.

conformable (*adj*) beds (p.80) are conformable when they lie on each other in a regular way. **conformity** (*n*).

unconformity (*n*) an unconformity is present between sedimentary rocks (p.80) and the rocks on which they rest if a period of non-deposition, i.e. a period during which no sediments were deposited (p.80), has taken place between the formation of the older rocks and the deposition (p.80) of the later sediments. An unconformity may be shown by a difference in dip (p.123) between the two series (*angular unconformity*) or by an irregular surface between them. In some cases the two series may be separated only by a period of non-deposition without later movement. There is then no difference in dip between the older and the newer sediments. This type of unconformity is called a *non-depositional unconformity*, a *non-sequence*, or a *diastem*. **unconformable** (*adj*).

non-sequence (*n*) a non-depositional unconformity (↑); an unconformity in which the beds above and below the plane of unconformity (↑) are parallel to each other. The time covered by a non-sequence is usually relatively short.

diastem = non-sequence (↑).

disconformity (*n*) more or less the same as a non-sequence (↑).

contemporaneous (*adj*) occurring at the same time. **contemporaneously** (*adv.*).

penecontemporaneous (*adj*) almost at the same time, e.g. the penecontemporaneous erosion of sediments shortly after their deposition.

intraformational (*adj*) occurring within one stratigraphical formation (p.113), e.g. an *intraformational conglomerate*, a conglomerate formed by contemporaneous (↑) erosion (p.20) and deposition (p.80).

標準地區，標準地點 被選作某一地層單位(第112頁)範例的地方。

外露層(名) 新岩層全為老岩層所包圍，因而較年青的地層與其主要露頭分隔開的露頭(第122頁)。

內露層(名) 老岩層全為新岩層所包圍的露頭(第122頁)。

整合的(形) 當地層(第80頁)以規則平整方式相互疊置時是整合的。(名詞為 conformity)

不整合(名) 不整合出現在沉積岩層(第80頁)與此岩層的上覆岩石之間，如果這其間在老岩層與後來沉積物沉積(第80頁)之間出現非沉積期，即沒有沉積物沉積(第80頁)的時期，那麼在沉積岩層與其上覆的岩石之間，就存在一個不整合。不整合可由該兩岩系間傾斜(第123頁)上的不同(角度不整合)或其間一不規則的面所顯現。在某些情況下，兩岩系可能僅由一個非沉積期隔開，而沒有後期的運動，而且較老和較新的沉積物之間的傾斜情況沒有區別，這種不整合類型叫做"非沉積不整合"、"小間斷"或"沉積停頓"。(形容詞為 unconformable)

小間斷(名) 非沉積不整合(↑)；不整合(↑)面上、下岩層相互平行的一種不整合。一個小間斷所經歷的時間一般比較短。

沉積停頓 同小間斷(↑)。

假整合(名) 大致相似於小間斷(↑)。

同時的(形) 發生在同一時間內。(副詞為 Contemporaneously)

準同時的(形) 幾乎同時的。例如沉積物在沉積之後的短時間內準同時發生侵蝕作用。

層內的(形) 出現在一個地層組(第113頁)內部的。例如"層內礫岩"，是由同時(↑)侵蝕(第20頁)和沉積(第80頁)形成的礫岩。

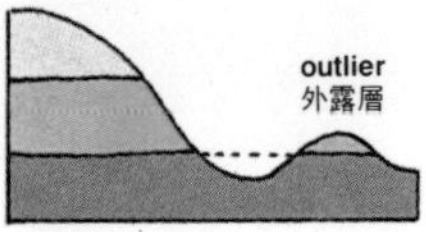

outlier
外露層

inlier
內露層

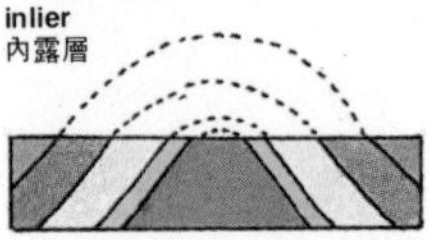

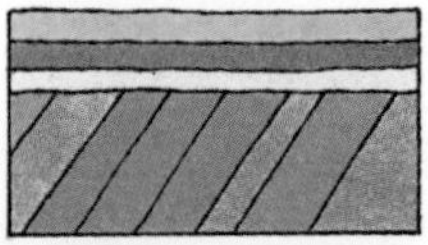

angular unconformity
角度不整合

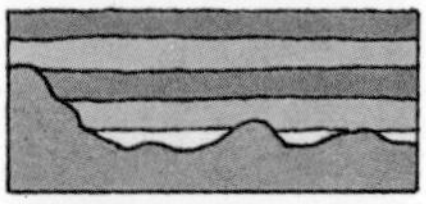

buried landscape
隱伏景觀

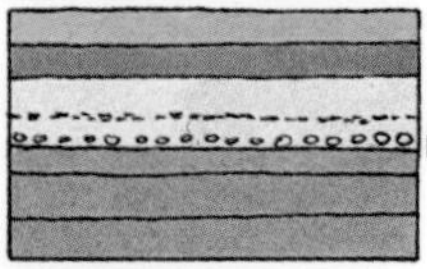

unconformity with basal conglomerate
帶有底礫岩的不整合

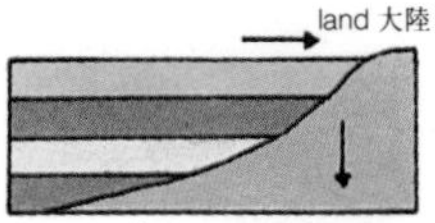

overlap 超覆

overstep (*n*) a relationship that is produced by unconformity (↑) in which the beds at the base of a younger series rest upon older and older beds of the series below the unconformity as they are followed across country. Overstep is the usual relationship produced by a marine transgression (↓). *See also* **overlap** (↓).

跨覆(名) 由不整合(↑)產生的一種接觸關係，當沿着接觸界綫追踪時，較新岩系底部的岩層覆于不整合下伏岩系之越來越老的岩層之上。跨覆是由海進(↓)所造成的常見接觸關係。參見**超覆**(↓)。

overlap (*n*) a relationship in which the members of a younger series of sediments spread in turn further and further over an older series below them. Overlap results when a marine transgression (↓) takes place while the surface on which the new rocks are being deposited (p.80) is sinking. *See also* **overstep** (↑).

超覆(名) 較新沉積岩系的各層自下而上一層層依次側向延伸並超蓋其下較老岩系的一種接觸關係。超覆是在海進(↓)時發生的，與此同時，沉積(第 80 頁)新岩石的沉積面下沉。參見**跨覆**(↑)。

offlap (*n*) a relationship in which younger members of a series of sediments cover a smaller and smaller area as the sequence is followed upwards. Offlap is produced by deposition (p.80) during a marine regression (↓). *See also* **overlap** (↑).

退覆(名) 沉積岩系較新的組成部分沿層序向上的覆蓋面積越來越小的一種接觸關係。退覆是海退(↓)過程中沉積(第 80 頁)所產生的，參見**超覆**(↑)。

transgression, marine (*n*) the spreading of the sea over the land in a relatively short period of geological time. A marine transgression results in overlap (↑). *See also* **regression** (↓).

海進(名) 在一個相對較短的地質年代中，海域向大陸的擴展，海進導致超覆(↑)。參見**海退**(↓)。

regression, marine (*n*) the opposite of transgression (↑): the movement of the sea away from the land in a relatively short period of geological time. A marine regression produces offlap (↑). *See also* **transgression** (↑).

海退(名) 海進(↑)的反面。即在一個較短的地質年代內，海域離開大陸的運動。海退導致退覆(↑)。參見**海進**(↑)。

interstratified (*adj*) strata (p.80) laid down or alternating with other strata. *See also* **interbedded** (↓), **intercalated** (↓). **interstratify** (*v*).

間層的；互層的(形) 指某些地層(第 80 頁)與其他一些地層一起交替沉積的。參見**夾層的**(↓)。**插入的**(↓)。(動詞為 interstratify)

interbedded (*adj*) deposited (p.80) in sequence (p.112) between one bed and another; used especially of lavas (p.70) between beds in a sedimentary (p.80) succession (p.112).

夾層的(形) 指按照層序(第 112 頁)沉積(第 80 頁)於兩個地層之間的形式，特別指一個沉積(第 80 頁)序列(第 112 頁)地層中的火山熔岩(第 70 頁)夾層。

intercalated (*adj*) (1) put into a series of bedded (p.80) rocks after their formation (e.g. a lava); (2) interstratified (↑). *See also* **interbedded** (↑). **intercalate** (*v*).

插入的(形) (1)在成層(第 80 頁)岩石系列形成之後的插入現象(例如一層熔岩)；(2)間層的、互層的(↑)。參見**夾層的**(↑)。(動詞為 intercalate)

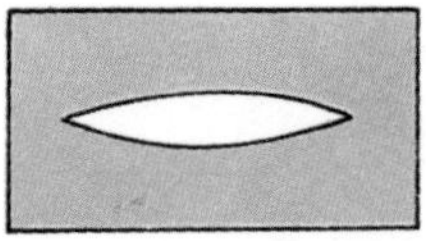

lens 透鏡體

lens (*n*) a mass of rock or other material that is thick in the centre and thin at the edges.

透鏡體(名) 中部厚而邊緣薄的一種岩石塊體或其他物質。

lenticular (*adj*) shaped like a lens (↑).

扁豆狀的；透鏡狀的(形) 狀似透鏡體(↑)的。

geochronology (*n*) the science of dating rocks and geological events in years. Radiometric dating (↓) and the counting of varves (↓) and tree-rings (↓) are the chief methods used.

地質紀年學；地質年代學(名) 測定岩石年齡和地質事件年代的科學。所用的主要方法有放射性年齡測定(↓)法、紋泥計齡法(↓)和樹木年輪年齡鑒定法(↓)。

radiometric dating a method of dating rocks and minerals by measuring the amounts of radioactive elements (p.19) in them and the daughter elements (p.19) into which they decay.

放射性年齡測定法 藉測定岩石和礦物中放射性元素(第 19 頁)的總量和由它衰變而成的子元素(第 19 頁)的總量，以求出岩石和礦物年齡的方法。

potassium–argon dating the isotope (p.19) of potassium ^{40}K is radioactive and decays in two ways to yield ^{40}A and ^{40}Ca, isotopes of argon and calcium. In the potassium–argon (K–Ar) method of radiometric dating (↑) the amounts of ^{40}K and ^{40}Ar are measured in order to calculate the age of the rock. The K–Ar method can be used for ages from 3400 Ma down to 30 000 years.

鉀氬年齡測定法 鉀的同位素(第 19 頁)^{40}K 是具放射性的元素，可以兩種方式衰變產生氬的同位素 ^{40}A 和鈣的同位素 ^{40}Ca。在鉀氬(K-Ar)放射性年齡測定(↑)法中，測定 ^{40}K 和 ^{40}A 的總量以計算出岩石的年齡。鉀氬法可用於測定 34 億年至 3 萬年範圍內的年齡。

K–Ar dating = potassium–argon dating (↑).

K-Ar 年齡測定法 同鉀氬年齡測定法(↑)。

rubidium–strontium dating the isotope of rubidium ^{87}Rb decays to give ^{87}Sr, an isotope of strontium. In the rubidium–strontium (Rb–Sr) method of radiometric dating (↑), the amounts of ^{87}Rb and ^{87}Sr are measured in order to calculate the age of the rock. The Rb–Sr method is used for Precambrian (p. 114) rocks and for igneous and metamorphic rocks (pp. 62, 90).

銣鍶年齡測定法 銣的同位素 ^{87}Rb 衰變為鍶的同位素 ^{87}Sr。在銣鍶的放射性年齡測定法(↑)中，測定 ^{87}Rb 和 ^{87}Sr 的總量以計算出岩石的年齡。銣鍶法適用於測定前寒武紀(第 114 頁)岩石、火成岩和變質岩(第 62、90 頁)。

Rb–Sr dating = rubidium–strontium dating (↑).

Rb-Sr 年齡測定法 同銣鍶年齡測定法(↑)。

uranium–lead, lead–lead, and thorium–lead dating the isotope of uranium ^{238}U decays to give ^{206}Pb; and ^{235}U, another uranium isotope, decays to give ^{207}Pb. The ratio of ^{207}Pb to ^{206}Pb provides a second method of measuring the age of a rock. The thorium–lead (Th–Pb) ratio provides a third method, but the results are less reliable. These three methods are used mainly for rocks and minerals with ages greater than 100 Ma.

鈾鉛、鉛鉛和釷鉛年齡測定法 鈾的同位素 ^{238}U 衰變為 ^{206}Pb，而鈾的另一種同位素 ^{235}U 衰變為 ^{207}Pb。將 ^{207}Pb 和 ^{206}Pb 之比作為第二個計算岩石年齡的方法。釷鉛(Th-Pb)之比可作為第三個方法，但測定結果的可靠性較差。這三種方法主要適用於年齡大於一億年的岩石和礦物。

		half-life 半衰期
^{40}K	^{40}Ar	1300 Ma
^{87}Rb	^{87}Sr	47000 Ma
^{238}U	^{206}Pb	4510 Ma
^{235}U	^{207}Pb	713 Ma
^{232}Th	^{208}Pb	13900 Ma
^{14}C	^{14}N	5570 years 年

half-life 半衰期 註：Ma 表示百萬年

carbon-14 dating the isotope of carbon ^{14}C decays to give ^{14}N, an isotope of nitrogen with a half-life (p.19) of 5570 years. In the ^{14}C method the ratio of these two isotopes is measured in order to find the age of the specimen. The ^{14}C method is used for dating events up to 70 000 years before the present.

碳 14 年齡測定法 碳的同位素 ^{14}C 衰變為氮的同位素 ^{14}N，^{14}N 的半衰期(第 19 頁)為 5570 年。在 ^{14}C 法中，計算出這兩個同位素之比以求出樣品的年齡。^{14}C 法用於測定距今 7 萬年以來的地質事件的年齡。

^{14}C dating = carbon-14 dating (↑).

^{14}C 年齡測定法 同碳 14 年齡測定法(↑)。

isotopic age an age (of a rock or mineral) measured by using radiometric methods (↑). The amounts of radioactive isotopes (p.19) are measured in order to calculate the age of the specimen.

同位素年齡 用放射性方法(↑)測定的岩石或礦物的年齡。測定放射性同位素(第 19 頁)的總量以計算標本的年齡。

absolute age an age (of a rock or mineral) expressed in numbers of years. 'Absolute age' is also used to mean 'isotopic age', but the term 'absolute age' is better avoided.

絕對年齡 以年數表示的岩石或礦物的年齡。"絕對年齡"也用來表示"同位素年齡"，但最好避免這樣用。

apparent age a radiometric age (↑) that is not the true age of the rock or mineral concerned.

視年齡；表觀年齡 放射性測定的年齡(↑)並非所涉及的岩石或礦物的真實年齡。

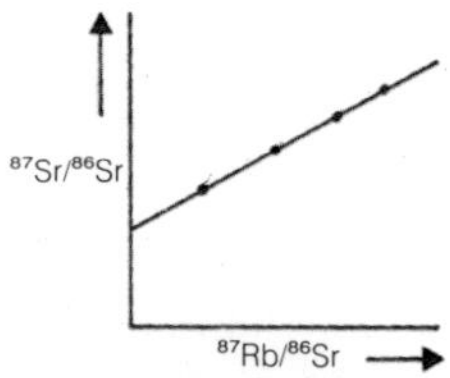

isochron 等時線

isochron (*n*) measurements of isotope ratios (p.19) of several minerals in a rock can be shown on a graph (p.157) in what is known as an *isochron plot*. The isochron is a straight line on the graph and its slope represents the age of the rock.

等時線(名) 岩石中幾個礦物的同位素比(第 19 頁)的測量值可以在圖(第 157 頁)上表示，此圖稱為"等時線圖"。在圖上，等時線是一條直線，其斜率表示岩石的年齡。

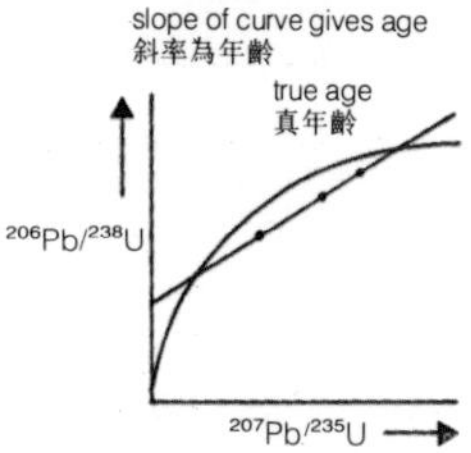

discordant age-pattern
不一致的年齡模式

concordant (*adj*) when more than one method of radiometric dating (↑) is used on a mineral or group of minerals in a rock and the dates found agree with each other within the limits of the method they are called *concordant*. *See also* **discordant** (↓).

一致的(形) 對一種岩石中的一種礦物或一組礦物使用一種以上的放射性年齡測定法(↑)時，如所測得的各個年齡在該方法的誤差範圍以內相互吻合，則稱這些年齡是"一致的"。參見**不一致的**(↓)。

discordant (*adj*) when more than one method of radiometric dating (↑) is used on a mineral or group of minerals in a rock and the dates found do not agree with each other they are called *discordant*. Discordant ages result when a rock has had a geological history that is not simple. They can provide useful geological information.

不一致的(形) 對一種岩石中的一種礦物或一組礦物使用一種以上的放射性年齡測定法(↑)時，如所測得的各個年齡在該方法的誤差範圍以內不相吻合，則稱這些年齡是"不一致的"。當一種岩石的地質史較複雜時，就會得出不一致的年齡結果。不一致年齡可提供有價值的地質資料。

varve-count (*n*) sediments (p.80) deposited in lakes formed by water flowing from glaciers (p.28) show a layered arrangement because the sediment deposited is fine in winter and coarse in summer. These layers are called *varves*. By counting them the ages of Pleistocene rocks (p.115) can be found.

紋泥計齡法(名) 沉積在冰川(第 28 頁)流水作用所形成湖泊內的沉積物(第 80 頁)，由於在冬季沉積時細而在夏季沉積時粗，因而顯示層狀排列。這些層就叫"紋泥"。計算紋泥層數可以求出更新世岩石(第 115 頁)的年齡。

tree-ring dating a tree usually adds a growth-layer to itself every year. By counting the rings in cross-sections of trees and by comparing the patterns of wider and narrower rings in the trees of a region it is possible to measure ages back to 7000 years BP (before the present).

樹木年輪年齡鑒定 樹木通常每年增長一層，計算樹木橫切面上的年輪數目，並比較這一地區樹木年輪寬窄型式，便可計算出距今 7000 年以來的年齡。

dendrochronology (*n*) = tree-ring dating (↑).

年輪測年學(名) 同樹木年輪鑒定(↑)。

structure (*n*) the shapes and positions of rock masses and their relationships to each other. **structural** (*adj*).

構造(名) 岩塊的形狀、位置及其相互關係。(形容詞為 structural)

tectonics (*n*) the study of the structure (↑) of the Earth's crust (p.9) or of a particular region. **tectonic** (*adj*).

大地構造學；構造地質學(名) 研究地殼(第 9 頁)或特定地區構造(↑)的學科。(形容詞為 tectonic)

outcrop (*n*) the area where a rock-unit occurs at the surface of the Earth. The rock-unit may be a stratum (p.80), an igneous intrusion (p.64), or any other body. It need not be exposed (↓) at the surface. **crop out** or **outcrop** (*v*).

露頭(名) 岩石地層單位在地表露出的區域。岩石地層單位可以是一個地層(第 80 頁)，一種火成侵入體(第 64 頁)或任何其他岩體。露頭不一定暴露(↓)在地面上。(動詞為 crop out 或 outcrop)

exposure (*n*) a place where rocks can be seen in their natural position and are not covered by vegetation or buildings. **expose** (*v*), **exposed** (*adj*). *See also* **outcrop** (↑).

露頭點(名) 岩石在其自然位置可見且未被植被或建築物覆蓋的地方。(動詞為 expose，形容詞為 exposed)，參見**露頭**(↑)。

isopachyte, isopach (*n*) a line on a map joining points at which a bed (p.80) has the same thickness.

等厚綫(名) 在圖上連結岩層(第 80 頁)厚度相等的各點的連綫。

stress (*n*) if a force (p.156) is applied to the surface of a body, such as a mass of rock, the force per unit area is called the *stress*. If the stress acts in a particular direction it is called a *directed stress*; if it acts equally in all directions it is called a *hydrostatic stress*. Stress may be of three types: *compressional* or *compressive*, when the forces are acting towards the centre of the body; *tensile*, when the forces tend to pull the body out; and *shear*, when two forces are acting that tend to turn the body round.

應力(名) 將一個力(第 156 頁)施於一物體(例如一塊岩石)的表面，則每單位面積受的力稱為"應力"。應力作用於特定方向則為"定向應力"；應力均勻作用於所有方向則稱"靜水應力"。應力有三個類型：壓應力，諸力的作用指向物體中心；張應力，諸力傾向於拉伸物體；剪應力，傾向於使物體轉動的一對作用力。

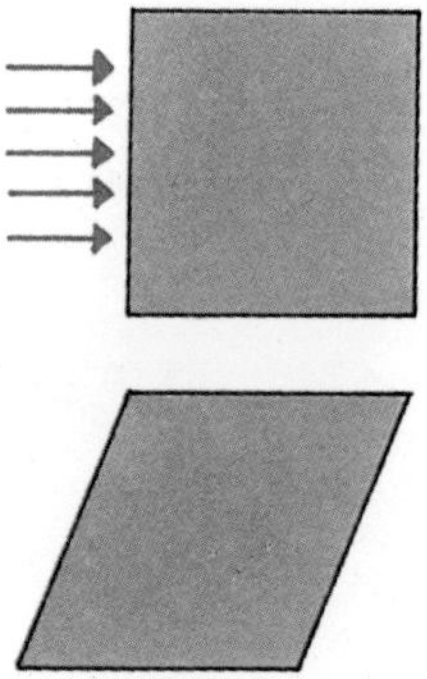

simple shear strain
簡單剪切應變

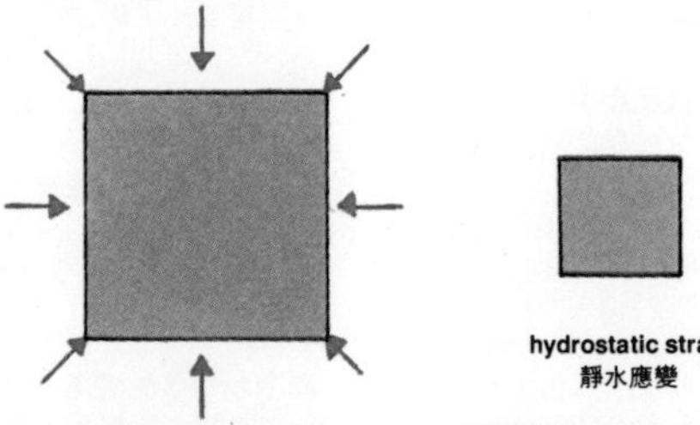

hydrostatic strain
靜水應變

strain (*n*) the changes in size and shape produced in rocks and other materials by stress (↑).

應變(名) 應力(↑)引起岩石或其他物體形狀或體積的變化。

deformation (*n*) any change in the shape of a mass of rock, whether large or small. **deform** (*v*).

形變(名) 岩塊形狀上任何或大或小的變化。(動詞為 deform)

fracture (*n*) a break in a rock caused by deformation (↑). **fracture** (*v*), **fractured** (*adj*).

破裂(名) 岩石因形變(↑)而致的破裂。(動詞為 fracture，形容詞為 fractured)

dip (*n*) the angle that a bedding-plane (p.80) or other surface on or in a rock makes with the horizontal. *True dip* is measured at 90° to the strike (↓); *apparent dip* is the angle as measured in any other direction (e.g. in a vertical section).

傾角；傾向（名） 層理面（第 80 頁）或岩石上（或岩石中）其他面與水平面的交角。真傾角是在與走向（↓）呈 90° 交角的方向上量度的；視傾角則是在其他任何方向上（例如在一垂直剖面上）量度得的。

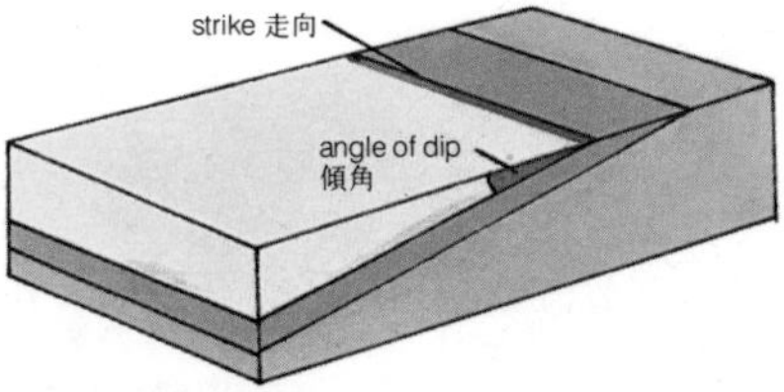

strike (*n*) the direction in which a horizontal line can be drawn on a bedding-plane (p.80) or other structural surface at any particular point. The strike is at 90° to the dip (↑).

走向（名） 層理面（第 80 頁）或其他構造面上可在任意點上畫出水平線的方向。走向與傾角（↑）呈 90° 相交。

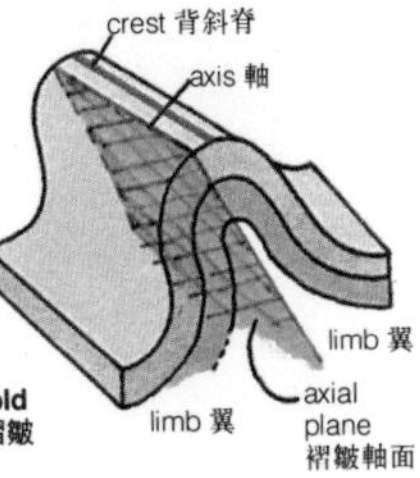

fold
褶皺

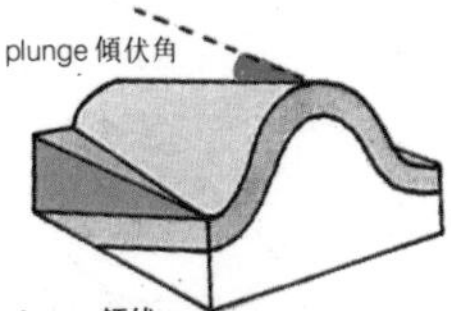

plunge 傾伏

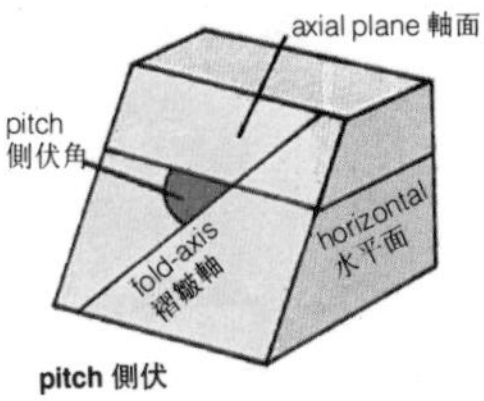

pitch 側伏

fold (*n*) a bend in a rock mass such as a bed in a series of strata (p.80). **fold** (*v*), **folded** (*adj*).

褶皺（名） 岩石塊體的彎曲，例如一系列層狀（第 80 頁）岩石中某一岩層的彎曲。（動詞為 fold，形容詞為 folded）

axial plane an imaginary plane that divides a fold (↑) into two more or less equal halves.

軸面 將褶皺（↑）劃分為近於相等的兩半的假想面。

fold-axis (*n*) an imaginary line that passes through the points where the axial plane (↑) of a fold (↑) cuts a bedding surface (p.80).

褶皺軸（名） 穿過褶皺（↑）軸面（↑）與某一層理面（第 80 頁）相交的各點的一條假想線。

limb (*n*) one of the two sides of a fold (↑).

翼（名） 一個褶皺（↑）的兩側之一。

hinge (*n*) the part of a fold (↑) where it is most sharply curved.

轉折端；樞紐（名） 褶皺（↑）彎曲最顯著的部分。

plunge (*n*) if the axis (↑) of a fold (↑) is not horizontal it is said to *plunge*. The amount of plunge is the angle between the fold-axis (↑) and the horizontal, as measured in the vertical plane. **plunging** (*adj*).

傾伏；傾伏角（名） 褶皺（↑）軸（↑）不是水平的則稱此褶皺為"傾伏"。在垂直面上測量時，傾伏角是褶皺軸（↑）和水平面的交角。（形容詞為 plunging）

pitch (*n*) the angle between a line (e.g. a fold-axis) in a plane and the horizontal as measured in the plane containing the line (e.g. the axial plane of the fold); (2) = plunge (↑). **pitching** (*adj*).

側伏角（名） (1)某一平面上的一條線（例如褶皺軸）與包含該線的平面（例如褶皺軸面）上的水平線之間的夾角。(2)同傾伏角（↑）。（形容詞為 pitching）

closure (*n*) the direction of closure of a fold (↑) is the direction in which the limbs (↑) become closer together. **close** (*v*).

閉合（名） 褶皺（↑）閉合的方向是兩翼（↑）會合一起的方向。（動詞為 close）

quaquaversal (*adj*) pointing away from a central point in all directions. A word used especially to describe dip (↑); e.g. a dome (p.125) has quaquaversal dip.

穹狀的（形） 由一個中心點向四周傾斜的。這個詞是專用於描述傾斜（↑）。例如，穹窿（第 125 頁）具穹狀傾斜。

anticline (*n*) a fold (p.123) shaped like an arch; a fold in which older rocks are in the centre and younger rocks are outside them. *See also* **syncline** (↓). **anticlinal** (*adj*).

syncline (*n*) a fold (p.123) shaped like the letter U; the opposite of an anticline (↑). **synclinal** (*adj*).

背斜（名） 一種形狀像拱門的褶皺（第 123 頁）：老地層在中心，新地層分佈於外面的一種褶皺。參見**向斜**（↓）。（形容詞為 anticlinal）

向斜（名） 形狀像字母 U 的褶皺（第 123 頁）：背斜（↑）的反意詞。（形容詞為 synclinal）

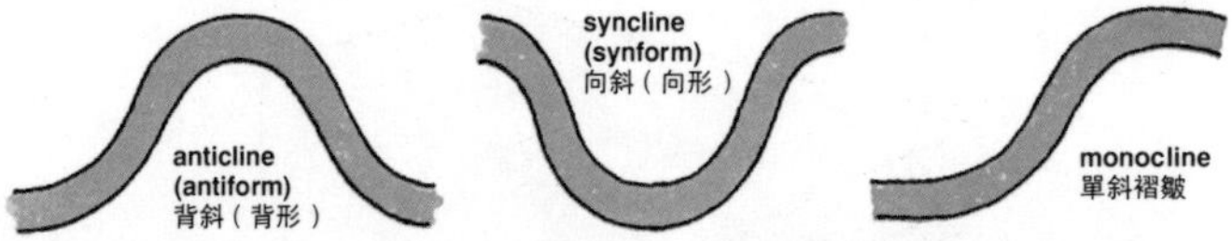

monocline (*n*) a sharp bend in a bed (p.113) that has the same dip on either side of the bend. **monoclinal** (*adj*).

anticlinorium (*n*) a large anticline (↑) which is made up of smaller folds (p.123).

synclinorium (*n*) a large syncline (↑) which is made up of smaller folds (p.123).

fan fold an anticlinal fold (↑) in which the limbs dip (p.123) towards each other or a synclinal fold (↑) in which the limbs dip away from each other.

antiform (*n*) a fold (p.123) of anticlinal shape (↑). The word 'antiform' can be used for a syncline (↑) that is upside down or where it is not certain whether the beds (p.113) are the right way up. *See also* **synform** (↓). **antiformal** (*adj*).

synform (*n*) a fold of synclinal (↑) shape. The word 'synform' can be used for an anticline that is upside down or where it is not certain whether the beds are the right way up. *See also* **antiform** (↑), **synformal** (*adj*).

crest (*n*) the highest parts of an antiformal (↑) fold (p.123).

trough (*n*) = syncline (↑). *See also* p.133.

concentric fold a fold (p.123) in which the bedding-planes (p.80) are a series of circles with the same centre. The thickness of the beds (p.80) (as measured at 90° to the bedding-planes) is the same in all parts of the fold.

parallel fold = concentric fold (↑).

similar fold a fold (p.123) in which the bedding-planes (p.80) are of the same shape. The beds (p.80) are thus thicker near the hinge (p.123) of the fold and thinner in its limbs (p.123).

單斜褶皺（名） 岩層（第 113 頁）的明顯彎曲，其兩側岩層的傾斜角相同。（形容詞為 monoclinal）

複背斜（名） 由許多小型褶皺（第 123 頁）構成的巨型背斜（↑）。

複向斜（名） 由許多小型褶皺（第 123 頁）構成的巨型向斜（↑）。

扇形褶皺 兩翼相向傾斜（第 123 頁）的背斜褶皺（↑）或兩翼相背傾斜的向斜褶皺（↑）。

背形（名） 一種具背斜（↑）形的褶皺（第 123 頁）。這個詞可指倒轉的向斜（↑）。也可指岩層（第 113 頁）層序不明的背斜形褶皺。參見**向形**（↓）。（形容詞為 antiformal）

向形（名） 一種具向斜（↑）形的褶皺。這個詞可指倒轉的背斜（↑），也可指岩層層序不明的向斜形褶皺。參見**背形**（↑）。（形容詞為 synformal）

背斜脊（名） 背形（↑）褶皺（第 123 頁）的最高部位。

褶皺槽（名） 同向斜（↑）。參見第 133 頁。

同心褶皺 諸層面（第 80 頁）為一系列同心圓的一種褶皺（第 123 頁）。在褶皺中岩層（第 80 頁）的厚度（垂直層理測量）處處相同。

平行褶皺 同同心褶皺（↑）。

相似褶皺 諸層面（第 80 頁）具有相同形態的一種褶皺（第 123 頁）。因而，岩層（第 80 頁）厚度在近轉折端（第 123 頁）加厚而在翼部（第 123 頁）變薄。

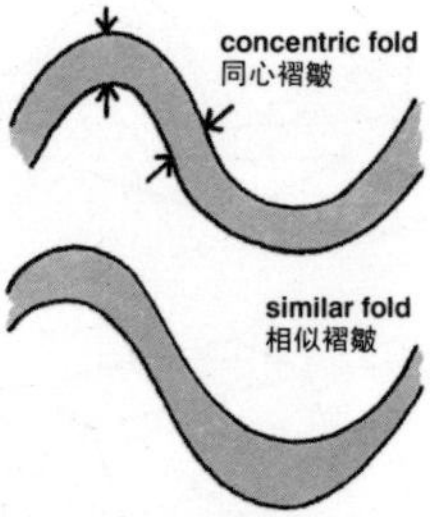

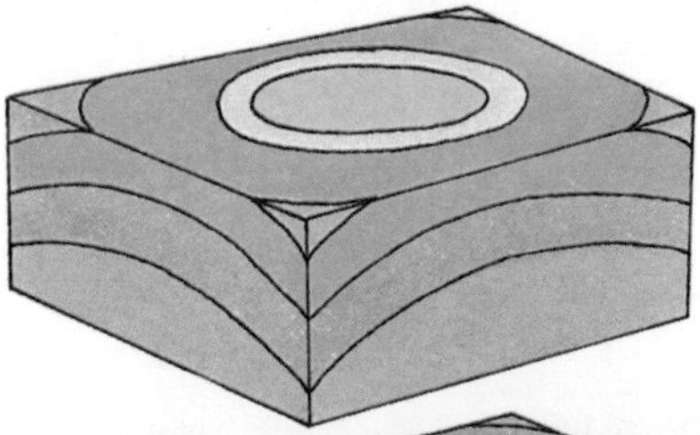

dome 穹窿

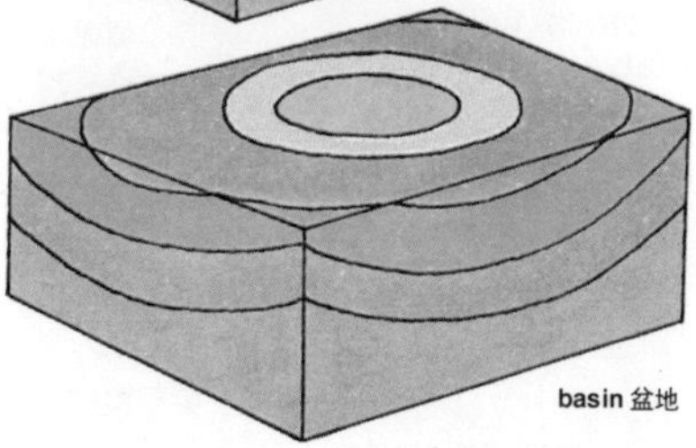

basin 盆地

basin, structural a structure of generally round shape in which the beds dip (p.123) inwards towards the centre, with younger beds in the centre.

dome (*n*) an anticlinal (↑) structure in which the beds dip (p.123) outwards in all directions.

pericline (*n*) a general term for domes (↑) and basins (↑). It is more generally used for domes.

uplift (*n*) upward movement of a large area of the Earth's crust (p.9). *See also* **subsidence** (↓). **uplift** (*v*).

subsidence (*n*) a sinking of a large area of the Earth's crust (p.9); the opposite of uplift (↑). **subside** (*v*).

warping (*n*) gentle bending of the Earth's crust (p.9). **warp** (*v*), **warp** (*n*).

upwarp (*n*) a large area of the Earth's crust (p.9) that has been uplifted (↑), usually as a broad anticline (↑). *See also* **downwarp** (↓).

downwarp (*n*) a large area of the Earth's crust (p.9) that has moved downward, usually as a broad syncline (↑); the opposite of upwarp (↑).

epeirogenic movements upward or downward movements of large areas of the Earth's crust (p.9) without folding (p.123).

構造盆地　諸岩層向內傾向(第 123 頁)中心，較新的地層則位於盆地中心的一種大致為圓形的構造。

穹窿(名)　諸岩層在各個方向上均向外傾斜(第 123 頁)的一種背斜(↑)構造。

圍斜構造(名)　穹窿(↑)和盆地(↑)的通稱，這個詞更常用於穹窿。

上升(名)　地殼(第 9 頁)大面積升起的運動。參見**沉陷**(↓)。(動詞為 uplift)

沉陷(名)　地殼(第 9 頁)大面積的下沉：上升(↑)的反意詞。(動詞為 subside)

翹曲(名)　地殼(第 9 頁)平緩的彎曲。(動詞、名詞均為 warp)

隆起(區)(名)　一大片已隆起(↑)的地殼(第 9 頁)區域，通常是一片開濶的背斜(↑)。參見**坳陷(區)**(↓)。

坳陷(區)(名)　一大片已下沉的地殼(第 9 頁)區域，通常是一片開濶的向斜(↑)。隆起(區)(↑)的反意詞。

造陸運動　沒有褶皺作用(第 123 頁)相伴隨的大面積地殼(第 9 頁)上升或下降運動。

zig-zag fold a fold (p.123) in which the limbs (p.123) are straight and the hinges (p.123) are sharp bends; its general shape is thus like a letter Z.
chevron fold = zig-zag fold (↑).
concertina fold = zig-zag fold (↑).
disharmonic fold a fold (p.123) in which the beds do not show a regular arrangement.
symmetrical fold a fold (p.123) in which the two limbs (p.123) dip (p.123) at about the same angle. *See also* **asymmetrical fold** (↓).
asymmetrical fold a fold (p.123) in which the two limbs do not dip at the same angle. *See also* **symmetrical fold** (↑).
cylindrical fold a fold (p.123) in which the bedding-planes (p.80) are shaped like a drum or a tube.
box fold a fold (p.123) that in cross-section has a square shape.
shear fold a fold (p.123) formed by small movements along closely spaced cleavage planes (p.95) or fractures (p.122).
slip fold = shear fold (↑).
flow fold a fold (p.123) formed in incompetent (↓) beds (p.80) that flow like a thick liquid.

鋸齒狀褶皺　翼(第 123 頁)平直而樞紐(第 123 頁)急彎的褶皺(第 123 頁)，其一般形狀像字母"Z"字形。
尖稜褶皺　同鋸齒狀褶皺(↑)。
手風琴褶皺　同鋸齒狀褶皺(↑)。
不協調褶皺　岩層不顯現規則排列的褶皺(第 123 頁)。
對稱褶皺　兩翼(第 123 頁)傾斜(第 123 頁)的角度大致相同的褶皺(第 123 頁)。參見**不對稱褶皺**(↓)。
不對稱褶皺　兩翼傾斜角度不相同的褶皺(第 123 頁)。參見**對稱褶皺**(↑)。
圓柱狀褶皺　層面(第 80 頁)呈鼓狀或管狀的褶皺(第 123 頁)。
箱狀褶皺　橫剖面呈正方形的褶皺(第 123 頁)。
剪切褶皺　沿密集劈理面(第 95 頁)或破裂面(第 122 頁)發生小規模運動所形成的褶皺(第 123 頁)。
滑褶皺　同剪切褶皺(↑)。
流狀褶皺　由軟(↓)岩層(第 80 頁)像黏稠液體般流動而形成的褶皺(第 123 頁)。

zig-zag fold 鋸齒狀褶皺

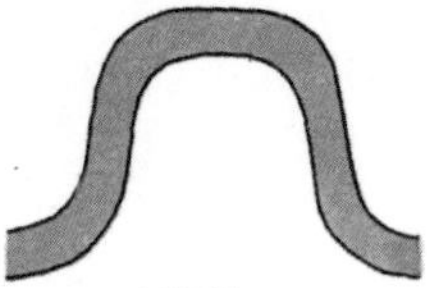
box fold 箱狀褶皺

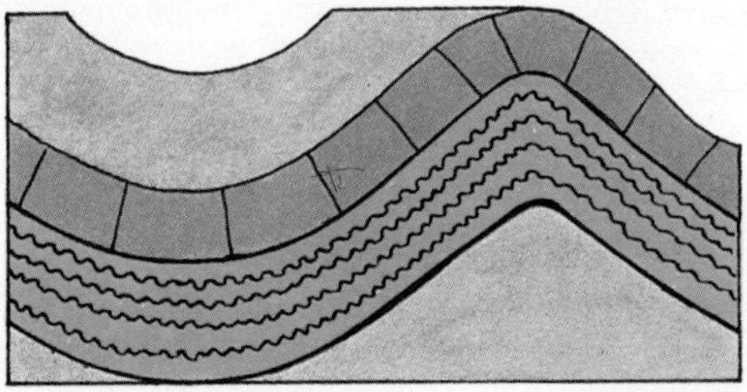

competence 硬性

competent bed a bed (p.80) whose thickness remains the same in all places when it is folded. *See also* **incompetent bed** (↓).
incompetent bed a bed (p.80) whose thickness varies from place to place when it is folded. *See also* **competent bed** (↑).
drag fold a small fold (p.123) formed in an incompetent bed (↑) by slip (p.128) along the bedding-planes (p.80).

硬岩層；強岩層　褶曲厚度處處保持相同的岩層(第 80 頁)。參見**軟岩層**(↓)。
軟岩層；弱岩層　褶曲時厚度處處不同的岩層(第 80 頁)。參見**硬岩層**(↑)。
拖曳褶皺　在軟岩層(↑)中沿層面(第 80 頁)滑移(第 128 頁)而形成的小褶皺(第 123 頁)。

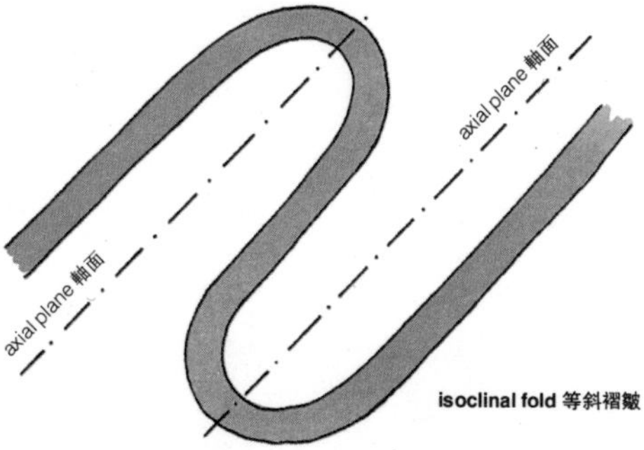

isoclinal fold 等斜褶皺

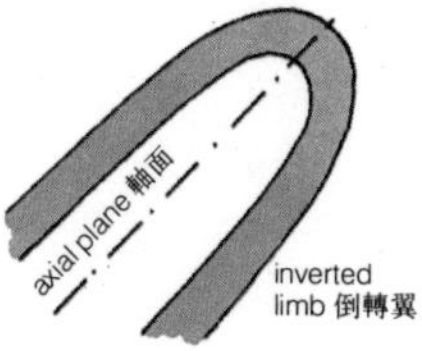

overturned fold 倒轉褶皺

isoclinal fold a fold (p.123) in which the two limbs (p.123) are parallel.
overturned fold a fold (p.123) in which the axial plane (p.123) is so far from the vertical that one limb (p.123) is over part of the other limb.
recumbent fold a fold (p.123) in which the axial plane (p.123) is horizontal or nearly horizontal. *See also* **nappe** (p.130).

等斜褶皺　兩翼(第 123 頁)平行的褶皺(第 123 頁)。
倒轉褶皺　軸面(第 123 頁)與垂直面相差甚遠，其一褶皺翼(第 123 頁)蓋在另一翼的一部分上的褶皺(第 123 頁)。
平臥褶皺　軸面(第 123 頁)水平或接近水平的褶皺(第 123 頁)。參見**推覆體**(第 130 頁)

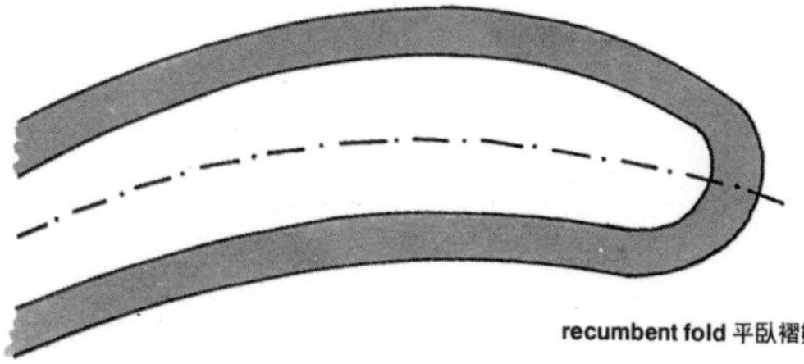
recumbent fold 平臥褶皺

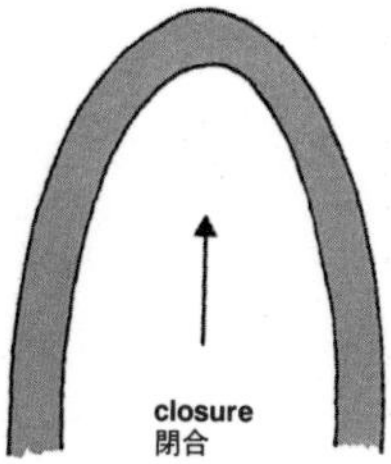

inverted (*adj*) upside down. **inversion** (*n*).
uninverted (*adj*) the right way up.
way up strata (p.80) are the right way up if the youngest beds (i.e. those deposited (p.80) last) are on top. Strata are inverted (↑) if they have been overturned (↑).
young (*v*) beds (p.80) are said to 'young' in the direction in which the youngest surface faces. The verb 'to young' is useful in describing the relationships of beds in areas of complex folding (p.123).
face (*v*) = young (↑).

倒轉的(形)　上下倒置的。(名詞為 inversion)
未倒轉的(形)　指層位頂面向上的。
層位頂面向上　如果最年輕的岩層(即最後沉積(第 80 頁)的岩層)位於頂部，則地層(第 80 頁)層位頂面向上的方向是正常的。如果地層倒轉(↑)，則層位頂面是倒轉的(↑)。
面向(動)　岩層(第 80 頁)所面向的方向就是最年輕的層面所面向的方向。動詞"to young(面向)"在描述複雜褶皺(第 123 頁)變動地區岩層之間的關係時是很有用的。
朝向(動)　同面向(↑)。

fault (*n*) a break in the rocks along which movement has taken place.
fault plane the surface along which fault movement (↑) has taken place. A fault-plane may be a smooth surface or a broad zone. *See also* **fault zone** (↓).
fault zone a broad area along which fault movement (↑) has taken place. A fault zone may be several hundred metres wide. *See also* **fault plane** (↑).
upthrow (*adj*) the side of a fault (↑) on which the throw (↓) is upward in relation to the other side. *See also* **downthrow** (↓).
downthrow (*adj*) the side of a fault on which the throw (↓) is downward in relation to the other side. *See also* **upthrow** (↑).
hanging wall the rocks that lie above the fault plane (↑) of a fault that is not vertical. *See also* **footwall** (↓).
footwall (*n*) the rocks that lie below the fault plane (↑) of a fault that is not vertical. *See also* **hanging wall** (↑).
throw (*n*) the amount of movement that has taken place along a fault (↑) as measured in the vertical direction. *See also* **heave** (↓).
heave (*n*) the amount of horizontal movement that has taken place between the two sides of a fault (↑). *See also* **throw** (↑).
hade (*n*) the angle between a fault plane (↑) and the vertical. It is equal to 90° minus the angle of dip (p.123) of the fault.
displacement (*n*) the distance between two points that were next to each other before fault (↑) movement took place.
offset (*n*) the horizontal displacement (↑) of a fault (↑), measured parallel to the strike (p.123) of the fault.
net slip = displacement (↑).
slip = net slip (↑).
normal fault a fault (↑) in which the hanging wall (↑) has moved downward in relation to the footwall (↑). *See also* **reverse fault** (↓).
reverse fault a fault (↑) in which the hanging wall (↑) has moved upward in relation to the footwall (↑). *See also* **normal fault** (↑).

斷層(名)　岩石中曾沿該岩石發生運動的一條裂縫。
斷層面　曾沿之發生斷層運動(↑)的面。斷層面可以是一個平滑的面也可以是一個寬濶的帶。參見**斷層帶**(↓)。
斷層帶　曾沿之發生斷層運動(↑)的一個寬濶區域。一個斷層帶可寬達幾百米。參見**斷層面**(↑)。
上昇的(形)　指斷層(↑)的一盤，其上的垂直斷距(↓)相對於另一盤是上昇的。參見**下落的**(↓)。
下落的(形)　指斷層的一盤，其上的垂直斷距(↓)相對於另一盤是下降的。參見**上昇的**(↑)。
上盤　位於非直立斷層的斷層面(↑)以上的岩石。參見**下盤**(↓)。
下盤(名)　位於非直立斷層的斷層面(↑)以下的岩石、參見**上盤**(↑)。
垂直斷距；落差(名)　在垂直方向上測得的沿一個斷層(↑)發生的運動量。參見**平錯**(↓)。
平錯；水平斷距(名)　斷層(↑)兩側之間發生的水平運動量。參見**垂直斷距**(↑)。
伸角；斷層餘角　斷層面(↑)與鉛垂線間的夾角。它等於 90° 減去斷層的傾角(第 123 頁)。
位移(名)　斷層(↑)發生運動之前彼此相鄰的兩點之間現時的距離。
水平斷錯(名)　平行於斷層(↑)走向(第 123 頁)所測得該斷層的水平位移(↑)。
總滑距　同位移(↑)。
滑距　同總滑距(↑)。
正斷層　上盤(↑)已相對於下盤(↑)向下運動的斷層(↑)。參見**逆斷層**(↓)。
逆斷層　上盤(↑)已相對於下盤(↑)向上運動的斷層(↑)。參見**正斷層**(↑)。

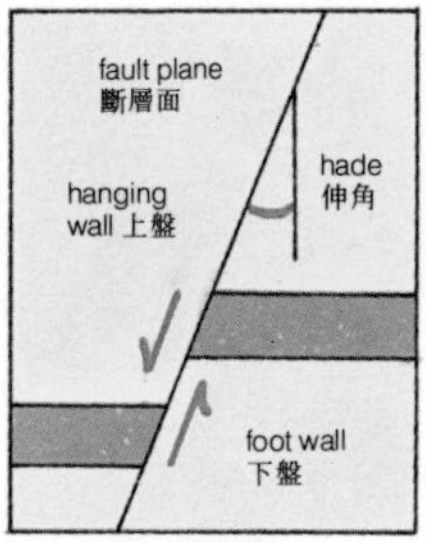

normal fault 正斷層

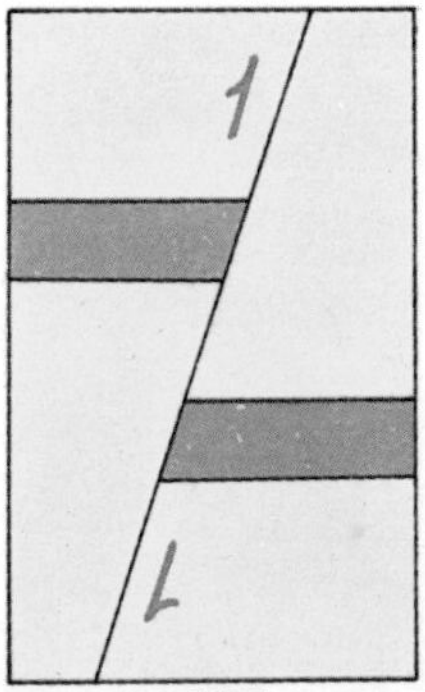
reverse fault 逆斷層

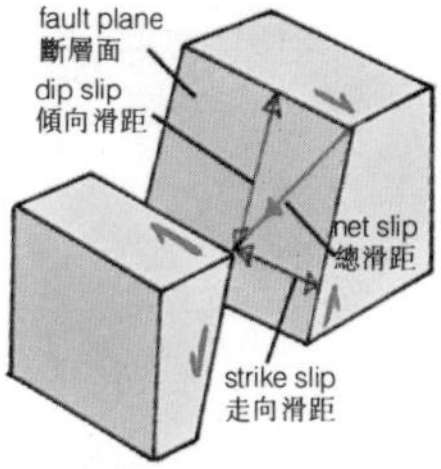

strike slip 走向滑距

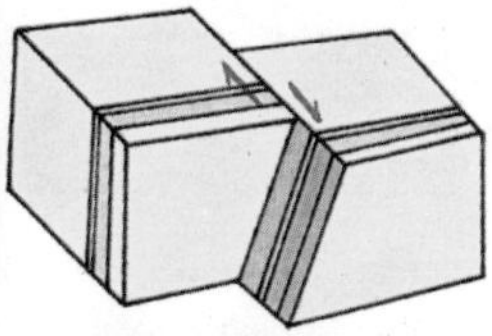
transcurrent fault (dextral) 橫推斷層（右行的）

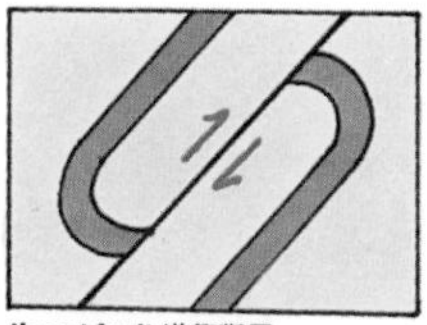
thrust fault 逆衝斷層

strike-slip fault a fault (↑) in which the movement on the fault plane (↑) is parallel to the strike (p.123) of the fault, i.e. sideways.

平移斷層；走向滑動斷層 斷層面(↑)上的運動與斷層走向(第 123 頁)，即側向平行的斷層(↑)。

wrench fault a steeply dipping (p.123) fault (↑) on which the movement has been horizontal, i.e. strike-slip (↑).

扭斷層；走向斷層 斷層上曾發生水平運動(即走向滑動(↑))的陡峭傾斜(第 123 頁)斷層(↑)。

tear fault (1) a wrench fault (↑); (2) a strike-slip fault (↑) that crosses the strike (p.123).

捩斷層 (1)一種扭斷層(↑)；(2)橫切走向(第 123 頁)的平移斷層(↑)。

transcurrent fault = tear fault (↑).

橫推斷層 同捩斷層(↑)。

dextral fault a strike-slip fault (↑) in which the relative displacement (↑) is to the right as seen across the fault plane (↑). *See also* **sinistral fault** (↓).

右行平移斷層 對着斷層面(↑)看時相對位移(↑)向右的平移斷層(↑)。參見**左行平移斷層**(↓)。

sinistral fault a strike-slip fault (↑) in which the relative displacement (↑) is to the left as seen across the fault plane (↑). *See also* **dextral fault** (↑).

左行平移斷層 對着斷層面(↑)看時相對位移(↑)向左的平移斷層(↑)。參見**右行平移斷層**(↑)。

oblique-slip fault a fault (↑) in which the displacement (↑) is not parallel to the strike (p.123) or to the dip (p.123) of the fault plane.

斜向滑動斷層 位移(↑)和斷層面的走向(第 123 頁)或傾向(第 123 頁)不平行的斷層(↑)。

rotational fault a fault (↑) in which one block (p.133) has turned about a point on the fault plane (↑). The displacement (↑) varies from place to place.

旋轉斷層 斷塊(第 133 頁)繞斷層面(↑)上的一點轉動的斷層(↑)，位移(↑)處處不同。

dip-slip fault a fault (↑) in which the movement on the fault plane (↑) is parallel to the dip (p.123) of the fault.

傾向滑動斷層 斷層面(↑)上的運動和斷層傾向(第 123 頁)平行的斷層(↑)。

slickenside (*n*) (*usually in the plural, slickensides*) fine parallel scratches or grooves on a fault (↑) surface that have been produced by the movement of the rocks on either side of the fault.

斷層擦痕(名) 斷層任一側的岩石運動在斷層(↑)面上所產生的細微的平行擦痕或槽溝。(英文常用其複數 slickensides)

fault breccia a breccia (p.87) made up of sharp pieces of rock broken up during movement along a fault (↑).

斷層角礫岩 由岩石沿一個斷層(↑)運動時所破裂的稜角狀岩石碎塊組成的角礫岩(第 87 頁)。

fault gouge finely divided rock material produced by movement along a fault (↑).

斷層泥 岩石沿一個斷層(↑)運動時所產生的碾碎岩石物質。

thrust fault a reverse fault (↑) in which the fault plane (↑) is at a low angle to the horizontal.

逆衝斷層 斷層面(↑)與水平面成低角度夾角的逆斷層(↑)。

thrust (*n*) = thrust fault (↑).

衝斷層(名) 同逆衝斷層(↑)。

thrust plane the plane of a thrust fault (↑).

逆衝斷層面 逆衝斷層(↑)的面。

overthrust (*n*) = thrust fault (↑).

逆掩斷層(名) 同逆衝斷層(↑)。

fault block a mass of rock with faults (↑) on two or more sides.

斷塊 兩邊或數邊帶有斷層(↑)的岩塊。

fault scarp an escarpment (p.33) produced by movement on a fault (↑).

斷層崖 由某個斷層(↑)運動所產生的懸崖(第 33 頁)。

listric fault a fault (p.128) that curves downwards, the fault plane (p.128) being steep at the surface and more nearly horizontal at depth. Listric faults are characteristic of continental margins (p.135).

lag (*n*) a thrust fault (p.129) in which the uninverted (p.127) limb (p.123) of a recumbent fold (p.127) has been cut out.

slide (*n*) a fault (p.128) that is nearly horizontal. A slide may be a thrust fault (p.129) or a lag (↑).

nappe (*n*) a large mass of rock that has been moved several kilometres or more. A nappe may be either the hanging wall (p.128) of an overthrust (p.129) or a recumbent fold (p.127).

decke (*n*) **(decken)** = nappe (↑).

thrust sheet = nappe (↑).

imbricate structure a structure in which reverse faults (p.128) are formed between thrust planes (p.129) that are more or less parallel.

schuppen structure = imbricate structure (↑).

décollement (*n*) 'unsticking'. A series of beds (p.80) may be folded (p.123) and slide (↑) over a lower series that is little folded or not folded at all; this is a décollement. It is necessary for there to be a bed at the base of the folded series that can slip easily over the rocks below, e.g. salt (p.17) or anhydrite (p.52) beds.

鏟狀斷層　斷層面（第 128 頁）在地表很陡，在深部則接近水平的向下彎曲的斷層（第 128 頁），鏟狀斷層是大陸邊緣（第 135 頁）的特徵。

滯後斷層（名）　平臥褶皺（第 127 頁）的未倒轉（第 127 頁）翼（第 123 頁）被削去的逆衝斷層（第 129 頁）。

滑移斷層（名）　接近水平的斷層（第 128 頁），可為逆衝斷層（第 129 頁）或滯後斷層（↑）。

納布（名）　已被運移數千米或更遠的巨大岩塊，可以是逆掩斷層（第 129 頁）的上盤（第 128 頁），或者是平臥褶皺（第 127 頁）。

推覆體（名）　同納布（↑）。

逆衝岩片　同納布（↑）。

疊瓦構造　在大致平行的逆衝斷層面（第 129 頁）之間形成一組逆斷層（第 128 頁）的構造。

疊置構造　同疊瓦構造（↑）。

滑脱（名）　"鬆開"。一組岩層（第 80 頁）在另一組輕微褶曲或完全不褶曲的下部岩層之上發生褶曲（第 123 頁）和滑移（↑），謂之滑脱。這必須是被褶曲的那組岩層底部有一個容易在下伏岩石之上滑動的岩層，例如鹽（第 17 頁）層或硬石膏（第 52 頁）層。

nappe 推覆體（納布）

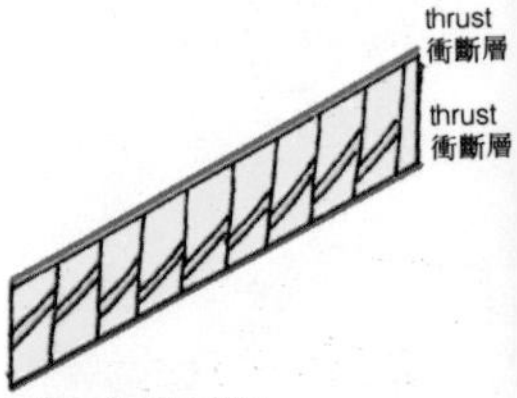

imbricate structure
疊瓦構造

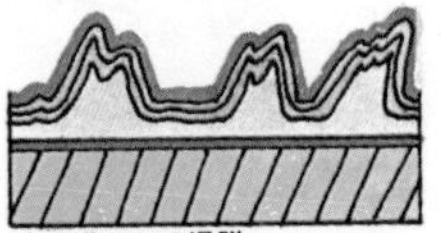

decollement 滑脱

klippe (*n*) (*klippen*) a piece of a nappe (↑) that stands apart and is separated from the other rocks of the nappe and rests on a thrust plane (p.129), thus forming a *tectonic outlier* (p.118).

window (*n*) an area in which the rocks above a thrust fault (p.129) have been eroded to expose (p.122) the rocks below.

fenster (*n*) = window.

gravity tectonics the movement of rocks under the force of gravity (p.11) to produce tectonic structures (p.122) such as faults (p.128) and folds (p.123).

飛來峯　推覆體（↑）的一部分，它脱離隔開推覆體的其他部分並覆於逆衝斷層面（第 129 頁）之上，形成一個"構造外露層"（第 118 頁）。

構造窗（名）　逆衝斷層（第 129 頁）上方的岩石被剝蝕後，露出（第 122 頁）下方岩石的區域。

蝕窗（名）　同構造窗。

重力構造　岩石在重力（第 11 頁）下運動所產生的地質構造（第 122 頁），例如斷層（第 128 頁）和褶皺（第 123 頁）。

diapir (*n*) a structure (p.122) in which the core (p.9) of an anticline (p.124) breaks through the rocks above. Salt (p.17) can form diapirs; so can igneous (p.62) rocks.

底闢構造（名） 背斜（第 124 頁）的核（第 9 頁）穿透上方岩石的構造（第 122 頁）。岩鹽（第 17 頁）、火成岩（第 62 頁）可形成底闢構造。

salt dome a diapir (↑) of salt (sodium chloride).

鹽丘 岩鹽（氯化鈉）的底闢構造（↑）。

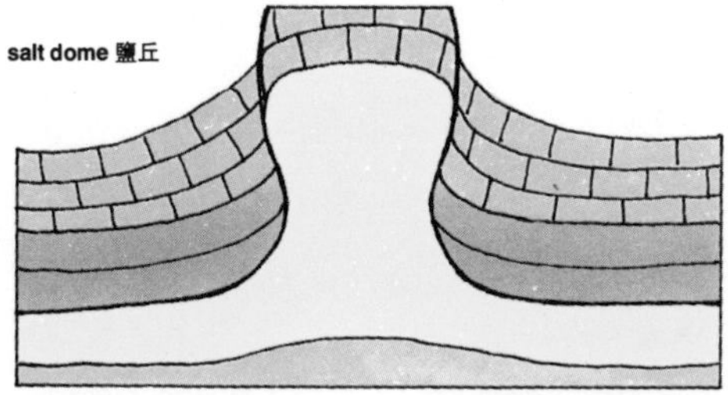

piercement dome a salt dome (↑) in which the mass of salt has broken through the rocks above it and has reached or nearly reached the Earth's surface.

刺穿穹丘 鹽體穿透上方岩石並達到或接近達到地表的一種鹽丘（↑）。

mantled gneiss dome a dome (p.125) of granite (p.76) surrounded by gneiss (p.97) and sediments (p.80). The foliation (p.95) of the granite is parallel to the bedding (p.80) of the sediments where they are in contact.

覆蓋的片麻岩穹丘 片麻岩（第 97 頁）和沉積岩（第 80 頁）所圍繞的花崗岩（第 76 頁）穹丘（第 125 頁）。在花崗岩與沉積岩接觸處，花崗岩的頁理（第 95 頁）與沉積岩的層理（第 80 頁）平行。

cryptoexplosion structure a term used to describe more or less circular structures that show intense rock deformation (p.122) but do not appear to have been caused by volcanic or tectonic activity. These structures range from 1.5 km to more than 50 km in diameter. They have evidently been formed by an explosive force but the term 'cryptoexplosion structure' does not imply any particular method of formation. *See also* **astrobleme** (↓).

隱爆構造 描述略帶圓形構造所用的術語，此構造顯示岩石強烈變形（第 122 頁），而此變形似非由火山或構造活動所造成。構造的直徑可從 1.5 千米到大於 50 千米，顯然是由某種爆炸力所形成，但"隱爆構造"這個術語並不指任何特定的形成方法。參見**古隕擊坑**（↓）。

astrobleme (*n*) an ancient mark on the surface of the Earth produced by the fall of a large body from space. Astroblemes are usually circular and the rocks show signs of great shock. *See also* **cryptoexplosion structure** (↑).

古隕擊坑（名） 由宇宙空間一個巨大物體墜落而在地球表面產生的古標誌。古隕擊坑通常是圓形的，岩石則顯示巨大衝擊的標誌。參見**隱爆構造**（↑）。

lineament (*n*) any geological structure of large size that appears as a geographical feature in the form of a straight line, e.g. a valley, a hill, or a coastline.

線狀構造（名） 地理特徵方面表現直線形狀的任何大型地質構造。例如河谷、山丘或海岸線。

diastrophism (*n*) the deformation (p.122) of large masses of the Earth's crust (p.9) to form mountains, etc. **diastrophic** (*adj*).

地殼變動(名) 地殼(第 9 頁)的大塊體形成山脈等的變形(第 122 頁)。(形容詞為 diastrophic)

orogeny, orogenesis (*n*) the process of mountain-building; a period of mountain-building. Deformation (p.122), folding (p.123), and thrusting (p.129) are characteristic features of orogenies and are usually accompanied by the intrusion (p.64) of igneous rocks (p.62). The period of time required for an orogeny may be hundreds of millions of years. **orogenic** (*adj*).

造山運動，造山作用(名) 造山過程：造山期。變形(第 122 頁)、褶皺作用(第 123 頁)和衝斷層作用(第 129 頁)都是造山運動的特徵，通常都伴有火成岩(第 62 頁)的侵入(第 64 頁)。一場造山運動所需的時間可長達幾億年。(形容詞為 orogenic)

orogen, orogenic belt a relatively narrow region, which may be thousands of kilometres long and hundreds of kilometres wide, that has been affected by an orogeny (↑). The deeper parts are affected by regional metamorphism (p.90) and the emplacement (p.64) of igneous rocks (p.62).

造山帶 受造山運動(↑)影響的一個相對狹窄的區域，可長達幾千公里、寬幾百公里。其較深部則受區域變質作用(第 90 頁)和火成岩(第 62 頁)侵位(第 64 頁)的影響。

mobile belt a long, narrow region of the Earth's crust (p.9) in which there is deformation (p.122), igneous activity (p.62), and metamorphism (p.90). A mobile belt will have stable blocks (↓) on either side of it.

活動帶 地殼(第 9 頁)中存在變形(第 122 頁)、火成活動(第 62 頁)和變質作用(第 90 頁)的一個狹長區域。一個活動帶的兩側都有穩定的地塊(↓)。

geosyncline (*n*) a long, narrow area of the Earth's crust (p.9) in which a great thickness of sediment (p.80) is deposited. Volcanic rocks (p.68) are also present with the sediments. The floor of the geosyncline subsides (p.125) as sedimentation continues. The sediments may later be deformed (p.122) to produce a mountain range. **geosynclinal** (*adj*).

地槽(名) 地殼(第 9 頁)中沉積着巨厚沉積物(第 80 頁)的一塊狹長地帶。火山岩(第 68 頁)也與沉積物一起出現。地槽底板隨着沉積作用的進行而下沉(第 125 頁)。而後，沉積物可能發生變形(第 122 頁)形成山脈。(形容詞為 geosynclinal)

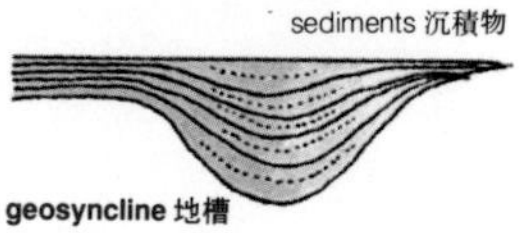

miogeosyncline (*n*) a geosyncline (↑) formed next to a craton (↓). The sediments (p.80) in a miogeosyncline are relatively thin and there are no volcanic rocks (p.68). **miogeosynclinal** (*adj*).

冒地槽(名) 在克拉通(↓)鄰近形成的地槽(↑)。冒地槽內的沉積物(第 80 頁)較薄，且沒有火山岩(第 68 頁)。(形容詞為 miogeosynclinal)

eugeosyncline (*n*) a geosyncline (↑) formed away from a craton (↓) and containing a great thickness of sediments (p.80), including greywackes (p.87) and volcanic rocks (p.68). **eugeosynclinal** (*adj*).

優地槽(名) 一種遠離克拉通(↓)形成的地槽(↑)，含有巨厚的沉積物(第 80 頁)，包括雜砂岩(第 87 頁)和火山岩(第 68 頁)。(形容詞為 eugeosynclinal)

geanticline (*n*) an uplift of anticlinal (p.124) form developed in a geosyncline (↑) as the sides of the geosyncline move closer together. **geanticlinal** (*adj*).

地背斜(名) 由於地槽(↑)兩邊相互移近而在地槽內發育起來的一種背斜(第 124 頁)形隆起。(形容詞為 geanticlinal)

cordillera (*n*) (1) a mountain range or a series of more or less parallel mountain ranges; (2) a row of islands in a geosyncline (↑), formed when the rocks of a geanticline (↑) reach sea level.

山系；雁列山脈（名） (1)山脈或一系列近於平行的山脈；(2)在地槽中(↑)，地背斜(↑)岩石達到海面時形成的一列島嶼。

block (*n*) in tectonics (p.122), a large mass of the Earth's crust, tens or hundreds of kilometres across, that behaves as a single rigid unit.

地塊（名） 大地構造學(第 122 頁)上，指一塊對徑幾十或幾百公里，相當於一塊單獨的剛性單元的一大塊地殼。

flexure (*n*) a bend in a strata (p.80), usually a gentle one; a fold (p.123). **flexural** (*adj*).

撓曲（名） 地層(第 80 頁)中通常較為平緩的彎曲；一種褶皺(第 123 頁)。(形容詞為 flexural)

taphrogenesis (*n*) vertical movements of large size that produce steeply dipping faults (p.128).

地裂作用（名） 產生陡傾斷層(第 128 頁)的大規模垂直運動。

shield (*n*) a large area of very old igneous (p.62) and metamorphic (p.90) rocks of Pre-Cambrian (p.114) age that have not been folded or deformed (p.122) since Pre-Cambrian times.

地盾（名） 由前寒武紀(第 114 頁)極古老火成岩(第 62 頁)和變質岩(第 90 頁)組成的廣濶區域，自前寒武紀年代以來未曾受到褶曲或變形(第 122 頁)。

platform (*n*) a large area of old, stable basement rocks (p.114) covered by younger horizontal or nearly horizontal strata (p.80) resting on the eroded (p.20) surface of the basement.

地臺（名） 由古老而穩定基底岩石(第 114 頁)構成的廣濶區域，其基底剝蝕(第 20 頁)面上覆蓋着較年輕的水平或近水平的地層(第 80 頁)。

foreland (*n*) a resistant block (↑), i.e. a block that is stable and rigid, on one side of a geosyncline (↑); the side of a folded mountain range towards which the overturned folds (p.123) are leaning.

前陸（名） 一塊穩定的地塊(↑)，即位於地槽(↑)一側的一塊穩定而剛性的地塊；倒轉褶皺(第 123 頁)所倚伏的褶皺山脈的一側。

craton (*n*) a stable area of the Earth's crust (p.9), usually large. More or less the same as a shield (↑).

克拉通（名） 地殼(第 9 頁)的一塊穩定區域，通常很巨大。與地盾(↑)大致相同。

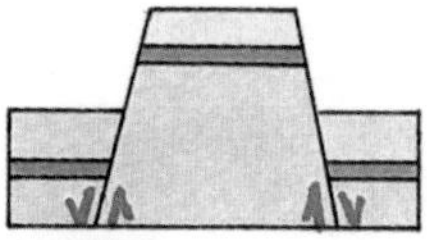

horst 地壘

horst (*n*) an area that has been lifted up as a block between normal faults (p.128) on either side of it.

地壘（名） 兩正斷層(第 128 頁)之間被抬昇起的一個斷塊區域。

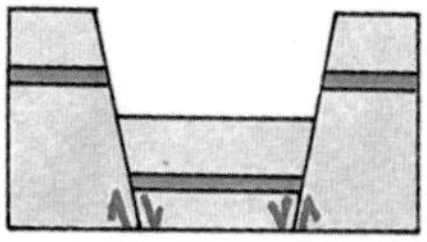

graben 地塹

graben (*n*) (*graben*) a narrow block of the Earth's crust (p.9) that has been moved down between normal faults (p.128) on either side. *See also* **rift valley** (↓).

地塹（名） 兩正斷層(第 128 頁)之間向下移動的地殼(第 9 頁)的一個狹窄地殼斷塊。參見**裂谷**(↓)。

rift (*n*) a structural feature in which the rocks between two faults (p.128) have been let down in relation to those on either side. The use of the term 'rift' does not necessarily mean that there is a valley in the geographical sense. *See also* **rift valley** (↓).

斷陷谷；裂陷（名） 兩條斷層(第 128 頁)之間的岩石已相對於其兩側的岩石下落的構造形跡。使用"斷陷谷"這一術語時未必表示存在一個地理意義上的谷地。參見**裂谷**(↓)。

trough faulting = rift (↑).

槽形斷層作用 同斷陷谷(↑)。

rift valley a valley formed between two more or less parallel faults (p.128).

裂谷 在兩條大致平行的斷層(第 128 頁)之間形成的谷地。

aulacogen (*n*) a trough (p.124) formed by a rift (↑) that has failed to develop.

塹溝（名） 由中途夭折的斷陷谷(↑)形成的褶皺槽(第 124 頁)。

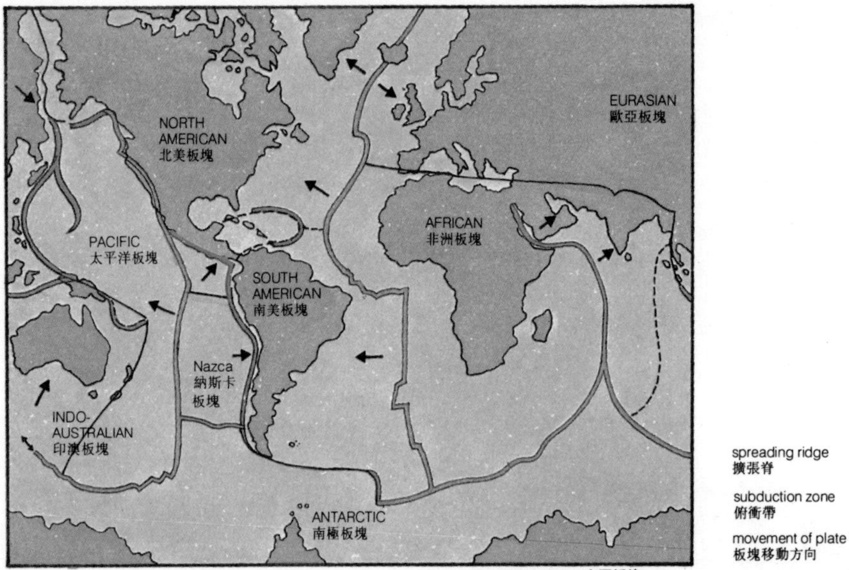

the major plates 主要板塊

plate tectonics the theory that the Earth's surface is covered by a number of relatively thin plates (↓), which move over the material below. Many geological facts that earlier appeared to be unrelated have been explained by this theory.

板塊構造 這理論認為地球表面為若干較薄的板塊(↓)所覆蓋，而這些板塊在下伏物質之上移動。此理論解釋了早先許多似乎不相關的地質事實。

plate (*n*) part of the Earth's surface that behaves as a single rigid unit. Plates are about 100 to 150 km thick. They may be made up of continental crust or oceanic crust (p.9), or both, on top of a layer of the upper mantle (p.9). Plates move in relation to the Earth's axis and to each other. There are seven large plates (the African, Eurasian, Indo-Australian, Pacific, North American, South American, and Antarctic plates) and several smaller ones.

板塊(名) 作用相當於一個單一的剛性單元的地球表面部分。板塊可以由上地幔(第 9 頁)層頂部的大陸地殼或海洋地殼(第 9 頁)單獨地或共同組成，厚約 100 到 150 公里。板塊相對於地球軸運動，彼此間亦相對運動。地球上有七個大板塊(非洲、歐亞、印澳、太平洋、北美、南美和南極板塊)和一些較小的板塊。

microplate (*n*) a small plate (↑), e.g. those round the Mediterranean Sea.

微板塊(名) 小板塊(↑)，例如地中海周圍的小板塊。

continental drift the theory that the present continents have been formed by the breaking up of one large continent and have since moved to their present positions.

大陸漂移 這理論認為地球上現今的大陸是由一個巨型大陸分裂後漂移到現今的位置形成的。

plate margin the edge of a plate (↑). It is at the plate margins that most seismic (p.12), volcanic (p.68), and tectonic (p.122) activity is found. There are three types: constructive margins, at which new crust (p.9) is being formed; destructive margins, at which one plate is moving down below another; and conservative margins, at which plates simply move past each other.

constructive margin see plate margin (↑).

destructive margin see plate margin (↑).

conservative margin see plate margin (↑).

plate boundary the line between two plates (↑) that touch each other. Plate boundaries are marked by seismic activity (p.12) and tectonic activity (p.122).

triple junction a point on the Earth's surface where three plate boundaries (↑) meet.

island arc a curved chain of islands with the convex (outer) side of the curve facing the open ocean. There is a deep oceanic trench (p.35) on the convex side of the arc and deep sea on the opposite side. Island arcs are regions where deep-focus earthquakes (p.12) occur and where gravity (p.11) and magnetic anomalies (p.14) are found. The islands may also show volcanic activity (p.68).

板塊邊緣 板塊(↑)的邊緣。地震(第12頁)、火山(第68頁)和構造(第122頁)活動大多數見於板塊邊緣。有三種類型：增生性邊緣，此處在正形成新地殼(第9頁)；破壞性邊緣，一塊板塊正在此處下移到另一板塊之下；保守性邊緣，此處的板塊只作相對運動。

增生性邊緣 見板塊邊緣(↑)。

破壞性邊緣 見板塊邊緣(↑)。

保守性邊緣 見板塊邊緣(↑)。

板塊邊界 兩個板塊(↑)之間相互接觸的接觸線。板塊邊界是以地震活動(第12頁)和構造活動(第122頁)為標誌。

三接合點；三重點 地球表面上三個板塊邊界(↑)相會之點。

島弧 弧的凸(外)邊面向開濶海洋的弧形列島。在弧的凸邊有深海溝(第35頁)，另一邊則為深海。島弧是深源地震(第12頁)發生的區域，在此處出現重力異常(第11頁)和磁異常(第14頁)。這些島也顯示有火山活動(第68頁)。

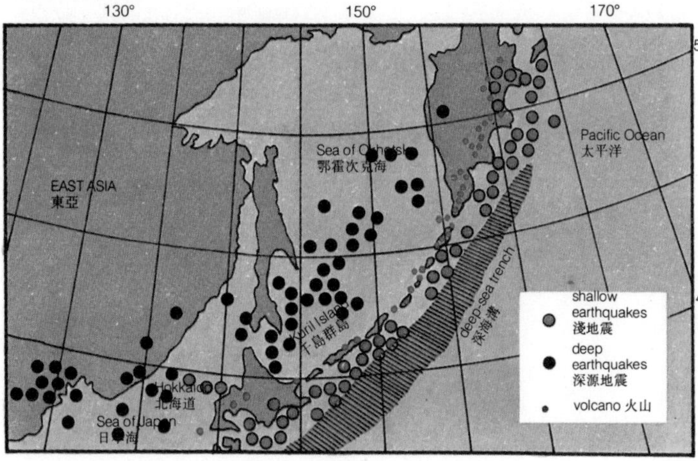

island arc: 島弧
the Kuril island arc, north of Japan
日本北部的千島島弧

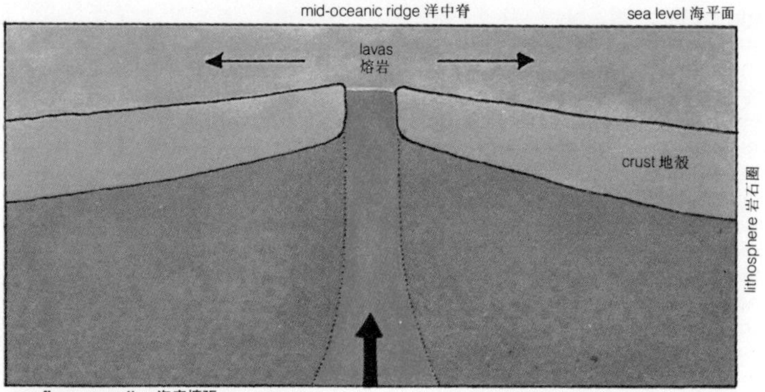

sea-floor spreading 海底擴張

sea-floor spreading the theory that growth of new crust (p.9) takes place at active mid-oceanic ridges (p.35). This takes place by the intrusion (p.64) of submarine lavas (p.70) at the mid-oceanic ridges. The rocks nearest to the ridge are thus the youngest and the age of the rocks on the sea floor increases with distance from the axis of the ridge. Magnetic anomalies (p.14) shown by the rocks on the two sides of the ridge are symmetrical about the ridge axis.

transform fault a fault (p.128) along which two plates (p.134) move past each other without lithosphere (p.9) being formed or destroyed. A typical transform fault is a strike-slip fault (p.129) that cuts across a mid-oceanic ridge (p.35), the ridge being offset (i.e. its two halves do not meet across the fault). There is seismic activity (p.12) at the transform fault between the two points where it meets the mid-oceanic ridge. There are also continental transform faults, e.g. the San Andreas fault in California and the North Anatolian transform fault south of the Black Sea.

zone of divergence a constructive margin (p.135); a region where two plates (p.134) are moving away from each other, e.g. the Mid-Atlantic Ridge. New lithospheric (p.9) material is formed in these regions.

pull-apart zone = zone of divergence (↑).

海底擴張　這理論認為新地殼(第 9 頁)的生長是在活動的洋中脊(第 35 頁)發生的。在洋中脊處由於海底熔岩(第 70 頁)的侵入(第 64 頁)而生成新地殼。因此，最靠近洋中脊的岩石最年輕，海底岩石的年齡則隨與洋中脊軸的距離的增加而增加。洋中脊兩邊的岩石所顯示的磁異常(第 14 頁)在脊軸兩側對稱分佈。

轉換斷層　兩個板塊(第 134 頁)沿斷層相互對向移動，且不生成或破壞岩石圈(第 9 頁)的斷層(第 128 頁)。切割洋中脊(第 35 頁)使之錯開(即斷層兩側的洋中脊不再重合一起)的平移斷層(第 129 頁)是一種典型的轉換斷層。在洋中脊與轉換斷層相交的兩個交點之間有地震活動(第 12 頁)。大陸也有轉換斷層。例如加利福尼亞州的聖安德烈斯斷層和黑海以南的北安納托利亞轉換斷層。

離散帶　一種增生性邊緣(第 135 頁)：兩板塊(第 134 頁)彼此離開的區域，例如大西洋中脊。這些區域中有新岩石圈(第 9 頁)物質生成。

拉開帶　同離散帶(↑)。

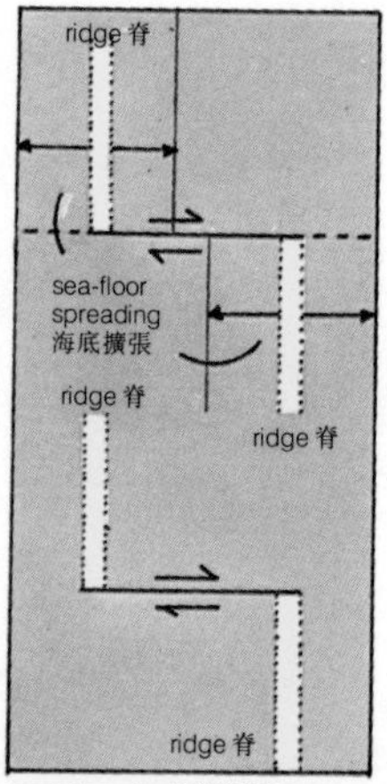

transform fault 轉換斷層

subduction zone a region in which a lithospheric plate (pp.9, 134) is forced down, or subducted, into the asthenosphere (p.9) and mesosphere (p.9). The movement of the lithospheric plate is thought to be the cause of the earthquakes (p.12) that occur in island arc regions (p.135). As it moves down into the mantle (p.9) the plate is heated, and at a depth of between 100 and 300 km it is partly melted. At 700 km depth it breaks up completely. *See also* **Benioff zone** (↓).

俯衝帶 岩石圈板塊(第 9、134 頁)被迫向下進入或消滅於軟流圈(第 9 頁)和中圈(第 9 頁)之中的一個區域。岩石圈板塊運動是島弧區域(第 135 頁)發生地震(第 12 頁)的起因。當它向下進入地幔(第 9 頁)時，板塊被加熱，並在 100 至 300 公里深度之間被部分熔化。在 700 公里深度處則完全熔化。參見**畢烏夫帶**(↓)。

Benioff zone a sloping surface of seismic activity (p.12) that is characteristic of island-arc (p.135) systems. The Benioff zone meets the Earth's surface close to an ocean trench (p.35) and dips below the island arc. The angle of dip is typically about 45° but may be between 30° and 80°.

畢烏夫帶 島弧(第 135 頁)系所特有的地震活動(第 12 頁)的傾斜面。畢烏夫帶與地球表面在海溝(第 35 頁)附近相交並傾斜於島弧之下，傾角一般約為 45°，但可在 30° 至 80° 之間變動。

zone of convergence = subduction zone (↑).

會聚帶 同俯衝帶(↑)。

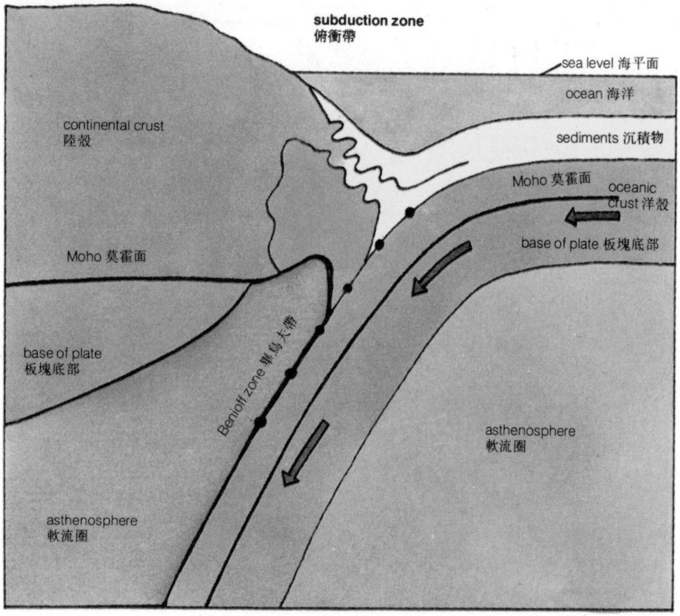

accretionary prism a mass of sediments (p.80) pushed together at a subduction zone (p.137). Older sediments are thrust (p.129) over younger sediments as one plate is driven under the other. (*See diagram on p.137.*)

back-arc upwelling island arc systems (p.135) tend to move away from the continent and into the ocean. This is thought to be caused by convection currents (p.142) in the asthenosphere (p.9) between the island arc and the continent, which in turn are caused by cooling of the mantle (p.9) by the descending mass of oceanic lithosphere (p.9).

加積稜柱體 在俯衝帶(第 137 頁)被推擠在一起的沉積物(第 80 頁)塊體。一個板塊被推入另一板塊之下時，較老的沉積物被衝斷(第 129 頁)而覆於較年輕的沉積物之上(見第 137 頁圖示)。

弧後上湧 島弧系(第 135 頁)傾向於背離大陸而移入海洋。這是島弧與大陸之間的軟流圈(第 9 頁)內的對流(第 142 頁)所引起的，這種對流則是地幔(第 9 頁)受下降的大洋岩石圈(第 9 頁)冷却所引起的。

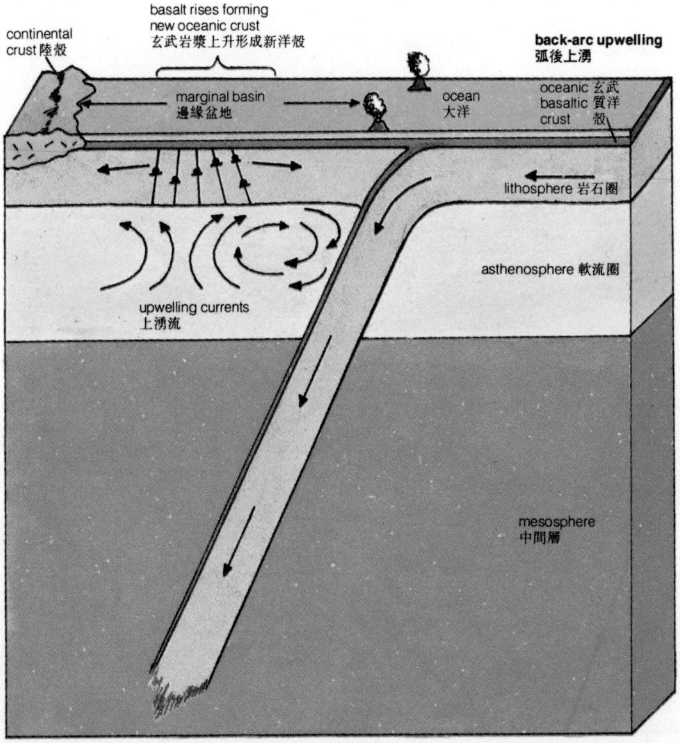

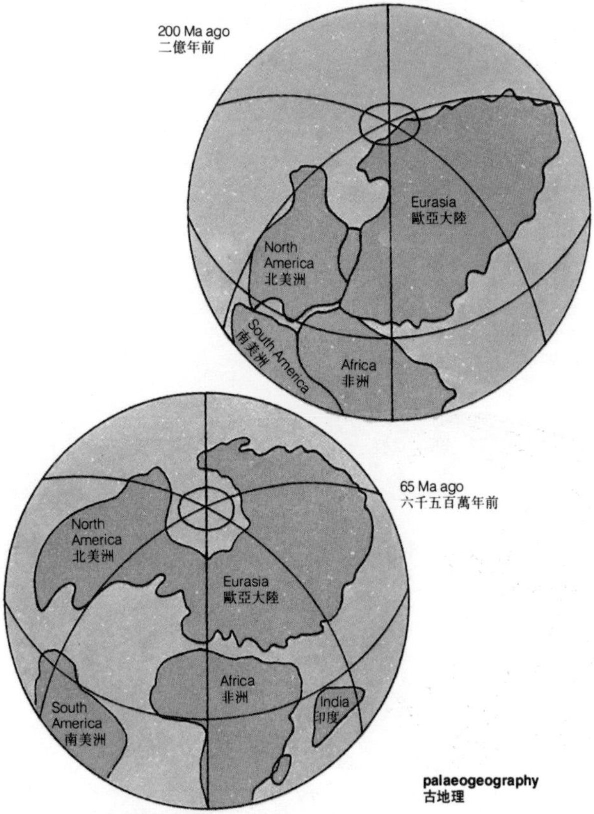

palaeogeography
古地理

palaeogeography (*n*) the study of the geography of past geological ages and especially of the positions of former continents and oceans. This is done by studying rocks of known ages to discover the environments (p.81) in which they were formed and the directions in which the rivers and the ocean currents were flowing at the time. The information is then put together, usually in the form of a *palaeogeographical map*. **palaeogeographical, palaeo-geographic** (*adj*).

古地理學(名) 研究遠古地質年代的地理，特別是從前大陸和海洋位置的學科。其做法是：研究已知年齡的岩石，查明這些岩石生成時的環境(第 81 頁)以及當時河流和洋流的流向，然後綜合這些資料，通常編成"古地理圖"。(形容詞為 palaeogeographical，palaeogeographic)

Gondwanaland (*n*) the 'supercontinent' that is thought to have existed in the southern hemisphere until the Cretaceous (p.115). It consisted of South America, Africa, Arabia, Madagascar, India, Sri Lanka, Australia, New Zealand, and Antarctica. *See also* **Pangaea** (↓).

岡瓦納古陸(名)　人們認為在白堊紀(第 115 頁)以前曾存在於南半球的一個"超級大陸"，此古陸由南美洲、非洲、阿拉伯、馬達加斯加、印度、斯里蘭卡、澳大利亞、紐西蘭和南極洲組成。參見**泛古陸**(↓)。

Laurasia (*n*) the 'supercontinent' that is thought to have existed in the northern hemisphere at some time before the Tertiary (p.115). It consisted of North America, Greenland, and Eurasia (Europe and Asia). *See also* **Pangaea** (↓).

勞亞古陸(名)　人們認為在第三紀(第 115 頁)以前的某個時期曾存在於北半球的一個"超級大陸"。此古陸由北美洲、格陵蘭和歐亞大陸(歐洲和亞洲)組成。參見**泛古陸**(↓)。

Pangaea (*n*) the 'supercontinent' formed by Gondwanaland (↑) and Laurasia (↑) together. Pangaea began to break up about 200 Ma ago in the Jurassic (p.115).

泛古陸(名)　由岡瓦納古陸(↑)和勞亞古陸(↑)一起形成的"超級大陸"。泛古陸在大約二億年前的侏羅紀(第 115 頁)開始解體。

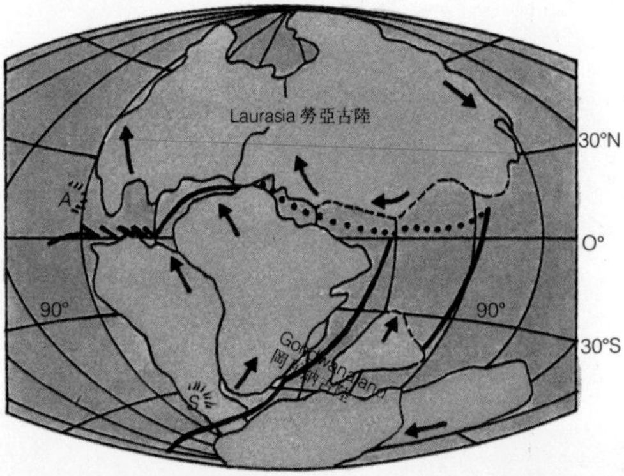

Laurasia and Gondwanaland 180 Ma ago
一億八千萬年前的勞亞古陸和岡瓦納古陸

divergent boundaries 離散邊界

convergent boundaries 會聚邊界

continental accretion the growth of continents by adding material to their margins. The central parts of the continents are very old: Pre-cambrian (p. 114). Younger rocks have, it is thought, been added to these central parts by accretion during mountain-building periods when the original continents have come together as a result of plate movements (p.134).

大陸增生　大陸由於其邊緣增加物質而生長。大陸中央部分是很古老的，屬前寒武紀(第 114 頁)。人們認為在造山時期，較年輕的岩石由於增生作用而被增加到這些中央部分上，當時的諸原始大陸已經由於板塊運動(第 134 頁)而滙合在一起。

suture (*n*) a belt of deformed rock (an orogenic belt, p.132), which marks the zone where two continents have come together and joined. The suture may be several hundred kilometres wide.

地縫合線(名)　是一個變形岩石帶(造山帶，第 132 頁)，它標誌着兩大陸會聚和結合的地帶。地縫合線可以寬達幾百公里。

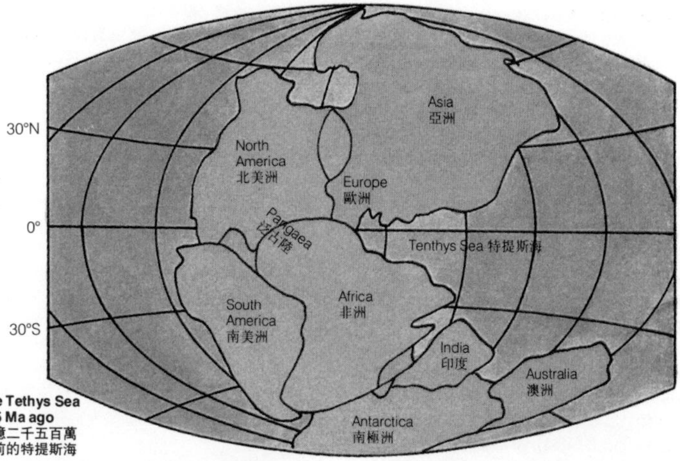

The Tethys Sea
225 Ma ago
二億二千五百萬
年前的特提斯海

Tethys (*n*) the ocean that is thought to have existed between the eastern ends of Gondwanaland (p.140) and Laurasia (p.140). The Tethys sea was formed when Gondwanaland moved south between 340 and 225 Ma ago. The eastern end of the Tethys sea was closed in the Tertiary (p.115), leaving the present Mediterranean Sea between Europe and Africa.

特提斯海(名)　人們認為是曾存在於岡瓦納古陸(第 140 頁)東端和勞亞古陸(第 140 頁)東端之間的古海洋。特提斯海是於三億四千萬年至二億二千五百萬年前之間，岡瓦納古陸南移時形成的。特斯提海的東端在第三紀(第 115 頁)時閉合，留下現今位於歐洲和非洲之間的地中海。

Iapetus sea a sea that is thought to have separated the English and European continent from Scotland and North America in Lower Palaeozoic times.

亞皮特斯海　人們認為是在早古生代時已將英吉利和歐洲大陸與蘇格蘭和北美洲分開的一個海。

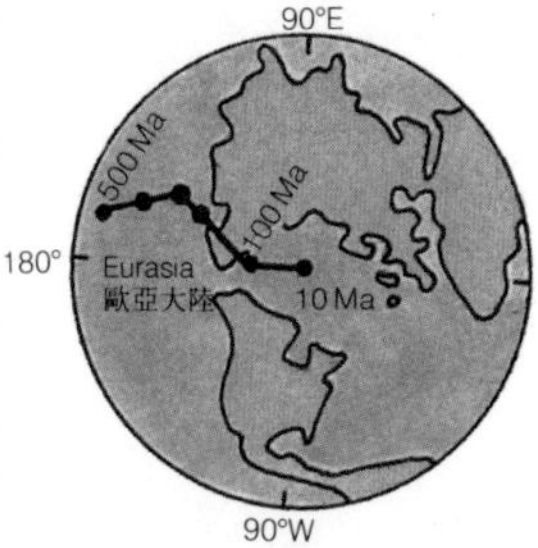

apparent polar wandering path for Eurasia
歐亞大陸的視地
極游移路徑

polar wander when the palaeomagnetism (p.14) of rocks of various ages is studied and their directions of magnetization (p.14) are measured, the positions of the Earth's magnetic poles (p.14), as shown by the magnetism of the older rocks, are not close to the Earth's geographical poles. Lines called *polar wandering curves* can be drawn on maps to show the apparent movements of the poles in the geological past. The apparent movement of the poles can be explained by movements of the continents, and this explanation fits in with the theory of continental drift (p.134).

地磁極游移　在研究不同年代的岩石的古地磁(第 14 頁)並測量它們的磁化(第 14 頁)方向時，較老岩石的磁性所顯示的地磁極(第 14 頁)位置並不接近於地球的地極。在地質史上不同時期地磁極的視運動可以用繪在地圖上的"地磁極游移曲線"來表示。地磁極的視運動可以用大陸的運動來解釋，而這一解釋與大陸漂移理論(第 134 頁)相吻合。

convection current convection could occur in the Earth's mantle (p.9) if the material in the upper part of the mantle were cooled. This material would then descend and its place would be taken by hotter material from below, thus setting up a convection current. Currents of this kind could provide one possible means of moving the lithospheric plates (pp.9, 134).

plume (*n*) it is thought that hotter material could move upward in the Earth's mantle (p.9) in a number of *thermal plumes*, each a few hundred kilometres across. On reaching the top of the mantle, the hotter material would spread out sideways in all directions and would be cooled. There would then be a downward flow of material in other regions to balance the upward flow in the plumes. The sideways movement of material could provide a means of moving the lithospheric plates (pp.9, 134), and the plumes could explain the occurrence of chains of volcanic islands (p.68) that have no relationship to plate boundaries (p.135). *See also* **hot spot** (↓).

hot spot an area of the lithosphere (p.9) that is heated by a plume (↑). Volcanic activity (p.68) and upward doming (p.125) of the lithosphere could be caused by a hot spot. A line of volcanoes (p.68) could be formed when a plate (p.134) moved over a fixed hot spot.

對流 地幔(第9頁)上部分所含的物質冷却時，可以在地幔中發生對流。被冷却的物質下沉，來自下面的較熱的物質則取代其位置，從而形成對流。這種對流成為岩石圈板塊(第9、134頁)運動的推動力。

地幔羽(名) 人們認為，地幔(第9頁)中較熱的物質可以以一些"熱羽"的形式向上運動，每一熱羽的橫斷面直徑為幾百公里。當其運動達到地幔頂部時，較熱物質向側旁各個方向擴散開並冷却下來。其他區域則有向下運動的物質流來平衡熱羽中向上移動的物質流。物質向側旁的運動可以成為推動岩石圈板塊(第9、134頁)的推動力。熱羽則可解釋與板塊邊界(第135頁)無關係的火山列島(第68頁)鏈如何出現。參見**熱點**(↓)。

熱點 岩石圈(第9頁)中被地幔羽(↑)加熱的地區。熱點可以引起火山活動(第68頁)以及岩石圈的向上窿起(第125頁)。當板塊(第134頁)在一固定熱點之上移動時就形成一列火山(第68頁)。

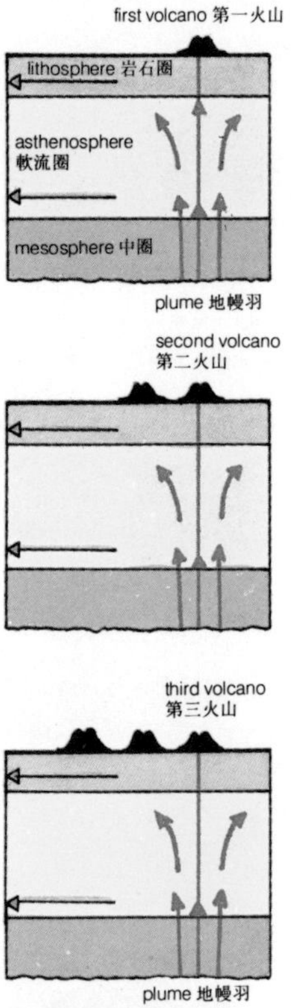

formation of a line of volcanoes by a fixed hot spot
一固定熱點所形成的一列火山

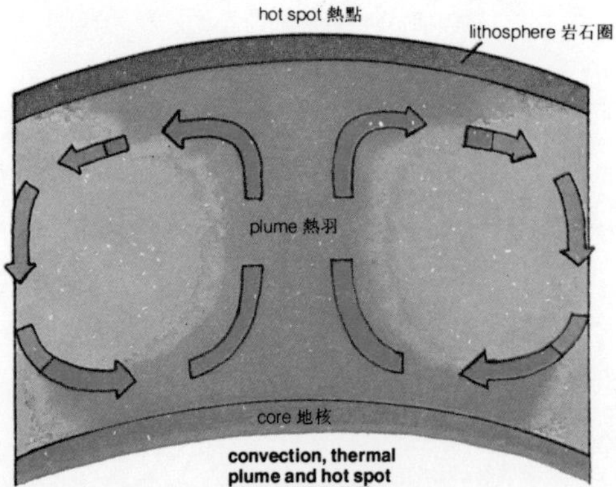

convection, thermal plume and hot spot
對流、熱羽和熱點

rock mechanics, soil mechanics the study of the physical properties of rocks and soils, especially those properties that affect their ability to support a load.

岩石力學，土壤力學 研究岩石和土壤的物理性質，特別是那些影響岩石和土壤承載力的物理性質的學科。

geotechnics = soil mechanics (↑).

土力學 同土壤力學(↑)。

site investigation the geological study of a piece of ground on which a building or some other structure (e.g. a dam) is to be placed.

場地勘察 對擬在其上建大樓或其他建築物(例如水壩)的一塊土地所作的地質調查。

soil creep the very slow movement of soil (p.23) at the surface down a slope.

土滑 地表土壤(第 23 頁)順坡向下的十分緩慢的移動。

earth flow movement of surface material that is faster than soil creep (↑) but slower than a mud flow (↓) or landslide (↓). An earth flow usually slides on a spoon-shaped surface. At its upper end is a curved cliff and at its lower end is a swelling shaped like a tongue.

土流 地面物質的運動，速度快於土滑(↑)而慢於泥流(↓)或滑坡(↓)。土流通常滑落在匙形地面上，其上端是彎曲的懸崖，下端則為舌狀隆起。

mud flow the rapid movement of a mixture of mud and water, which flows like a liquid. A typical mud flow can carry large rocks and boulders. Mud flows are characteristic of desert regions and alpine regions (i.e., the higher regions of mountain systems).

泥流 泥水混合物像液體流動般的快速運動。典型的泥流可以携帶大量石塊和漂礫。泥流是沙漠區域和高山區域(即山系的較高區域)所具的特徵。

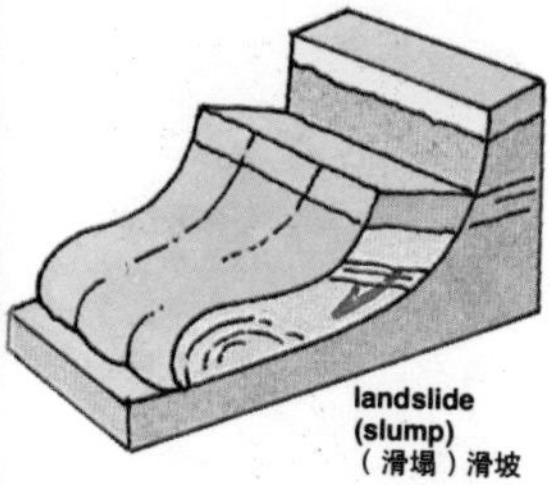

landslide (slump)
(滑塌) 滑坡

landslide (*n*) the movement of a mass of rock or soil, or both, down a slope. A landslide differs from an earth flow (↑) in that the mass of rock remains more or less in one piece.

滑坡(名) 岩石、土壤或兩者的混合物整體沿斜坡向下的運動。滑坡與土流(↑)不同之處在於岩塊下滑時或多或少保持其整體性。

landslip (*n*) = landslide (↑).

地滑(名) 同滑坡(↑)。

slump (*n*) a landslide (↑) in which a mass of rock – typically clay – moves on a curved surface (a *shear surface*).

滑塌(名) 岩體(特別是黏土)在弧形面(剪切面)上移動的滑坡(↑)。

glide (*n*) a landslide (↑) in which a mass of rock moves downward along the surface of a sloping bedding plane (p.80).

滑移(名) 岩體沿傾斜層面(第 80 頁)向下滑動的滑坡(↑)。

rock fall the free fall of pieces of rock from a cliff or steep slope. Freezing and thawing are a common cause of rock falls. The fallen rocks may pile up as talus cones (p.21) at the foot of the cliff.

岩崩 岩塊從懸崖或陡坡自由崩落。凍結而後解凍是岩崩的一般成因。崩落的岩石可以在懸崖脚下堆積成岩錐(第 21 頁)。

quick clay a special type of clay (p.88) containing a large amount of water (often more than 50 per cent by weight). Such clays are normally solid but a shock can cause them to turn liquid. They can thus form sudden earth flows. Quick clays were originally deposited on the sea floor close to glaciers (p.28).

不穩黏土；超靈敏黏土 含大量水(其重量往往佔一半以上)的特殊類型黏土(第 88 頁)。這種黏土通常是固體，但受到衝擊或震動即可變為液態，因此可以突然形成土流。不穩黏土原先是沉積在靠近冰川(第 28 頁)的海底上。

petroleum (*n*) oil (↓) occurring naturally in the Earth's crust; natural gas (↓), oil (↓), and solid bitumens (p.89).

oil (*n*) the oil that occurs naturally in the Earth's crust is a mixture of compounds of carbon and hydrogen (hydrocarbons). It is generally thought to have formed from the remains of plants and animals, or both, but we have no direct knowledge of the way in which oil is formed in the Earth's crust.

crude oil, crude oil (↑) as it occurs naturally in the Earth's crust.

natural gas gas occurring naturally in the Earth's crust that consists of hydrocarbons (compounds of carbon and hydrogen). Natural gas may be found alone or with oil (↑).

oil shale an argillaceous rock (p.85) containing a solid material that when distilled gives off oil. The oil cannot be obtained without distillation.

migration (*n*) the upward movement of oil (↑) from the rocks in which it was originally formed to other rocks that are porous (p.84) and permeable (p.145). Pressure (p.93) from the weight of the beds on top of the beds containing the oil can cause it to start to migrate. **migrate** (*v*), **migrating** (*adj*).

trap (*n*) something that stops the upward migration (↑) of oil and causes it to accumulate. Traps are of two types: structural (p.122) and stratigraphical (p.112). Examples of structural traps are anticlines, synclines, domes, and faults (pp.116, 117, 120). Stratigraphical traps include unconformities (p.118) and variations in lithology (p.85).

cap rock the impermeable (p.146) rock(s) that make a trap (↑) effective in preventing oil from migrating (↑) further.

reservoir (*n*) beds in which oil accumulates.

borehole (*n*) a hole drilled into the Earth for oil (↑), gas, water, etc. or to gain information about the rocks below the surface.

well-logging (*n*) the use of physical measurements from instruments lowered down boreholes (↑) to obtain information about the rocks below the surface.

石油(名)　地殼中天然出產的油(↓)：天然氣(↓)、油(↓)和固體瀝青(第 89 頁)。

油(名)　地殼中天然出產的油是碳和氫的各種化合物(烴)的混合物。一般認為，油是從植物、動物或二者的遺體形成的，但我們對油在地殼中形成的方式並沒有直接的知識。

原油　天然產於地殼中的油(↑)。

天然氣　天然產於地殼中，由各種烴(碳氫化合物)組成的氣體。天然氣可以單獨產出，也可以與油(↑)一起產出。

油頁岩　可從所含固體物質中泥質岩石(第 85 頁)蒸餾出油。如不進行蒸餾就不能獲得油。

運移(名)　油(↑)從生油的母岩中向上遷移至其他多孔狀(第 84 頁)透水性(第 145 頁)岩石的運動。含油層頂部岩層的重量賦予油開始運移的壓力(第 93 頁)。(動詞為 Migrate，形容詞為 migrating)

圈閉(名)　某些能阻止油向上運移(↑)並使油積聚的空間。圈閉有構造的(第 122 頁)和地層的(第 112 頁)兩種類型。構造圈閉的例子有背斜、向斜、穹窿和斷層(第 116、117、120 頁)。地層圈閉包括不整合(第 118 頁)和岩性(第 85 頁)變化。

蓋層　形成圈閉(↑)的不透水(第 146 頁)岩石，它能有效地防止油進一步運移(↑)。

儲集層(名)　油在其中積聚的岩層。

鑽孔(名)　為找尋油(↑)、氣、水等，或為獲得地下岩石資料而向地下鑽的孔洞。

測井(名)　將儀器放入鑽孔(↑)中，以物理測量法獲得有關地下岩石的資料。

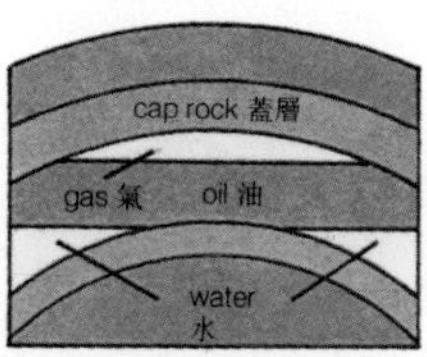

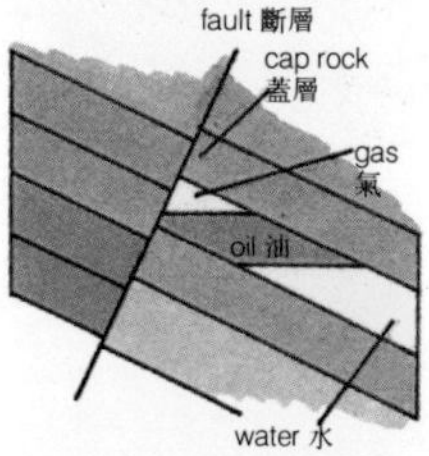

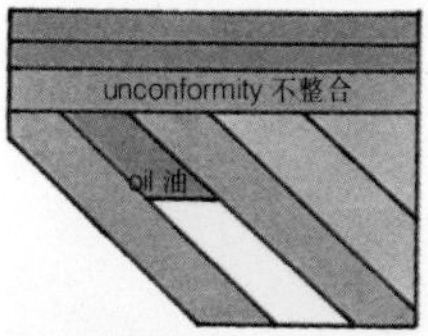

traps 圈閉

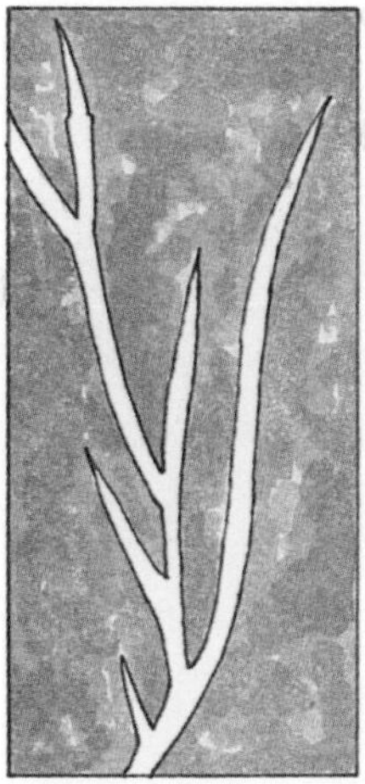

mineral veins 礦物脈

opencast working 露天開採

mine (*n*) a hole dug in the Earth for the purpose of obtaining ores (↓) or minerals, etc. **mine** (*v*).
mineral deposit a mass of ore (↓) that is of value.
ore (*n*) a mineral from which a metal that is of value can be obtained at a cost that makes the work worth while.
ore body a mass of ore (↑). It may be of hydrothermal (↓) origin or intrusive (p.64): a dyke, sill (p.66), or vein (↓).
mineralization (*n*) the formation of new minerals, especially ore minerals, in an existing rock – usually as veins (↓) or masses.
metallogenetic province, metallogenic province a region in which there is a series of mineral deposits (↑) having characters in common.
hydrothermal deposit a mineral deposit (↑) produced by liquids coming from a magma (p.62) that contain a large proportion of hot water.
replacement deposit a mineral deposit (↑) in which the mineral takes the place of a rock that was there earlier, e.g. the hydrothermal (↑) replacement of limestone (p.86) by galena (p.50).
placer deposit a deposit, usually at the surface, containing a mineral of value such as gold.
vein (*n*) a thin mass of rock or a mineral, especially a thin ore body (↑).
wall rock the country-rock (p.65) on either side of a vein (↑).
lode (*n*) more or less the same as a vein (↑).
reef (*n*) in mining geology, a vein (↑) of quartz (p.55) containing gold.
gangue (*n*) the part of an ore (↑) body that does not contain the metal(s) that are being worked.
stockwork (*n*) a large mass of rock that is cut across by many small veins (↑).
opencast (*adj*) applies to a method of mining (↑) a bed (p.80) or vein (↑) near the surface by cutting into it from above rather than by digging a mine (↑) under the ground.
overburden (*n*) rock or other material of no value that lies over a deposit of useful material.
quarry (*n*) a place where rock is dug out in the open. **quarry** (*v*), **quarried** (*adj*).

礦井；礦坑（名）　為獲得礦石（↓）或礦物等而掘入地下的坑洞。（動詞為 mine）
礦床　有開採價值的礦石（↓）體。
礦石（名）　能以合算的成本提煉出有價值的金屬的礦物。
礦體　礦石（↑）的塊體。礦體可以是熱液（↓）成因的或侵入的（第 64 頁）：如岩牆、岩床（第 66 頁）或礦脈（↓）。
礦化（名）　在現存的岩石中形成新礦物、特別是形成礦石礦物的過程。這些新礦物通常呈脈（↓）狀或塊體狀。
成礦區　具備一系列有共同特徵的礦床（↑）的區域。
熱液礦床　由含有大量熱水岩漿（第 62 頁）的液體所生成的礦床（↑）。
交代礦床　指一種礦床（↑），其中的礦物置換早先存在的岩石。例如方鉛礦（第 50 頁）對石灰岩（第 86 頁）的熱液（↑）交代。
砂礦床　通常存在於地表，含有價值礦物如金的礦床。
脈（名）　薄的岩體或礦物體，特別是薄礦體（↑）。
圍岩　礦脈（↑）兩側的原岩（第 65 頁）。
礦脈（名）　大致與脈（↑）同意。
含金石英脈（名）　採礦地質學上指含金的石英（第 55 頁）脈（↑）。
脈石（名）　不含所需提取金屬的那部分礦石（↑）體。
網狀脈（名）　被許多細脈（↑）縱橫交切的大岩石塊體。
露天開採的（形）　描述用於近地表的礦層（第 80 頁）或礦脈（↑）的採礦（↑）方法，即自上而下挖採礦體而非在地下挖掘礦坑（↑）。
剝離層（名）　有用礦物礦床上方所覆蓋着的岩石或其他無價值的物質。
採石場（名）　露天採掘岩石的場地。（動詞為 quarry，形容詞為 quarried）。

hydrology (*n*) the study of water as it occurs on the Earth; in streams, as runoff (↓), from springs, etc.

水文學(名)　研究水在地球上的產出狀態，例如成河流、呈徑流(↓)、來自泉水等的學科。

hydrogeology (*n*) the geology of water supplies that are obtained from under the ground.

水文地質學(名)　地下水供水的地質學。

precipitation (*n*) water falling as liquid (rain, dew) or solid (snow, hail, frost, etc.) on the surface of the Earth from the atmosphere.

降水(名)　以液態(雨、露)或固態(雪、雹、霜等)從大氣降落地面的水。

runoff (*n*) the water falling on the Earth's surface that reaches the streams and rivers.

徑流(名)　降落到地面並沿地面流到溪流和河流的水。

catchment area the area from which the rain-water or other precipitation (↑) enters a particular river or stream. Outside the catch-ment area the water flows in another direction.

集水區；集水面積　雨水或其他降水(↑)流入特定溪流或河流的區域。集水區以外的水則向其他方向流去。

groundwater (*n*) water that is present in the pore spaces (p.84) and other spaces in the rocks below the Earth's surface.

地下水(名)　地面下的岩石孔隙(第 84 頁)或其他空間中存在的水。

water table the upper surface of the groundwater (↑); the surface below which the pore spaces (p.84) of the rocks are filled with water.

地下水位　地下水(↑)的上界面，此界面以下的岩石孔隙空間(第 84 頁)中充滿水。

spring (*n*) water from under the ground coming out at the surface of the Earth. Springs occur where the water table (↑) meets the ground surface or where water under pressure reaches the surface.

泉(名)　從地下冒出地面的水。泉出現於地下水位(↑)與地面交會之處，或地下水受壓力而昇達地面之處。

aquifer (*n*) a stratum (p.80) of rock below the Earth's surface that holds water and through which water can move.

含水層(名)　地面下含有的水能在其中流動的岩層(第 80 頁)。

artesian (*adj*) refers to an aquifer (↑) with impermeable (↓) beds above it and in which water is under a high enough pressure for it to rise above the aquifer.

自流的；承壓的(形)　形容上覆有不透水(↓)層的含水層(↑)，其中的水承受着足够高的壓力因而能上昇到含水層上方。

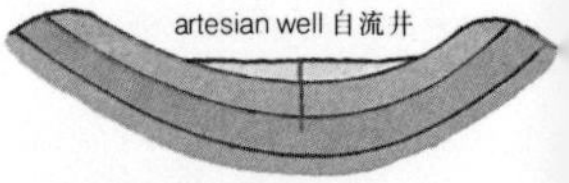

artesian basin 自流盆地

meteoric water water at and below the surface of the Earth that has come from the atmosphere, i.e. from precipitation (↑).

大氣降水　來自大氣圈，即來自降水(↑)過程的地表水和地下水。

juvenile water water in the Earth's crust that has come from magma (p.62).

初生水；岩漿水　來自岩漿(第 62 頁)且留在地殼內的水。

connate water water in a sedimentary rock (p.80) that is believed to have been trapped in the sediment at the time it was formed.

原生水　在沉積岩生成的同時被圈閉在沉積岩(第 80 頁)中的水。

permeability (*n*) in hydrology (↑), permeability is a measure of the ease with which liquids and gases can pass through a rock. **permeable** (*adj*).

滲透率(名)　在水文學(↑)上，滲透率是指液體和氣體通過一種岩石的難易程度的一個尺度。(形容詞為 permeable)

impermeable (*adj*) not permeable (↑).

不透水的(形)　非透水的(↑)。

specimen (*n*) a thing or part of a thing – e.g. a rock or a fossil (p.98) – that is taken as an example.

hand specimen a piece of a rock of a size that is suited to examination in the laboratory (↓).

laboratory (*n*) a room or rooms set aside for scientific work.

lens (*n*) a piece of glass with curved surfaces that can be used to make an object (e.g. a rock specimen) appear larger.

microscope (*n*) an instrument that uses lenses (↑) to give a view of a small object in which its size appears to be greatly increased.

標本(名) 被採集用作示例的一件物體或物體的某部分。例如一塊石頭或化石(第 98 頁)。

手標本 尺寸適合於在實驗室(↓)作觀察研究用的一塊岩石。

實驗室(名) 專供科學研究工作用的一間或多間房室。

透鏡；放大鏡(名) 具彎曲表面的一塊玻璃片，用於使物體(例如岩石標本)看來較大。

顯微鏡(名) 一種使用透鏡(↑)使細小物體看來像放大許多倍的儀器。

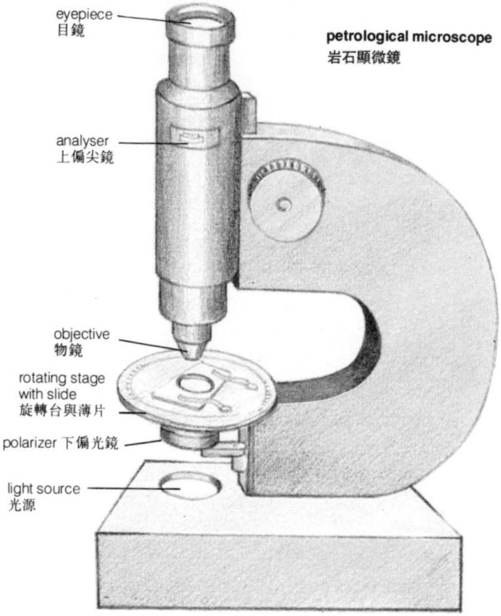

petrological microscope
岩石顯微鏡

petrological microscope a microscope (↑) with special fittings for studying rocks and minerals by using polarized light (p.158).

thin section a very thin piece of a rock or mineral fixed to a piece of glass so that it can be viewed under a microscope (↑).

岩石顯微鏡 一種帶有特殊配件，利用偏振光(第 158 頁)研究岩石和礦物的顯微鏡(↑)。

薄片 固定在玻璃片上以供在顯微鏡(↑)下觀察用的一片極薄的岩石或礦物。

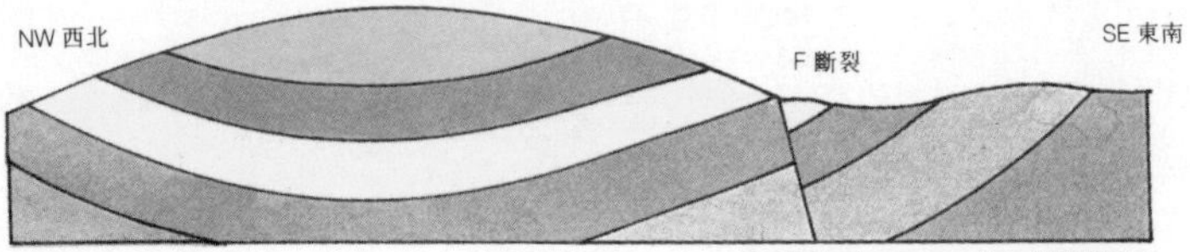

geological cross section 地質橫切面

section (1) a vertical exposure (p.122) or a series of exposures of rocks, e.g. in a cliff; (2) also **cross-section**: a drawing made to show the geological structure (p.122) along a chosen line; (3) also **cross-section**: a cut made across a specimen (p.147) to show what is inside it; (4) a thin section (p.147).

cross-section (*n*) *see* **section** (2), (3) (↑).

in situ (*Latin*) in place, e.g. of a rock or fossil that has not been moved from the place where it was formed.

contact (*n*) the surface at which two different kinds of rock come together; especially between an igneous rock (p.62) and the country-rock (p.65).

clinometer (*n*) a simple instrument for measuring angles from the horizontal, such as angles of dip (p.123).

剖面　(1)直立的露頭(第 122 頁)或一系列岩石露頭，例如在懸崖上的露頭；(2)又稱**橫切面**：沿一條選定線繪製，用以表示地質構造(第 122 頁)的圖；(3)又稱**橫斷面**：橫切一塊標本(第 147 頁)以顯示其內部的切面；(4)薄片(第 147 頁)。

橫切面(名)　見**剖面**(2)、(3)(↑)。

在原地　在原位，例如一塊岩石或化石在原地生成後沒有受到移動。

接觸面(名)　兩種不同類型岩石緊挨在一起的面；尤指火成岩(第 62 頁)與原岩(第 65 頁)間的面。

測斜儀(名)　用以測量與水平面所成角度，例如測傾角(第 123 頁)的簡單儀器。

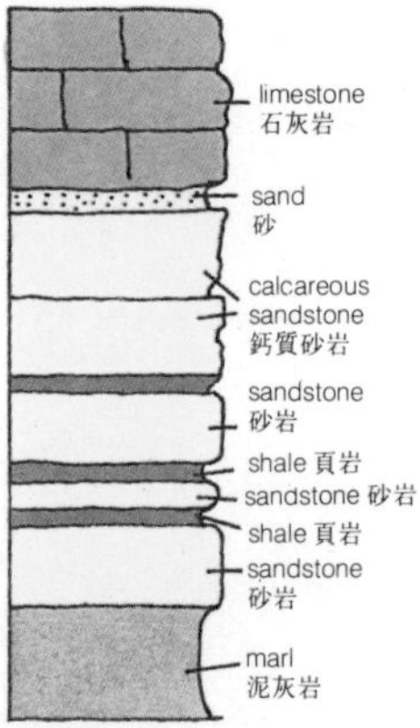

geological section
地質剖面

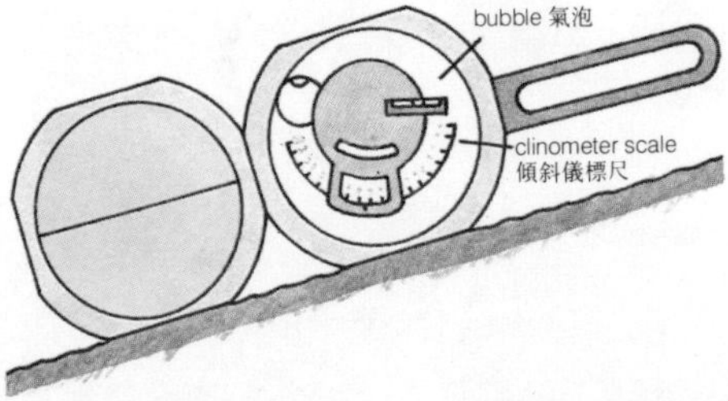

Burton pocket transit
when it is used as a clinometer the bubble is centred by moving a lever (not shown here) on the back of the case; the angle of dip (p.123) is then read off from the scale.

伯頓袖珍經緯儀(羅盤)
當用作測斜儀時，藉移動羅盤盒背後的旋轉桿(圖中未繪出)使氣泡對中，然後從標尺上讀出傾角(第 123 頁)

prismatic compass a hand compass fitted with sights so that the user can measure angles to chosen points.

field work, field geology geological work done in the open air outside the laboratory (p.147).

survey (*v*) to make measurements of an area of land and draw a map of it.

geological survey the work of studying the geology of part of the Earth's crust (p.9) and drawing a geological map of it.

稜鏡羅盤　一種裝有瞄準器的手持羅盤，可用於測量所選各點的角度。

野外工作，野外地質學　在實驗室(第 147 頁)外露天進行的地質工作。

測量(動)　在陸地區域進行測量並繪製地圖。

地質調查　研究地殼(第 9 頁)某一部分的地質情況並繪製其地質圖的工作。

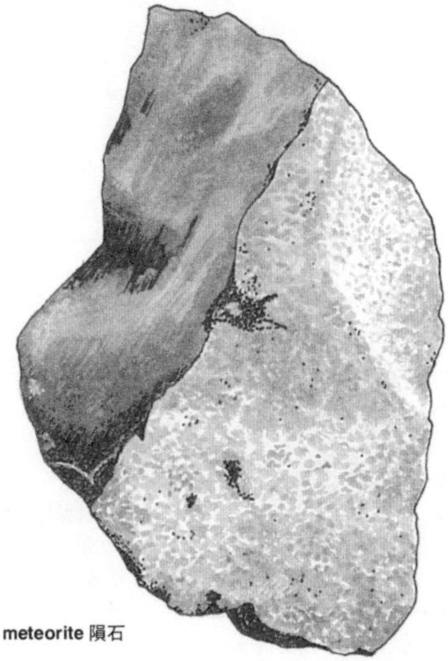

meteorite 隕石

meteorite (*n*) a small solid body from space that has fallen on the Earth's surface.
stony meteorite a meteorite made up chiefly of silicate minerals (pp.16, 53) such as olivine (p.58), pyroxenes (p.57), and feldspars (p.56).
stone (*n*) = stony meteorite (↑).
aerolite (*n*) = stony meteorite (↑).
chondrite (*n*) a stony meteorite (↑) containing *chondrules*: small round masses about 1 mm in diameter made up of olivine (p.58) and pyroxene (p.57). *See also* **achondrite** (↓).
achondrite (*n*) a stony meteorite (↑) that does not contain the *chondrules* that are found in chondrites (↑).
siderite (*n*) a meteorite (↑) consisting of metals: about 90% of iron (Fe) and 6 to 9% of nickel (Ni).
iron (*n*) = siderite (↑).
siderolite (*n*) a meteorite (↑) containing more or less equal amounts of metals and silicates (p.16).
stony-iron (*n*) = siderolite (↑).
tektite (*n*) a rounded, flat, glassy meteorite (↑).

隕石(名)　從太空落到地球表面上的小固體。

石隕石　主要由矽酸鹽礦物(第 16、53 頁)如橄欖石(第 58 頁)、輝石(第 57 頁)和長石(第 56 頁)組成的一種隕石。

石隕石(↑)英文亦稱stone。

石隕石(↑)英文亦稱aerolite。

球粒隕石(名)　含"隕石球粒"的石隕石(↑)。球粒的直徑約為 1 mm，是由橄欖石(第 58 頁)和輝石(第 57 頁)組成的圓形小團塊。參見**無球粒隕石**(↓)。

無球粒隕石(名)　不含球粒隕石(↑)中所見的"隕石球粒"的石隕石(↑)。

隕鐵(名)　含金屬的隕石(↑)，約 90% 為鐵(Fe)和 6 至 9% 為鎳(Ni)。

鐵隕石(名)　同隕鐵(↑)。

石鐵隕石(名)　金屬和矽酸鹽(第 16 頁)含量大致相等的隕石(↑)。

石鐵隕石(↑)英文亦稱stony-iron。

玻隕石(名)　圓形、扁平的玻璃質隕石(↑)。

lunar (*adj*) of the moon.
mare (*n*) (*maria*) one of the large and generally flat areas of the moon which appear dark in colour as seen from the Earth. These areas are made up of mafic (p.75) and ultramafic (p.75) rocks.
lunar highlands areas of the Moon's surface that are higher than the maria (↑), with many craters (↓). The lunar highlands appear bright as seen from the Earth.
terrae (*n.pl.*) = lunar highlands (↑).
crater (*n*) a circular hollow with steep slopes formed either by volcanic action (p.68) or by the fall of a meteor (p.149).
mascon (*n*) an area on the Moon where the density (p.154) of the rocks below the surface is especially high.
rille (*n*) one of the long, narrow valleys on the surface of the Moon. Rilles are up to several hundred kilometres long and one to two kilometres wide. The walls of the valley are steep and its bottom is flat. A rille may be straight (a *normal rille*) or winding (a *sinuous rille*).
mare ridge a long, narrow hill in a lunar mare (↑). Mare ridges are up to a few hundred kilometres long and several tens of metres high.
wrinkle ridge = mare ridge (↑).
lunar regolith a thin layer of grey material on the surface of the Moon. It consists of loose or partly cemented (p.84) fragments ranging from very fine dust to large blocks.
lunar soil = lunar regolith (↑).

月球的（形） 月的。
月海（名） 月球上的開闊而通常平坦的區域之一，在地球上看此區域顯得顏色深暗，此區域是由鎂鐵質（第 75 頁）和超鎂鐵質（第 75 頁）岩組成。（複數為 maria）
月球高地 月面上高出於月海（↑）的區域，此區域有許多月坑（↓）。在地球上看月球高地顯得很明亮。
月陸（名） 同月球高地（↑）。
月坑（名） 月球上有陡坡的圓形凹坑，是由火山作用（第 68 頁）或由隕石（第 149 頁）撞擊所形成。
質量瘤（名） 月面下岩石密度（第 154 頁）特別高的一個區域。
月溪；月面溝紋（名） 月面上的狹長溝谷之一。月溪長達幾百公里，寬 1 至 2 公里。溝谷壁陡而底平。月溪可以是平直的（直月溪），也可以是彎曲的（曲月溪）。
月海脊 月海（↑）中狹長的山丘。月海脊長達幾百公里，高達幾十米。
月面皺脊 同月海脊（↑）。
月壤 月面上一層薄的灰色物質。它由鬆散的或部分膠結的（第 84 頁）碎屑組成，大小從細塵狀到大的塊狀都有。
月土 同月壤（↑）。

Additional definitions 補充釋義

The following list gives definitions of words that are not in the basic word list (the defining vocabulary) that is used for most of the definitions in the main part of this dictionary. These additional words are needed to explain some of the geological terms that are included in the dictionary. Some of them are words in everyday use; others are scientific terms that are only indirectly related to geology.

For ease of reference the words in this appendix are listed in alphabetical order.

詞典正文中用於釋義的詞彙，大部分沒有列入基本詞彙表(釋義詞彙)中，以下補充列出這些詞的意義。這些補充詞彙對於解釋本詞典中的某些地質術語是必需的，其中有些是常用詞；有些則是與地質學只有間接關係的科學術語。

為方便讀者查閱，本附錄所列的詞均按英文字母順序排列。

acceleration (*n*) the increase in velocity (p.153) per unit time, i.e. the increase in velocity divided by the time taken to increase it. If the velocity of a motor-car increases from 20 metres per second (m/s) to 30 m/s in 5 seconds, then the acceleration = (increase in velocity) ÷ (time) = (30 – 20) m/s ÷ 5 s = 10 m/s ÷ 5 s = 2 m/s² (metres per second per second) **accelerate** (*v*), **accelerated, accelerating** (*adj*).

加速度(名) 單位時間內速度(第 153 頁)的增量，即速度的增量除以增加此速度所需的時間。例如一輛汽車在 5 秒鐘內其速度從每秒 20 米(m/s)增加到 30 m/s，則其加速度 = (速度增量) ÷ (時間) = (30 – 20)m/s ÷ 5 s = 10 m/s ÷ 5 s = 2 m/s²(米/秒²)。(動詞為 accelerate，形容詞為 accelerated，accelerating)

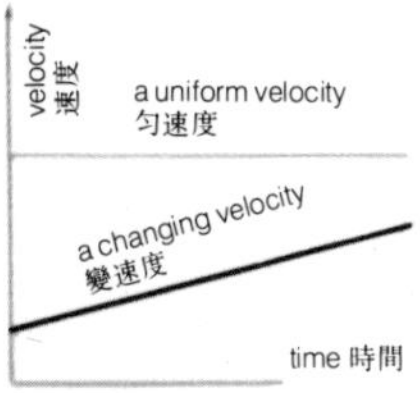

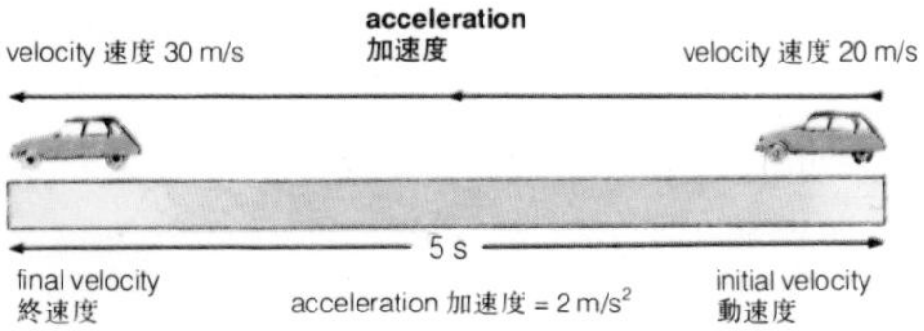

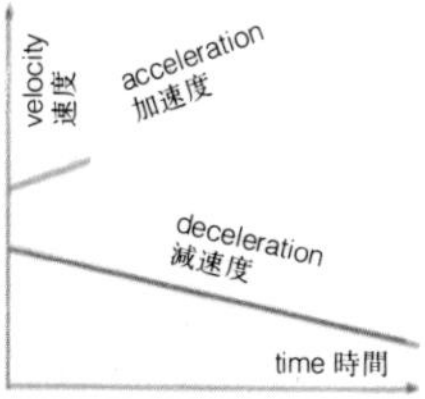

accumulate (*v*) to build up in one place, to heap up in a mass. **accumulation** (*n*).

aggregate (*v*) to gather together into one whole; to mass together. **aggregate** (*n*).

alter (*v*) to make something different in some way without changing the thing itself. **alteration** (*n*).

alternate (*adj*) (of two things) arranged or coming one after the other by turns. **alternately** (*adv*), **alternation** (*n*).

聚集(動) 聚合於一處並堆成一堆。(名詞為 accumulation)

集合(動) 集在一起成為一個整體；集中一起。(名詞 aggregate 意為集合體)

改變(動) 使某物在某些方面與原本有異而不改變該物自身。(名詞為 alteration)

交替的(形) (二物)輪流地一個接一個排列或一個跟隨在另一個之後。(副詞為 alternately，名詞為 alternation)

atom (*n*) the smallest particle of an element (p.15) that has the properties of that element and takes part in chemical reactions (p.17). **atomic** (*adj*).

原子(名) 具有一種元素(第 15 頁)本身的性質並能參與化學反應(第 17 頁)的最小質點。(形容詞為 atomic)

atomic number the number of protons (p.151) in the nucleus (p.150) of an atom (↑). It determines the chemical nature of the atom, i.e. to which element an atom belongs.

原子序數 一個原子(↑)的核(第 150 頁)中的質子(第 151 頁)數目。原子序數決定原子的化學性質，即決定該原子屬何種元素。

atomic weight the average weight of the atoms (↑) of an element in relation to the oxygen atom, which for this purpose is taken to be 16.

原子量 以氧原子量等於 16 為標準所測算出的某一元素原子(↑)的平均重量。

average (*n*) an average is the sum of variable quantities divided by the number of the quantities, e.g. the average of 10 m, 16 m, 8 m, 12 m, is: (10 m + 16 m + 8 m + 12 m) ÷ 4 = 46 m ÷ 4 = 11.5 m. **average** (*adj*).

平均值(名) 變量的總和除以變量的數目之值。例如，10 m、16 m、8 m、12 m 的平均值是：(10 m + 16 m + 8 M + 12 m) ÷ 4 = 46 m ÷ 4 = 11.5 m。(形容詞為 average)

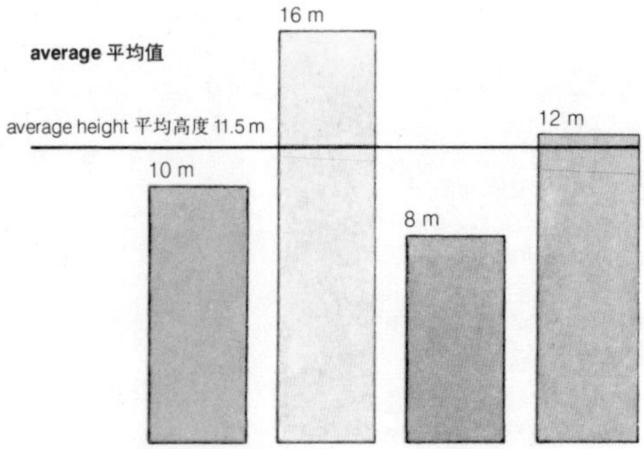

balance (*n*) a state in which two or more things are equal to each other in weight or some other respect so that there is no tendency for them to move up or down. **balanced** (*adj*).

平衡(名) 二件或多件物體的重量或其他方面彼此處於相等狀態，因而沒有向上或向下運動作調整的傾向。(形容詞為 balanced)

base (*n*) a substance which reacts with an acid to form a salt and water only, generally an oxide or a hydroxide of a metal. **basic** (*adj*).

鹼(名) 與酸反應只生成鹽和水的物質，一般為金屬氧化物或金屬的氫氧化物。(形容詞為 basic)

boundary (*n*) the edge or border of a thing.

邊界(名) 物體的邊緣或周界。

cartilage (*n*) a material in the body of a vertebrate (p.109) that holds other parts of the body together (e.g. bones). Some animals, e.g. sharks, have cartilage instead of bone in their skeletons (p.159).

軟骨(名) 脊椎動物(第 109 頁)體內用以將其他部分(例如硬骨)聯結在一起的物質。某些動物，例如鯊魚，其骨骼(第 159 頁)由軟骨而不是由硬骨組成。

cell (*n*) the smallest part of a plant or animal. The simplest living things consist of only one cell; others, e.g. mammals (p.109), contain many millions of cells. A cell has the ability to take in chemical substances and use them to make the various substances it needs in order to live. **cellular** (*adj*).

細胞（名） 植物或動物體的最小部分。最簡單的生物只由一個細胞組成；另一些生物例如哺乳動物（第 109 頁），其細胞則數以百萬計。細胞能攝取化學物質用以製造所需的各種物質以供生存。（形容詞為 cellular）

channel (*n*) a hollow bed in which water runs; a long narrow hollow.

水道（名） 水在其中流動的凹溝；狹長的溝谷。

charge (*n*) an electric charge cannot be explained but a body that carries a charge will be drawn towards another body carrying a charge of opposite sign (i.e. a body with a positive charge will be drawn towards a body with a negative charge); and bodies with charges of the same sign (both positive or both negative) will tend to be pushed away from each other.

電荷（名） 帶電荷的物體被吸引向另一個帶相反電荷的物體（即帶正電荷的物體被吸引向帶負電荷的物體），除此之外電荷是難以解釋清楚的；帶相同符號（二者均為正或二者均為負）電荷的物體會互相排斥。

chemistry (*n*) the study of the elements and their compounds (p.15), their nature and the ways in which they act upon each other. **chemical** (*adj*).

化學（名） 研究元素及其化合物（第 15 頁）的性質以及它們之間相互作用方式的一門學科。（形容詞為 chemical）

classify (*v*) to arrange in classes. **classification** (*n*), **classified** (*adj*).

分類（動） 把…排列成各類。（名詞為 classification，形容詞為 classified）

coarse (*adj*) made up of large pieces or particles, etc.; the opposite of 'fine'.

粗的（形） 描述由粗大的碎塊或粒子等組成；"細的"的反義詞。

combine (*v*) to join together. **combined** (*adj*).

使結合（動） 使聯結在一起。（形容詞為 combined）

complex (*adj*) made up of many parts; not simple. **complexity** (*n*).

複雜的（形） 指由許多部分組成的；不簡單的。（名詞為 complexity）

compose (*v*) to form by being together, e.g. minerals may compose a rock. **composed** (*adj*).

組成（動） 以在一起的方式形成。例如一些礦物可以組成一種岩石。（形容詞為 composed）

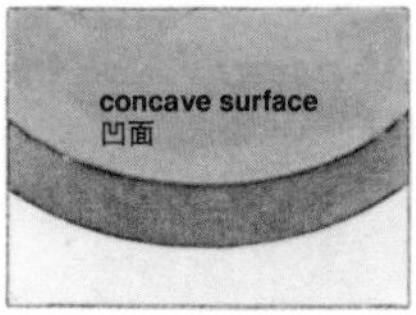

concave (*adj*) curved inwards; hollow. See also **convex** (p.154).

凹的（形） 向內彎曲；凹陷的。參見**凸的**（第 154 頁）。

concentric (*adj*) (of circles) having the same centre. **concentrically** (*adv*).

同心的（形） 指（多個圓）具有同一個中心點。（副詞為 concentrically）

conductor (*n*) a material through which heat or an electric current can flow. All metals are good conductors of heat and electricity. **conduct** (*v*).

導體（名） 熱或電流能在其中流過的材料。一切金屬都是熱和電的良導體。（動詞為 conduct）

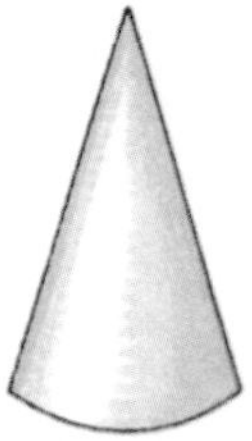

cone
錐體

cone (*n*) a solid figure produced by a straight line passing through a fixed point and a circle. **conical** (*adj*).

錐體（名） 由一條直線通過一個固定點和一個圓所產生的一種立體圖形（形容詞為 conical）。

continent (*n*) one of the larger unbroken land masses of the Earth's surface, e.g. Africa.

大陸（名） 地球表面上較大的完整的陸塊。例如非洲大陸。

contour (*n*) a line joining points on a map or diagram that have the same value; (usually) points on a map that have the same height.

等值線（名） 將地圖上或圖表上具有相同值（在地圖中通常是具有相同高度）的各點連結起來的線。

contract (*v*) to decrease in length, area, or volume of a solid, or in volume of a fluid. Contraction can be caused by a fall in temperature. **contraction** (*n*).

收縮(動) 固體長度、面積或體積縮小，或流體容積縮小。收縮可因溫度降低而引起。(名詞為 contraction)

convection (*n*) the transfer of heat in a fluid by the rising of hotter fluid and the sinking of colder fluid to take the place of the hotter fluid. A **convection current** is formed by the movement of the fluid.

對流(名) 流體中的熱流體上升、冷流體下沉取代熱流體位置的傳熱方式。**對流**是靠流體運動形成的。

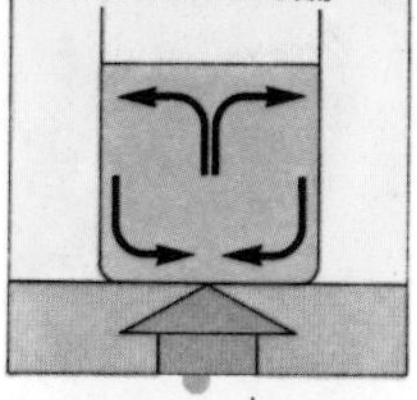
convection current 對流

convex (*adj*) of round shape on the outside; the opposite of **concave** (p.153).

凸的(形) 圓形凸向外邊：**凹的**(第 153 頁)的反義詞。

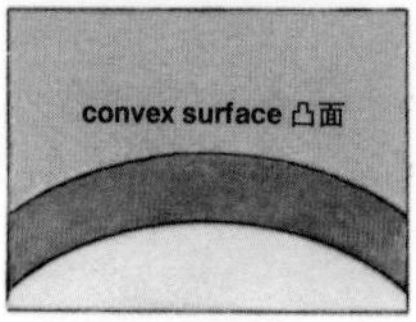

curved mirror 曲面鏡

correspond (to) (*v*) to be similar to in some character of working. **corresponding** (*adj*).

相當於(動) 某些行為特性上相似於。(形容詞為 corresponding)

decompose (*v*) to break down something into the parts of which it is made. **decomposition** (*n*), **decomposed** (*adj*).

分解(動) 使某物分裂為其組成的各部分。(名詞為 decomposition，形容詞為 decomposed)

density (*n*) mass per unit volume, i.e. the mass of a material divided by its volume, e.g. 128.2 cm^3 of iron has a mass of 1 kg: density of iron = 7.8 g/cm^3 = 7800 kg/m^3 (kilograms per cubic metre). Each material has a particular density, e.g. the density of water is 1 g/cm^3. Density is important in the identification (p.93) of materials. **dense** (*adj*).

密度(名) 單位體積的質量，即物質的質量除以其體積。例如體積為 128.2 cm^3 的鐵質量為 1 kg，則鐵的密度 = 7.8 g/cm^3 = 7800 kg/m^3 (千克每立方米)。每種物質都有其特有的密度。例如，水的密度是 1 g/cm^3。密度對於鑑定(第 93 頁)物質極為重要。(形容詞為 dense)

densities 密度

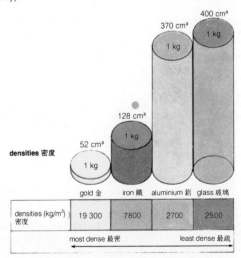

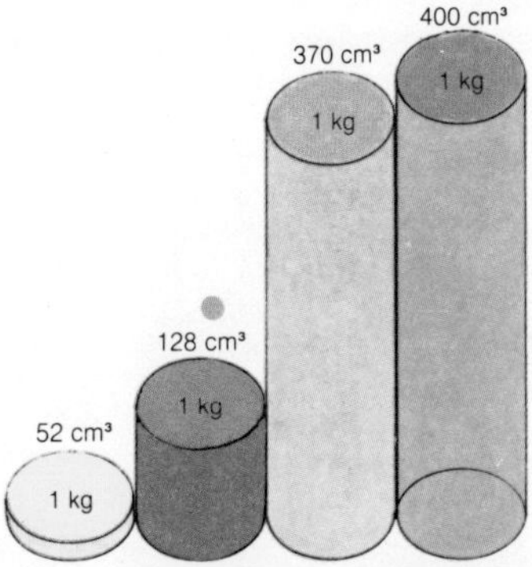

	gold 金	iron 鐵	aluminium 鋁	glass 玻璃
densities (kg/m^3) 密度	19 300	7800	2700	2500

most dense 最密 ⟷ least dense 最疏

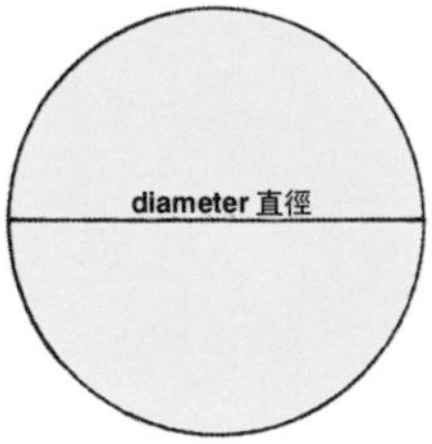

diagram (*n*) a drawing or figure that is intended to explain something or to show the relationships between two or more things.

圖解；圖表(名) 用以解釋某些事物，或用以表示兩種或多種事物之間的關係的一種繪圖或圖形。

diameter (*n*) a straight line passing from side to side through the centre of a circle or a sphere, or the length of that line.

直徑(名) 從圓或球的一邊通過圓心或球心延伸到另一邊的一條直線，或此直線的長度。

dimension (*n*) the dimensions of a solid are its length, its breadth, and its height. Liquids and gases do not have dimensions. A solid has three dimensions: a flat surface has two dimensions (length and breadth); and a line has only one dimension (length). **in three dimensions** in the three directions of space. **dimensional** (*adj*).

維(名) 固體的維是指其長、寬和高。液體和氣體沒有維。立體有三維；平面有二維(長和寬)；直線只有一維(長)。**在三維中**是指在空間的三個方向上。(形容詞為 dimensional)

dissolve (*v*) to put into solution (p.159). **dissolved** *(adj)*

溶解(動) 使進入溶液(第 159 頁)中。(形容詞為 dissolved)

distillation (*n*) a physical process for separating liquids from mixtures. The mixture is heated and the liquid is turned to gas; the gas is then cooled in a tube, where it turns to liquid and can be collected. Liquids with different boiling points can be separated by this process. **distil** (*v*).

蒸餾(名) 從混合物中分離出各種液體的物理過程。混合物被加熱之後其中的液體變為氣體，將氣體引入試管中冷却使之變回液體然後收集之。此法可用於分離沸點不同的各種液體。(動詞為 distil)

distribute (*v*) to spread out or scatter in space or over a surface. **distribution** (*n*).

分佈(動) 散佈在空間中或散佈於一個平面上。(名詞為 distribution)

division (*n*) the act of dividing; the state of being divided.

劃分(名) 分割的行動；被劃分的狀態。

downwards (*adv*) from higher to lower.

向下(副) 從高到低。

drill (*v*) to make a hole through or into something, e.g. to drill a borehole (p.139) into the Earth.

鑽(動) 在某物上作一個孔或打通一個孔。例如向地下鑽一個鑽孔(第 139 頁)。

electron (*n*) a very small particle with a mass about 1/1840 that of a hydrogen atom (↑) and a very small negative electrical charge (p.153).

電子(名) 一種很微小的質點，質量約為氫原子(↑)質量的 1/1840，而所帶的負電荷(第 153 頁)很小。

en echelon (*adj, French*) used to describe structures (e.g. folds) that are parallel to each other but are not opposite to each other.

雁列狀(形) 描述相互平行但不齊頭排列的構造(例如：褶皺)用的詞。

energy (*n*) the ability to do work. There are different forms of energy: potential energy (stored energy); kinetic energy (energy from motion); heat energy; light energy; electrical energy; chemical energy; nuclear energy. One form of energy can be transformed into another.

能(名) 做功的能力。能有不同的形式：勢能(潛能)；動能(運動的能)；熱能；光能；電能；化學能；核能。能可從一種形式轉變為另一種形式。

equilibrium (equilibria) (*n*) a state of balance (p.146) between opposing forces or effects. Forces that balance each other are *in equilibrium*. An object is equilibrium if the forces acting upon it are in equilibrium.

均衡(名) 反向力或效應之間的一種平衡(第 146 頁)狀態。相互平衡的力，處於平衡狀態。如果作用於某一物體上的諸力處於"平衡狀態"，則該物體是均衡的。(複數為 equilibria)

equivalent (*n*) equal in value, power, meaning, etc., to something else.

等價物(名) 數值、能力、意義等的相等物。

evaporation (*n*) the change of a liquid to a vapour at a temperature below, or at, its boiling point, e.g. the evaporation of rain water without the water boiling; the evaporation of a salt solution (p.159) leaving the salt. **evaporate** (*v*).

蒸發(名) 液體在溫度等於或低於其沸點時變為蒸汽的過程。例如雨水在水不沸騰情況下的蒸發；鹽的溶液(第 159 頁)蒸發，遺留下鹽。(動詞為 evaporate)

expand (*v*) to increase in length, area, or volume of a solid, or in volume of a liquid or gas. **expansion** (*n*).

膨脹(動) 固體長度、面積或體積的增大，或液體、氣體體積的增大。(名詞為 expansion)

external (*adj*) lying outside.

外部的(形) 位於外面的。

fine (*adj*) made up of small pieces or particles, etc. The opposite of 'coarse' (p.153).

細的(形) 指由微細的小塊、或粒子等組成的。"粗的"(第 153 頁)的反義詞。

force (*n*) a push or pull that causes: (1) an acceleration, or (2) a change in the shape of an object, or (3) a reaction (the opposite effect to an action, equal to it but in the opposite direction). A force can be measured by: (1) the amount it stretches a spring; (2) the acceleration it gives to a mass.

力(名) 推或拉引致：(1)產生加速度；(2)使物體改變形狀；(3)產生反作用力(與作用力的效應相反，二者大小相等而方向相反)。力的測量可以根據：(1)使彈簧伸長的長度；(2)使一定質量的物體具有加速度。

formula (*n*) chemical symbols written together to show the atoms in a molecule of a compound or in an ion; e.g. (a) the formula MgO stands for a molecule of magnesium oxide and shows that it is composed of one atom of magnesium combined with one atom of oxygen; (b) the formula NO_3^- stands for a nitrate ion. The formula gives the composition of a substance.

化學式(名) 寫在一起表示某一化合物分子或某一離子中的諸原子的化學符號。例如：(a)化學式 MgO 代表一個氧化鎂分子，並表示它由一個鎂原子與一個氧原子結合而成；(b)化學式 NO_3^- 代表一個硝酸根離子。化學式表示出一種物質的組成。

fragment (*n*) a piece broken off. **fragmented** (*adj*).

碎屑(名) 碎裂的一片。(形容詞為 fragmented)

frequency (*n*) the number of times an event is regularly repeated in unit time, e.g. the number of vibrations in 1 second. **frequent** (*adj*).

頻率(名) 單位時間內某一事件規律地重複出現的次數。例如一秒鐘內振動的次數。(形容詞為 frequent)

gas (*n*) a state of matter (p.159) in which the molecules (p.15) are free to move about, there being no forces to hold them together. A gas has no definite volume and no shape. **gaseous** (*adj*).

氣體(名) 物質(第 159 頁)的一態，其中的分子(第 15 頁)自由地運動，不存在使分子保持在一起的力，故氣體無確定的體積和形狀。(形容詞為 gaseous)

gene (*n*) a short length of a chromosome (a thread-like body in the part of a cell that controls the activities of a cell) which controls a characteristic of a living thing. A gene can be passed on from parent to the next generation.

基因(名) 長度短的一種染色體(細胞某部分中能控制細胞活動的一種線狀體)，它控制生物的特徵。基因可以從母體遺傳給下一代。

geography (*n*) the study of the Earth's surface; its physical features, etc. **geographical** (*adj*).

地理(名) 研究地球表面及其自然特徵等的一門學科。(形容詞為 geographical)

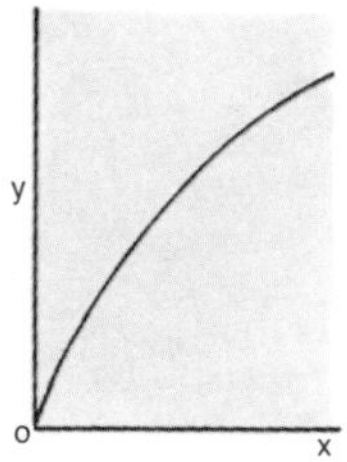

graph 曲線圖

graph (*n*) a diagram that shows the variation of a quantity in relation to another quantity, e.g. a graph of density against temperature for a given substance. **graphical** (*adj*).

曲線圖(名) 表示一個量相對於另一個量而變化的圖解。例如某一物質的密度相對於溫度變化的曲線圖。(形容詞為 graphical)

horizontal (*adj*) level; parallel with the horizon (the line where the Earth and sky appear to meet).

水平的(形) 平的；和地平線(看似地球與天空的交界線)平行的。

hydration (*n*) a chemical process in which a compound (p.15) absorbs and combines with water. **hydrated** (*adj*).

水合作用(名) 某一種化合物(第 15 頁)吸收水並與之結合的化學過程。(形容詞為 hydrated)

hydrous (*adj*) containing water.

含水的(形) 含有水的。

intense (*adj*) to a high degree; very strong, violent.

強烈的(形) 高度的；很強的、激烈的。

landscape (*n*) the scene presented by a piece of country.

景觀(名) 一個地帶所呈現的自然景色。

layer (*n*) a thickness of a substance spread over a surface, especially one of a series. **layered** (*adj*).

層(名) 分佈在一個表面上並具一定厚度的物質，特別是一個層系中的一層。(形容詞為 layered)

magnet (*n*) a solid object that attracts iron and attracts or pushes away (repels) other magnets. When free to turn, it points in a north – south direction. Magnets possess the property of magnetism (*n*). **magnetic** (*adj*), **magnetize** (*v*).

磁鐵(名) 可以吸引鐵，並可吸引或推開(排斥)另一磁鐵的一種固體。任其自由轉動時，它指向南北方向，具有磁性。(形容詞為 magnetic，動詞為 magnetize)

magnetic field the space round a magnet of an electric current in which a magnetic material experiences a magnetic force of attraction, or a magnet sets in the direction of the magnetic force from the magnet.

磁場 磁性物質在磁鐵或電流周圍的空間可感受到磁的吸引力的空間或在磁鐵周圍另一塊磁鐵可按磁力線的方向定位的空間。

margin (*n*) the part of a surface that lies just inside its border; an edge, a border line. **marginal** (*adj*).

邊緣(名) 正好在邊界以內的表面部分；邊界、邊界線。(形容詞為 marginal)

marine (*adj*) (1) of the seas and oceans; (2) inhabiting the sea, found in or formed by the sea.

海洋的(形) (1)海的，大洋的；(2)海棲的，出現於海中或形成於海中的。

medium (*adj*) between the highest and lowest levels; in the middle.

中等的(形) 最高和最低水平之間的；居中間的。

method (*n*) a means of doing something.

方法(名) 行事的方式。

nucleus (*atomic*) (*n*) the central part of an atom. It consists of one or more protons and has a positive charge (p.146) which is almost exactly equal to the total charge on the electrons that surround it in the atom. Nearly the whole of the mass of the atom is in the nucleus. **nuclear** (*adj*).

(原子的)核(名) 原子的中心部分。核含有一個或多個質子，並帶正電荷(第 146 頁)，電量和環繞核周圍的諸電子的總電荷幾乎相等。原子的質量幾乎全集中在核中。(形容詞為 nuclear)

occur (*v*) to be found, to happen.

出現(動) 發現，發生。

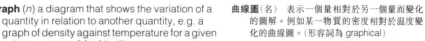

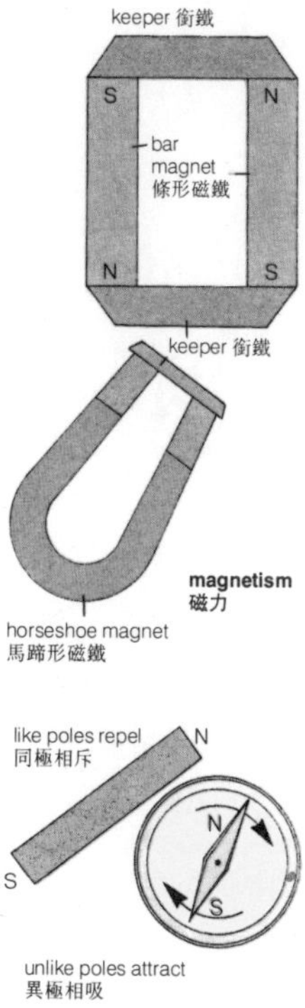

magnetism 磁力

optical interference under certain conditions two sets of light waves can act on each other, producing alternate light and dark bands (if the light is of one wavelength) or colours like those of the rainbow (if white light is used). (The colours seen in thin films of oil on water are an example.) Interference can occur in an anisotropic (p.47) mineral, and the colours that result can be used in identifying the mineral.

光波干涉 兩組相同波長的光波在一定條件下相互作用產生交替的亮帶和暗帶；若使用白光則產生像彩虹那樣的顏色。在水面上的油膜中所見到的顏色就是一個例子。干涉可以發生在非均質（第 47 頁）礦物中，可利用所產生的顏色來鑑定礦物。

origin (*n*) the first state of something; the first place from which something has come. **original** (*adj*).

起源（名） 事物的初始狀態；某物最先來自之處。（形容詞為 original）

oscil|ate (*v*) to move regularly to and fro or up and down. **oscillation** (*n*), **oscillating** (*adj*).

擺動（動） 有規律地來回或上下移動。（名詞為 oscillation，形容詞為 oscillating）

parallel (*adj*) (of two lines or planes) running in the same direction and everywhere at the same distance from each other; never meeting.

平行的（形） 指兩條直線或兩個平面向同一方向延伸，保持距離處處相等，永不相交的。

particle (*n*) a small or very small piece of something.

粒子（名） 小或極小的某物的微粒。

physics (*n*) the study of the properties of matter (except for their chemical properties) and of energy (p.148). **physical** (*adj*).

物理學（名） 研究物質性質（不包括化學性質）和能量（第 148 頁）的學科。（形容詞為 physical）

plan (*n*) a view or a drawing of something as seen from above.

平面圖（名） 某物的俯視圖。

plane (*n*) a surface that is flat. In geology the word plane is also used for surfaces that are not quite flat. **planar** (*adj*).

平面（名） 平坦的表面。在地質學上，"平面"這個詞亦表示不很平的面。（形容詞為 planar）

plastic (*adj*) able to be shaped into any form and of remaining in that shape. Clays (p.88) are plastic.

塑性的（形） 形容可以塑造成任何形狀並保持這種形狀。黏土（第 88 頁）是塑性的。

polarized light in ordinary light, vibrations take place in all possible directions at right angles to the path of the light beam. In plane polarized light the vibrations take place in only one plane. Polarized light is used in the petrological microscope for the identification of minerals.

偏振光；偏光 在尋常光中，振動發生在與光束路徑垂直的所有可能方向上。在平面偏振光中，振動只發生在一個平面內。岩石顯微鏡是使用偏振光鑑定礦物。

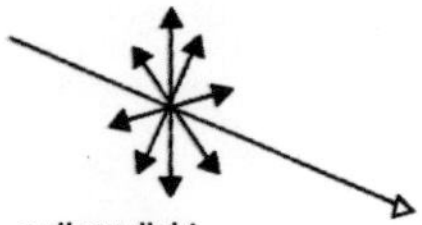

ordinary light
尋常光

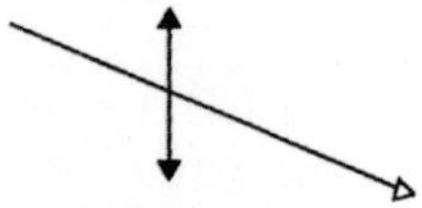

plane polarized light
平面偏振光

position (*n*) the place in which something is found.

位置（名） 某物出現之處。

process (*n*) a set of actions or events taking place one after the other.

過程（名） 一個接一個依次進行的一組作用或依次發生的一組事件。

property (*n*) something that a thing always, or usually, has or shows; a characteristic.

性質（名） 某物常常或通常具有的或顯示的某些東西；一種特性。

proportion (*n*) the relationship of two or more things to each other in their sizes, quantities, or numbers, etc. **proportional** (*adj*).

比例（名） 二物或多物在大小、數量或數目等方面的相互關係。（形容詞為 proportional）

proton (*n*) a particle with electric charge equal to that of the electron (p.155) but of opposite sign, and mass about 1 836 times that of the electron. A proton is a hydrogen ion.

質子（名） 電荷與電子（第 155 頁）的電荷相等但符號相反的一種粒子，其質量約為電子的 1836 倍。氫離子就是質子。

ratio (*n*) the relationship between two quantities expressed in numbers, e.g. if a mixture contains 2 parts of A to 5 parts of B then the ratio of A to B in it is 2:5.

比（名） 以數字表達的兩個量之間的關係。例如一種混合物含 2 份 A 和 5 份 B，則 A 與 B 之比為 2：5。

refer (*v*) to relate; to direct attention; to turn for facts or other knowledge. **reference** (*n*).

談及；涉及；參閱（動） 講述；引起注意；查閱某些事實或知識。（名詞為 reference）

refraction 折射

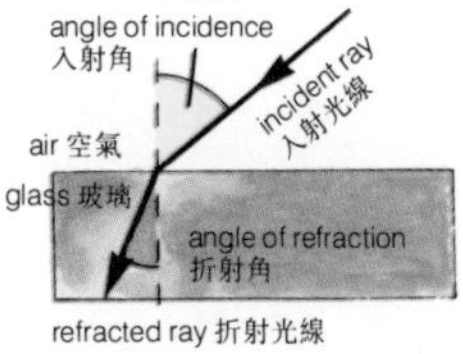

refractive index a measure of the ability of a material to bend a beam of light. For a particular material it is equal to the speed of light in a vacuum divided by the speed of light in the material. It is also measured by: {sine (angle of incidence)} ÷ {sine (angle of refraction)}: see diagram.

折射率 物質使光束彎曲的能力的一種尺度。某一特定物質的折射率，等於光在真空中的速度除以光在該物質中的速度。折射率也等於：（入射角的正弦）÷（折射角的正弦）（見圖）。

replace (*v*) to take the place of. **replacement** (*n*), **replaceable** (*adj*).

交代（動） 取代。（名詞為 replacement，形容詞為 replaceable）

ridge (*n*) a long narrow area of high ground; the edge where two slopes meet at the top.

脊（名） 高地的一個狹長的區域；兩個坡面在頂部相交所成的稜。

rigid (*adj*) describes a solid that does not change in shape when a force acts upon it.

剛性的（形） 描述一種固體受力時不改變形狀。

series (*n*) a set of things in a line or following one after the other, and having something in common.

系列（名） 一組具有某些共同性質並排列成行或一個接一個的事物。

skeleton (*n*) the frame of bone or other hard material that supports or contains an animal. **skeletal** (*adj*).

骨骼（名） 用以支撐或包容動物軀體的骨架或其他硬質材料的框架。（形容詞為 skeletal）

solution (*n*) the change of matter from the solid or gaseous state to the liquid state by putting it in a liquid.

溶解（名） 物質放入一種液體中而從固態或氣態轉變為液態的變化。

stable (*adj*) not easily changed or moved.

穩定的（形） 不易於變化或移動的。

states of matter all materials are solids, liquids, or gases. These are the three states of matter.

物態 物質有三態，即固態、液態和氣態。

subsurface (*adj*) below the surface, e.g. subsurface geology, the geology of the region below the Earth's surface.

地下的（形） 在地面以下。例如地下地質學，即地表以下地域的地質學。

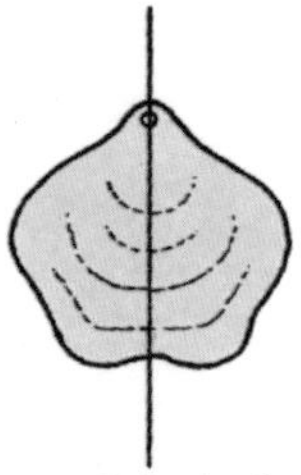

symmetry in a fossil Brachiopod
腕足類化石的對稱

symmetry (*n*) having parts that correspond (p.155) on either side of a plane, a straight line, or a point. Parts on opposite sides of the plane, line, or point are similar in size, shape, and position. In crystallography the word 'symmetry' is used with a special meaning: see p.42. **symmetrical** (*adj*).

對稱（名） 在一個平面、一條直線或一點的兩側具有相應的兩部分。在這一平面、直線或點兩側的兩部分，其大小、形狀和位置是相同的。在結晶學中，"對稱"這個詞有特殊含意，（見第 42 頁）。（形容詞為 symmetrical）

temperature (*n*) a measure, using a scale, of how hot, or how cold, an object, an organism (p.98), or the atmosphere is.

溫度（名） 用一種標度表示物體、有機體（第 98 頁）或大氣冷、熱程度的尺度。

term (*n*) a word used with a special meaning, as in science. Thus 'rock' used as a term in geology has a different meaning from the meaning it has in ordinary speech.

術語（名） 科學上作為具特殊含義使用的詞。例如"岩石"在地質學中作為一個術語使用，其含義與日常用語的含義有所不同。

theory (*n*) a supposition that is put forward to explain a body of facts; e.g. the theory of isostacy (p.11); the theory of plate tectonics (p.134). **theoretical** (*adj*).

學説；理論（名） 用以解釋一些事實的全體而提出的一種假設。例如地殼均衡説（第 11 頁）；板塊構造（第 134 頁）理論。（形容詞為 theoretical）

tissue (*n*) (in living things) a mass of cells (p.146) and the material between them that together have the same purpose.

組織（名） （生物體中）一群細胞（第 146 頁）以及細胞之間的物質，它們共同去完成同一種機能。

unit (*n*) a thing or a group of things that can be regarded as the smallest part into which something larger can be divided; a quantity chosen as a standard measurement for other quantities.

單元；單位（名） 可看成是某件較大之物所能分成的最小部分的一物或多物；選作測量其他量之標準的一個量。

upward (*adv*) moving towards a higher place.

向上（副） 向較高的位置運動。

variation (*n*) varying (↓) from a normal or earlier state or amount or standard.

變化（名） 從正常的或更早的狀態、量或標準向其他狀態、量或標準改變（↓）。

vary (*v*) to change or make (or become) different. **varied** (*adj*).

改變（動） 使變，或使不同。（形容詞為 varied）

velocity (*n*) speed in a given direction. A motor-car travelling along a straight road with a speed of 70 km/h has a velocity of 70 km/h. A motor-car travelling at 70 km/h round a bend in a road has a velocity that is changing all the time, because it is not travelling in a straight line.

速度（名） 在給定方向上的速率。一輛以 70 km/h 速率沿直路行駛的汽車，其速度為 70 km/h。一輛以 70 km/h 速率沿彎路行駛的汽車，其速度則時時在改變，因為它不是沿直線行駛。

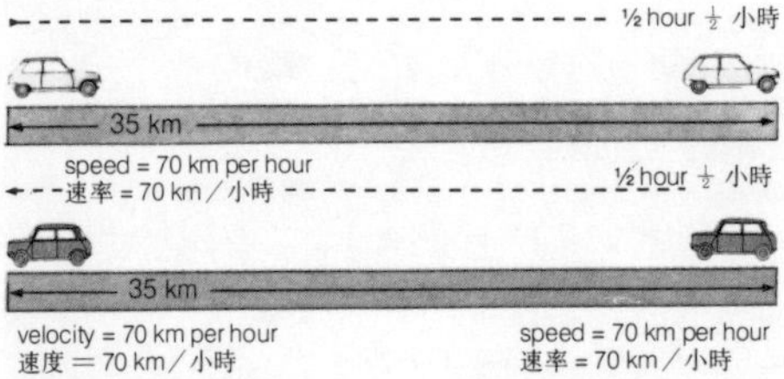

vertical (*adj*) upright; at 90° to the plane of the horizon (the line where the Earth and the sky appear to meet).

垂直的（形） 直立的；與地平綫（看似地球與天空的交會線）的面成 90°。

vibrate (*v*) to move to and fro, especially rapidly. **vibration** (*n*), **vibrating** (*adj*).

振動（動） 往返運動，特別是迅速地往返運動。（名詞為 vibration，形容詞為 vibrating）

longitudinal wave motion
縱波運動

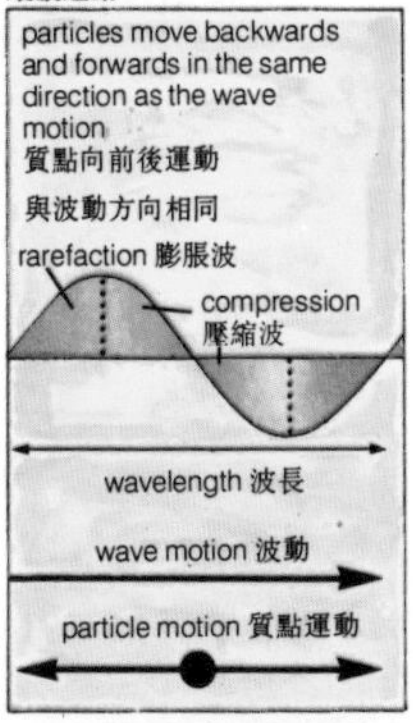

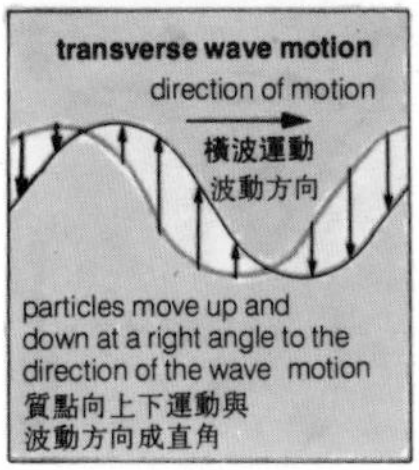

volume (*n*) the amount of space filled by a body or a substance. Volume is usually measured in cubic metres (m^3) or a related unit (mm^3, km^3 for example).

體積（名） 為一種物體或物質所填滿的空間的量。體積通常以立方米（m^3）或相關的單位（如 mm^3，km^3）來測度。

wave motion the sending of energy by a regular movement, in the form of a wave (see diagram). When a wave passes through a material, each particle moves about a central point to produce the wave motion. Waves are of two kinds: transverse, in which the particles move at 90° to the direction in which the wave is travelling; and longitudinal, in which the particles move backwards and forwards in the same direction as the wave is travelling.

波動 以波（見圖）的形式作規律性運動傳送能量。當一個波通過一種物質時，每個質點都在一個中心點周圍運動，從而產生波動。波有兩種：橫波，其中質點運動的方向與波傳播的方向成 90°；縱波，其中質點向前後運動，與波傳播的方向相同。

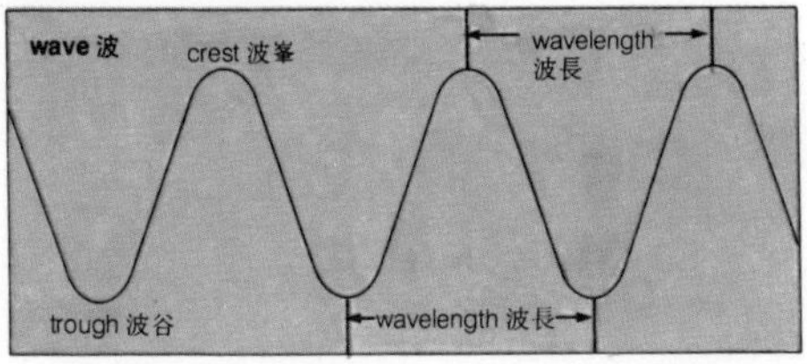

wavelength (*n*) the distance between a point in a wave and the next point at the corresponding place moving in the same direction; i.e. the distance between one crest (see diagram) and the next crest or between one trough and the next.

波長（名） 波中的一個點與在相應位置上向相同方向運動的鄰點之間的距離：即一個波峯（見圖）與相鄰波峯或一個波谷與相鄰波谷之間的距離。

zone (*n*) in the general sense, a long narrow area shaped like a band, especially one that stretches like a ring round something. **zonal** (*adj*). 'Zone' is used in a special sense in palaeontology and stratigraphy: see pp.111, 117.

帶（名） 一般是指帶狀的狹長地帶，特殊的情況是環繞某物伸展而成的環狀地帶。"帶"在古生物學和地層學中有特殊含義：見第 111、117 頁（形容詞為 zonal）。

Common abbreviations in geology 地質學上常用的縮寫詞

aff.	affinity	親合性；親和力
anh	anhydrous	無水的
approx.	approximate(ly)	近似（地）
aq.	aqueous	水的；水成的
av.	average	平均的；平均值
b.p.	boiling point	沸點
BP	before the present	距今
c.	circa (about)	約
corr.	corrected	修正的
cryst.	crystalline	結晶的；晶質的
cy	clay	黏土
exptl	experimental	實驗的
f.p.	freezing point	冰點；凝固點
gen.	genus	屬
H	hardness (Mohs' scale)	莫氏硬度
L.	lower	下；早期的
liq.	liquid	液態；液體
lst	limestone	灰岩
M.	middle	中
Ma	milion years	百萬年
max.	maximum	最大值（的）
min.	minimum	最小值（的）
m.p.	melting point	熔點

NRM	natural remnant magnetism	天然剩磁
pH	hydrogen ion concentration	氫離子濃度
ppm	parts per million	百萬分之一
ppt	precipitate	沉澱
R.I.	refractive index	折射率
s.l.	sea level	海平面
s.l.	*sensu lato* (in the broad sense)	在廣義上；廣義而言
sol.	soluble	可溶的
soln	solution	溶液
sp., spp.	species (sing., plural)	種
s.s	*sensu stricto* (in the strict sense)	嚴格地説；精確地説
sst	sandstone	砂岩
s.t.p.	standard temperature and pressure	標準溫度和壓力
temp.	temperature	溫度
U.	upper	上；晚期的
vol.	volume	體積
v.p.	vapour pressure	蒸汽壓
wt%	weight per cent	重量百分比

International System of Units (SI)
國際單位制

PREFIXES 詞頭

PREFIX 詞頭	FACTOR 因數	SIGN 符號	PREFIX 詞頭	FACTOR 因數	SIGN 符號
milli- 毫	$\times 10^{-3}$	m	kilo- 千	$\times 10^{3}$	k
micro- 微	$\times 10^{-6}$	μ	mega- 兆	$\times 10^{6}$	M
nano- 納(毫微)	$\times 10^{-9}$	n	giga- 吉	$\times 10^{9}$	G
pico- 皮(微微)	$\times 10^{-12}$	p	tera- 太	$\times 10^{12}$	T

BASIC UNITS 基本單位

UNIT	單位	SYMBOL 符號	MEASUREMENT	度量
metre	米	m	length	長度
kilogramme	千克(公斤)	kg	mass	質量
second	秒	s	time	時間
ampere	安培	A	electric current	電流
kelvin	開爾文	K	temperature	溫度
mole	摩爾	mol	amount of substance	物質的量

DERIVED UNITS 導出單位

UNIT	單 位	SYMBOL 符號	MEASUREMENT	度量
newton	牛 頓	N	force	力
joule	焦 耳	J	energy, work	能、功
hertz	赫 兹	Hz	frequency	頻率
pascal	帕斯卡	Pa	pressure	壓強、壓力
coulomb	庫 侖	C	quantity of electric charge	電量
volt	伏 特	V	electrical potential	電勢
ohm	歐 姆	Ω	electrical resistance	電阻

SOME MULTIPLES OF SI UNITS HAVING SPECIAL NAMES
一些有專門名稱的國際單位倍數

UNIT 單位	SYMBOL 符號	DEFINITION 定義	MEASUREMENT 量
ångstrom 埃	Å	10^{-10} m = 10^{-1} nm	length 長度
micron 微米	µm	10^{-6} m	length 長度
litre 升	l	10^{-3} m^3 = dm^3	volume 容積
tonne 噸	t	10^3 kg	mass 質量
dyne 達因	dyn	10^{-5} N	force 力
bar 巴	bar	10^5 Pa	pressure 壓力

SOME NON-SI UNITS
一些非國際單位制的單位

UNIT 單位	SYMBOL 符號	DEFINITION 定義	MEASUREMENT 量
atm 大氣壓	atm	101325 Pa, 1.01325 bar	pressure 壓力
degree Celsius 攝氏度	°C	K($t_c = t_k - 273$)	temperature 温度
million years 百萬年	Ma, m.y.	10^6 years 年	time 時間
billon(us) years 十億年	Ga	10^9 years 年	time 時間

Understanding scientific words 理解意義的科學用詞彙

New words can be made by adding **prefixes** or **suffixes** to a shorter word. Prefixes are put at the front of the shorter word and suffixes are put at the back of the word. Words can also be broken into parts, each of which can have a meaning, but cannot be used alone.

(i) correct → *in*correct (adding a prefix)
correct → correct*ness* (adding a suffix)
correct → *in*correct*ness* (adding a prefix and a suffix)

(ii) **isomorphism** is broken into iso-morph-ism
iso- is a prefix which means 'identical in structure'
morph is a word part which means 'form or shape'
-ism is a suffix which means 'a condition'

Hence *isomorphism* means the condition of having identical forms or shapes; it describes the condition of two crystalline substances.

Prefixes describing numbers or quantities are taken from Greek or Latin words. The following table shows the common prefixes from these two languages. Prefixes, suffixes, and word parts are listed alphabetically in separate sections after the table.

詞頭或**詞尾**加於簡短的單詞上即成新詞。詞頭是加於該簡短單詞之前，而詞尾則加於其後。單詞可斷分為幾部分，各個部分都各有其意義但不能獨立使用。

(i) correct ⟶ *in*correct （加詞頭）
correct ⟶ correct*ness* （加詞尾）
correct ⟶ *in*correct*ness* （加詞頭和詞尾）

(ii) **isomorphism** 可斷分為 iso-morph-ism
iso- 為詞頭，其含義是"構造上同一"
morph- 為詞部，其含義是"形式或形狀"
-ism 為詞尾，其含義是"某種狀態"

因此，*isomorphism* 的含義是具有同一形式或形狀之狀態；它描述兩種結晶物質的狀態。

描述數或量的詞頭都源自希臘詞或拉丁詞，下表列出源自這兩種語言的常用詞頭。此表之後，將按字母順序分別列出這些詞頭、詞尾和構詞成分。

	GREEK PREFIX 希臘詞頭	LATIN PREFIX 拉丁詞頭	PREFIX 詞頭	MEANING 意義	詞源
1 一	mono-	uni-	hemi-	half 半	Gr 希臘
2 二	di-	bi-	semi-	half 半	L 拉丁
3 三	tri-	ter-	poly-	many 多；聚	Gr 希臘
4 四	tetra-	quad-	multi-	many 多	L 拉丁
5 五	penta-	quinq-	omni-	all 全；總	L 拉丁
6 六	hexa-	sex-	dupli-	twice 兩倍；兩次	L 拉丁
7 七	hepta-	sept-	tripli-	three times 三倍；三次	Gr 希臘
8 八	octo-	oct-	hypo-	less, under 次；較少；在下	Gr 希臘
9 九	nona-	novem-	hyper-	more, over 超；過；高；在上	Gr 希臘
10 十；分	deca-	deci-	sub-	under 亞；次；在下、較低	L 拉丁
100 百；厘	hecta-	centi-	super-	over 過、超、高於	L 拉丁
1000 千；毫	kilo-	milli-	iso-	same, equal, identical 相同、相等、同一	Gr 希臘

PREFIXES 詞頭

a-	without, lacking, lacking in, e.g. *a*morphous, being without shape; *a*symmetrical, without symmetry, or lacking in symmetry.	無；非；缺。例如 amorphous 無定形的；asymmetrical 不對稱的或缺對稱的。
allo-	different, or different kinds, e.g. *allo*tropy, the existence of an element in two or more different forms.	異；異類。例如 *allo*tropy 同素異形現象，指一種元素存在兩種或多種不同的形態。
amphi-	on both sides, e.g. *ampho*teric, having the nature of both an acid and a base.	兩；雙；兩側。例如 *ampho*teric 兩性的，指兼具酸和鹼的性質的。
an-	the same prefix as **a-**, used in front of words beginning with a vowel, or the letter *h*, e.g. *an*isotropic, not having the same properties in all directions; *an*hydrous, being without, or lacking, water, in a crystal.	同詞頭 *a-*，加在以元音字母或以字母 h 開頭的詞之前。例如 *an*isotropic 各向異性的，指各方向均不具相同的性質；*an*hydrous 無水的，指晶體中無水或缺水。
anti-	opposite in direction, or in position, e.g. *anti*catalyst, a catalyst which slows down a chemical reaction, i.e. works in the opposite direction to a catalyst.	方向或位置相反。例如 *anti*catalyst 反催化劑，指降低化學反應(即與催化劑起相反作用)速度的催化劑。
auto-	caused by itself, e.g. *auto*xidation, reaction of a substance with atmospheric oxygen at room temperature, the substance oxidizes itself; *auto*catalysis, a chemical reaction in which the products act as catalysts for the reaction.	自；自動。例如 *auto*xidation 自氧化，指物質在室溫下與大氣的氧起反應，而本身起氧化之作用。*auto*catalysis 自催化，指化學反應的產物對該化學反應起催化劑作用。
cis-	on the same side, e.g. *cis*-compound, an isomer in which two like groups are on the same side of the double bond in the compound. See *trans-*.	在同側(順式)。例如 *cis*-compound 順式化合物，指兩個相同基團位於雙鍵同一側的同分異構體。見 *trans-*。
co-	acting together, with, e.g. *co*hesion, the force holding two or more objects together.	共；共同(作用)；和。例如 *co*hesion 內聚力，使兩個或多個物體保持在一起的力。
counter-	acting against, acting in the opposite direction, e.g. *counter*act, to act against, such as a mild alkali counteracts the effect of acid on skin; *counter*clockwise, turning in the opposite direction to the hands of a clock.	反抗；逆；向相反方向起作用。例如 *counter*act 反作用，意為抵銷，如弱鹼抵銷酸對皮膚的影響；*counter*clockwise 逆時針方向轉。
de-	opposite action, e.g. *de*compression, the lessening of a pressure, it is the opposite action to compression; *de*activate, to make less active, it is the opposite of activate.	倒轉；反動作。例如 *de*compression 減壓，與壓縮的作用相反；*de*activate 鈍化，與活化相反。
dia-	through, across, e.g. *dia*meter, the line going across a circle, through the centre.	通過；橫越。例如 *dia*meter 直徑，橫截圓且通過圓心的直線。
dis-	opposite action, e.g. *dis*charge, to take an electric charge away from a charged body, the opposite of charge; *dis*connect, to break, or open, a connection, the opposite of connect.	作用相反。例如 *dis*charge 放電，排除帶電體之電荷，與充電相反；*dis*connect 使連接斷開或脫開，與連接相反。
equi-	having the same number, equal, e.g. *equi*molecular, having the same number of molecules; *equi*librium, the condition of two rates of reaction being equal and opposite, so that there is no further change in a reversible reaction.	同數；相等。例如 *equi*molecular 等分子數的；*equi*librium 平衡，指可逆反應中兩個反應的速度相等，方向相反，因而不再發生變化之情況。
im-	the opposite, not. (Used with words beginning with b, m, p.) For example, *im*perfect, not perfect, the opposite of perfect; *im*permeable, not permeable.	不；非等否定含義(加在以 b、m、p 開頭的詞前)。例如，*im*perfect 不完美的，與"完美的"相反；*im*permeable 不可滲透的。
in-	the opposite, not. (Used with all words other than those beginning with b, m, p.) For example, *in*active, the opposite of active; *in*adequate, not adequate.	不；非；無等否定含義(加在非以 b、m、p 開頭的詞之前)。例如 *in*active 不活潑的、與活潑的相反；*in*adequate 不適當的。
infra-	below, e.g. *infra*molecular, having a size smaller than a molecule, so the size is below molecular size.	在下；較低。例如 *infra*molecular 亞分子的，即尺寸小於一個分子的。

inter- between, among, e.g. *inter*face, a common surface between two liquids or two solids; *inter*stice, a narrow space between two solid objects.
之間；之中。例如 *inter*face 界面，為兩液體或兩固體間的公共面；*inter*stice 間隙，兩固體之間的狹窄空間。

macro- great, large, e.g. *macro*molecule, a large molecule composed of many smaller molecules, as in a polymer.
大量的；巨大的。例如 *macro*molecule 大分子，由許多較小分子組成的一個大分子，例如聚合物。

micro- small, especially if too small to be seen by the human eye alone, e.g. *micro*balance, a balance used for measuring masses of less than 1 mg; *micro*analysis, analysis using very small amounts of substances.
微；微小；尤指微小得不能單靠人眼觀察。例如 *micro*balance 微量天平，測量少於 1 mg 質量用的天平；*micro*analysis 微量分析，用很少量物質進行的分析。

non- not, e.g. *non*-electrolyte, a substance which is not an electrolyte; *non*-ferrous, any metal other than iron.
非；不；無等否定含義。例如 *non*-electrolyte 非電解質；*non*-ferrous 非鐵的，即除鐵以外的任何其他金屬。

ortho- straight, right-angled, upright, e.g. *ortho*gonal, with parts at right-angles; *ortho*rhombic, a crystal system with three unequal axes at right-angles.
直的；直角的；豎直的。例如 *ortho*gonal 直交的，各部分成直角相交的；*ortho*rhombic 正交的，指三根不相等軸互成直角的晶系。

pan- all, complete, every, e.g. *pan*chromatic, covering all wavelengths of light in the spectrum.
總；全；泛。例如 *pan*chromatic 全色的，指包含光譜中全部光的波長。

para- at the side of, by, e.g. *para*casein, an insoluble form of casein, formed when soluble casein coagulates.
側；副。例如 *para*casein 副酪 ，為不溶解形酪 ，由可溶的酪 凝固所生成。

pseudo- has the same appearance, but is false, e.g. *pseudo*alum, a substance which has the appearance of an alum, but is not an alum.
貌同實假。例如 *pseu*doalum 假明礬，一種具礬外貌却非明礬的物質。

re- again, e.g. *re*activate, to make something activated again; *re*crystallize, to crystallize again.
再；重。例如 *re*activate 再活化，使某物再次活動；*re*crystallize 再結晶。

syn, sym- joined together, united, e.g. *syn*thesis, combining elements or compounds to make new compounds.
連在一起；合。例如 *syn*thesis 合成，使諸元素或諸化合物結合形成新的化合物。

trans- across, on the opposite side of, e.g. *trans*-compound, an isomer in which two like groups are on opposite sides of the double bond in the compound. See *cis-*.
越；在不同側(反式)。例如，*trans*-compound 反式化合物，化合物中兩個相同基團在雙鍵兩側的同分異構體。見 *cis-*。

ultra- beyond, e.g. *ultra*filter, a filter which has holes so small it filters out colloids; it thus has uses beyond those of the ordinary filter.
超。例如 *ultra*filter 超濾器，一種有許多微孔，可濾出膠體的濾器；用處大於普通濾器。

un- not, the opposite, e.g. *un*saturated, means not saturated; *un*stable, means not stable; *un*paired, means not in a pair, and so by itself.
不；反等否定含義。例如 *un*saturated 不飽和的；*un*stable 不穩定的；*un*paired 不成對的，即單獨的。

SUFFIXES 詞尾

-able forms an adjective which shows an action can possibly take place, e.g. change*able*, something which can change; transform*able*, something which it is possible to transform.
構成形容詞，表示某種作用有可能發生。例如 change*able* 可改變；transform*able* 可變換。

-al of, or to do with; forms a general adjective, e.g. experiment*al*, of, or to do with, experiment; fraction*al*, of, or to do with, fractions; therm*al*, of, or to do with, heat.
構成普通形容詞，表示"屬……的"；"與……有關的"。例如 experiment*al* 實驗的，與實驗有關的；fraction*al* 分數的或用分數表示；therm*al* 熱的；熱力的。

-ed forms the past participle of a verb, can be used as an adjective; it shows an action under the control of an experimenter, e.g. vari*ed*, describes a quantity changed by an experimenter; dehydrat*ed*, describes a substance from which water has been removed under the control of an observer.
構成動詞的過去分詞，可作形容詞，表示一種作用受實驗者控制。例如 vari*ed* 已改變的，形容一個量受實驗者改變；dehydrat*ed* 已脫水的，形容一種物質已在觀察者控制下除去所含水分。

-er (-or) forms a noun from a verb and describes an agent, e.g. mix*er*, a device which mixes; desiccat*or*, a device that desiccates; generat*or*, a device that generates a gas.
加於動詞構成名詞，描述動作者或描述能進行這種動作的工具。例如 mix*er* 混合裝置；desiccat*or* 乾燥裝置；generat*or* 氣體發生器。

-gram forms a noun describing a record which is written or drawn, e.g. chromato*gram*, the recorded result from an experiment on chromatography; tele*gram*, the written message recorded by telegraph.

構成名詞，描述一種寫成文字或繪成圖表的記錄。例如 chromato*gram* 色層譜，指從色層法實驗所得到的記錄：tele*gram* 電報。

-graph forms a noun describing an instrument or device that records variation in a quantity, or other information, e.g. thermo*graph*, a kind of thermometer which records changes of temperature over a period of time; tele*graph*, a device which records information in words.

構成名詞，描述一種記錄一個量的變化或記錄其他信息的工具或裝置。例如 thermo*graph* 溫度記錄儀，指在一段時間內紀錄溫度變化的一種溫度計：tele*graph* 電報機，一種記錄文字信息的裝置。

-ic of, or to do with; forms a general adjective, e.g. bas*ic*, of, or to do with, a base; cycl*ic*, of, or to do with, a cycle; ion*ic*, of, or to do with, ions.

表示屬於：與……有關的：構成一般形容詞。例如 bas*ic* 鹼性的，與鹼有關的：cycl*ic* 循環的，與循環有關的：ion*ic* 離子的，與離子有關的。

-ify forms a verb which is causative in action, e.g. pur*ify*, to cause to become pure; solid*ify*, to cause to become solid.

構成一個表示引起作用的動詞。例如 pur*ify* 淨化：solid*ify* 固化。

-ing forms the present participle of a verb, can be used as an adjective; it shows an action not under the control of an experimenter, e.g. fluctuat*ing*, describes a quantity varying above and below an average value, which cannot be controlled by an observer; disintegrat*ing*, describes a radioactive substance undergoing disintegration, as the process cannot be controlled by an observer.

構成動詞的現在分詞，可作形容詞用，表示不受制於實驗者的一種作用。例如 fluctuat*ing* 波動的，形容某個量變動於一個平均值上下，且不受觀察者所控制的：disintegrat*ing* 衰變的，描述一種放射性物質經受衰變，而這一過程不受觀察者控制。

-ity forms a noun of a state or quality, e.g. pur*ity*, the quality or state of being pure; acid*ity*, the quality of being acid.

構成一個與狀態或品質有關的名詞。例如 pur*ity* 純度：acid*ity* 酸性。

-ive forms an adjective by replacing *-ion* in nouns; the adjective describes an agent producing the effect described by the noun, e.g. inhibit*ion* → inhibit*ive*, describes an agent causing inhibition; oxidat*ion* → oxidat*ive*, describes a process causing oxidation; explos*ion* → explos*ive*, describes an agent causing an explosion.

取代名詞中的 *-ion* 構成形容詞：形容產生該名詞所描述的效應的動因。例如 inhibit*ion* → inhibit*ive* 抑制的：oxidat*ion* → oxidat*ive* 氧化的：explos*ion* → eplos*ive* 爆炸的。

-ize forms a verb which is causative in the formation of something, e.g. ion*ize*, to cause ions to be formed; polymer*ize*, to cause polymers to be formed.

構成動詞，表示形成某物。例如 ion*ize* 使電離：polymer*ize* 聚合。

-lysis forms a noun describing the action of breaking down into simpler parts, e.g. hydro*lysis*, the decomposition of a compound by the action of water; electro*lysis*, the decomposition of a substance by an electric current.

構成名詞，描述分裂為更簡單部分的作用。例如 hydro*lysis* 水解，指藉水的作用使化合物分解：electro*lysis* 電解，利用電流分解一種物質。

-meter forms a noun describing an instrument which measures quantitatively, e.g. thermo*meter*, an instrument which measures temperature accurately; volt*meter*, an instrument which measured electric potential in volts.

構成名詞，記述定量測量用的一種儀器。例如 thermo*meter* 溫度計，準確測量溫度的一種儀器：volt*meter* 電壓表，一種測量電位(以伏特為單位)的儀器。

-metry forms a noun describing a particular science of accurate measurement, e.g. thermo*metry*, the science of measuring temperature; hydro*metry*, the science of measuring the density of liquids.

構成名詞，記述準確測量的專門學科。例如 thermo*metry* 測溫學：hydro*metry* 液體密度測量學。

-ness forms an abstract noun of state or quality, e.g. sweet*ness*, the quality of being sweet; soft*ness*, the quality of being soft.

構成表示"狀態或性質"的抽象名詞。例如 sweet*ness* 甜性：soft*ness* 柔性。

-ous forms an adjective showing possession, or describing a state, e.g. anhydr*ous*, being in the state of not possessing water, homolog*ous*, in the state of being a homologue; homogen*ous*, in the state of having the same properties throughout a substance.

構成形容詞，表示"具有"之意，或描述一種狀態。例如 anhydr*ous* 無水的，指不具水的狀態：homolog*ous* 同系的，指以同系物的狀態存在：homogen*ous* 均勻的，均相的，指全部物質都具相同性質的狀態。

-philic forms an adjective describing a liking for something, e.g. proto*philic*, describes a substance which accepts protons.
構成形容詞，描述親近某物。例如 proto*philic* 親質子的，描述一種物質接受質子。

-phobic forms an adjective describing a dislike for something, e.g. lyo*phobic*, describes a colloid which does not go readily into solution.
構成形容詞，描述疏遠某物。例如 lyo*phobic* 疏液的，描述一種膠體難溶於溶液中。

-scope forms a noun describing an instrument which measures qualitatively, e.g. spectro*scope*, an instrument by which spectra can be observed qualitatively; hygro*scope*, an instrument which measures qualitatively the humidity of the atmosphere.
構成名詞，描述一種定性測量用的儀器。例如 spectro*scope* 分光鏡，用以定性觀察光譜的儀器；hygro*scope* 濕度器，用於定性測量大氣濕度的儀器。

-scopy forms a noun describing the use of instruments for observation in science, e.g. micro*scopy*, the use of microscopes for scientific observation.
構成名詞，描述應用儀器於科學觀察。例如 micro*scopy* 顯微鏡檢查法，指應用顯微鏡於科學觀察。

-stat forms a noun describing a device which keeps a quantity constant, e.g. hydro*stat*, a device which keeps water in a boiler at a constant level; thermo*stat*, a device which keeps a liquid, or an object, at a constant temperature.
構成名詞，描述一種能使某一量保持恆定的裝置。例如 hydro*stat* 防爆裝置，使鍋爐保持恆定水位的裝置；thermo*stat* 恆器器，係一種液體或物體保持恆定溫度的裝置。

-tion forms an abstract noun. With *-ation*, it forms a noun of action, e.g. pollu*tion*, the result of polluting; concentra*tion*, the degree to which a solution is concentrated; distill*ation*, the noun of action from distil; precipit*ation*, the noun of action from precipitate.
構成抽象名詞。用 *-ation* 構成表示動作或作用的名詞。例如，pollu*tion* 污染；concentra*tion* 濃度；distill*ation* 蒸餾作用；precipita*tion* 沉澱作用。

WORD PARTS 詞部

aqua water, to do with water, e.g. *aqu*eous, a solution containing water; *aqu*oion, an ion with molecules of water associated with it.
水：與水有關。例如 *aqu*eous 水的，指含水的溶液；*aquoion* 水合離子，與水分子締合的離子。

chrom colour, to do with colour, e.g. pan*chrom*atic, all the colours, and hence all the wavelengths of the visible spectrum; *chrom*atography, the analysis of complex substances in which a coloured record of the analysis is produced.
顏色：與顏色有關。例如 pan*chrom*atic 全色的，即可見光譜的全部波長；*chrom*atography 色層法，色譜法，係複雜物質的色譜分析法。

gen to produce, e.g. homo*gen*ize, to make a mixture of solid and liquid substances into a viscous liquid of the same texture throughout; *gen*erate, to produce energy or a flow of gas.
引起；產生。例如 homo*gen*ize 使均勻，將一種固體和液體物質的混合物製成結構處處相同的黏性液體；*gen*erate 產生能量或氣流。

hydr water or liquids, e.g. de*hydr*ate, to remove water; an*hydr*ous, describes a substance without water.
水或液體。例如 de*hydr*ate 脫水；an*hydr*ous 無水的。

hygro damp or humid, e.g. *hygro*scopic, attracting water from the atmosphere to become damp; *hygro*meter, an instrument that measures the relative humidity of the atmosphere.
潮濕；濕的。例如 *hygro*scopic 吸濕的；*hygro*meter 濕度表，測量大氣相對濕度的儀器。

morph shape or form, e.g. a*morph*ous, describes a substance which is without a crystalline form; poly*morph*ism, existing in different forms.
形狀或形式。例如 a*morph*ous 非晶質的，描述一種不具晶形的物質；poly*morph*ism 同質多晶形現象。

photo light, e.g. *photo*lysis, decomposition caused by light; *photo*halide, any halide which is decomposed by light.
光。例如 *photo*lysis 光解作用；*photo*halide 感光性鹵化物，見光能自行分解的任何鹵化物。

pneumo air or gas, e.g. *pneum*atic trough, a trough for the collection of gases.
空氣或氣體。例如 *pneum*atic trough 集氣槽。

pyro great heat, e.g. *pyro*lysis, decomposition caused by heating; *pyro*meter, a special kind of thermometer for measuring very high temperatures.
高溫。例如 *pyro*lysis 高溫分解；*pyro*meter 高溫計，用以測量極高溫度的一種專用溫度計。

therm heat, e.g. *therm*ostable, stable when heated; *therm*al, of, or to do with heat; *therm*ometer, an instrument for the quantitative measure of temperature.
熱。例如 *therm*ostable 耐熱的；*therm*al 熱的，屬於熱的，或與熱有關的；*therm*ometer 溫度計，用以定量測量溫度的儀器。

Acknowledgements 致謝

There are many standard source books in geology but the author would like to acknowledge his especial indebtedness to the authors of the following books:

地質學的權威性參攷書很多，其中作者特別感謝下列各書的作者

BIRKELAND, P.W. and LARSEN, E.E.: *Putnam's geology*, Oxford University Press, New York, 1978

FYFE, W.S.: *Geochemistry*, (Oxford chemistry series), Oxford University Press, Oxford, 1974

GASS, I.G., SMITH, P.J. and WILSON, R.C.L. (editors): *Understanding the Earth*, Open University Press, Milton Keynes, 1971

GUTENBERG, B.: *Internal constitution of the Earth*, Dover Books, New York, 1951

McKERROW, W.S.: *The ecology of fossils: an illustrated guide*, Duckworth, London, 1978

SPENCER, E.W.: *The dynamics of the Earth*, Thomas Crowell, New York, 1972

WELLS, A.F.: *Structural inorganic chemistry*, Oxford University Press, Oxford, 1962 (fourth edition)

WYLLIE, P.J.: *The way the Earth works*, John Wiley, New York, 1976

The story of the Earth, Institute of Geological Sciences, London, 1972

INDEX 索引

A horizon /e həˈraɪzn̩/ A 層 23
A 2 horizon /e tu həˈraɪzn̩/ A 2 層 23
á á (lava) /ˈa a/ 塊熔岩 70
abrasion /əˈbreʒən/ 磨蝕 20
absolute age /ˈæbsəˌlut edʒ/ 絕對年齡 121
abundances, cosmic /əˈbʌndənsɪz, ˈkɑzmɪk/ 宇宙豐度 18
  terrestrial /təˈrɛstrɪəl/ 地球豐度 18
abyssal /əˈbɪsl̩/ 深海的 34
abyssal hills /əˈbɪsl̩ hɪlz/ 深海丘陵 34
abyssal plain /əˈbɪsl̩ plen/ 深海平原 34
abyssal zone /əˈbɪsl̩ zon/ 深海帶 36
Acadian /əˈkedɪən/ 阿卡德期的 116
acceleration /ækˌsɛləˈreʃən/ 加速度 151
accessory mineral /ækˈsɛsərɪ ˈmɪnərəl/ 副礦物；附生礦物 75
accretionary prism /əˈkriʃənˌɛrɪ ˈprɪzəm/ 加積稜柱體 138
accretion, continental /əˈkriʃən, ˌkɑntəˈnɛntl̩/ 大陸增長 140
accumulate /əˈkjumjəˌlet/ 聚集 151
achondrite /eˈkɑnˌdraɪt/ 無球粒隕石 149
acicular /əˈsɪkjulɚ/ 針狀 45
acid (rocks) /ˈæsɪd/ 酸性的(岩) 74
actinolite /ækˈtɪnəˌlaɪt/ 陽起石 57
adaptation (evolutionary) /ˌædəpˈteʃən/ 適應(演化的) 102
adaptive radiation /əˈdæptɪv, redɪˈeʃən/ 適應輻射 102
adularia /ˌædʒəˈlɛrɪə/ 冰長石 56
aeolian weathering /iˈolɪən ˈwɛðərɪŋ/ 風力風化 22
aeon /ˈiɑn/ 宙 113
aerobic /ˌeəˈrobɪk/ 喜氧的 101
aerolite /ˈeəroˌlaɪt/ 石隕石 149
after-shock /ˈæftɚˌʃɑk/ 餘震 13
age (geological time-division) /edʒ/ 期(地質年代的劃分單位) 113
age, absolute, /edʒ, ˈæbsəˌlut/ 絕對年齡
  ***see* absolute age** 121
  apparent /əˈpærənt/ 視年齡；表觀年齡
  ***see* apparent age** 121
agglomerate /əˈglɑmərɪt/ 集塊岩 69
aggradiation /ˌægrəˈdeʃən/ 加積作用 32
aggrade /əˈgred/ 加積 32
aggregate /ˈægrɪgɪt/ 集合 151
Agnatha /ˈægnəθə/ 無頜超綱 109
albite /ˈælbaɪt/ 鈉長石 56
algae /ˈældʒi/ 藻類 110
alkali /ˈælkəˌlaɪ/ (強)鹼 16
alkali feldspars /ˈælkəˌlaɪ ˈfɛldˌspɑrz/ 鹼性長石 56
alkali metals /ˈælkəˌlaɪ ˈmɛtl̩z/ 鹼金屬 17
alkali pyroxenes /ˈælkəˌlaɪ ˈpaɪrɑkˌsinz/ 鹼性輝石 57
alkaline (rocks) /ˈælkəˌlaɪn/ 鹼性的(岩) 75
allanite /ˈælənaɪt/ 褐簾石 60
allochemical /ˌæləˈkɛmɪkl̩/ 異源化學的 85
allogenic /ˌælədʒɛˈnɪk/ 外源的 81
allothigenic, /əˌlɔθɪdʒɛnɪk/ 外源的
  ***see* allothigenous** 81
allothigenous /ˌæləθɪdʒɪnəs/ 他生的 81
allotriomorphic /əˌlɔtrɪrˈmɔfɪk/ 他形的 72
alloy /ˈælɔɪ/ 合金 17
alluvial /əˈluvɪəl/ 沖積的 24
alluvial fan /əˈluvɪəl fæn/ 沖積扇 24
alluvium /əˈluvɪəm/ 沖積層；沖積物 24
almandine /ˈælmənˌdin/ 鐵鋁榴石(貴榴石) 58
alpha quartz /ˈælfə kwɔrts/ α 石英；低温石英 55
Alpine (movements) /ˈælpaɪn/ 阿爾卑斯期(運動)的 116
alpine glacier /ˈælpaɪn ˈgleʃɚ/ 阿爾卑期型冰川 28
alter /ˈɔltɚ/ 改變 151
alternate /ˈɔltɚnɪt/ 交替的 151
alumina /əˈlumɪnə/ 礬土；鋁氧；氧化鋁 16
aluminium silicate minerals /ˌæljəˈmɪnɪəm ˈsɪlɪkɪt ˈmɪnərəlz/ 矽酸鋁礦物 59
ammonites, /ˈæməˌnaɪts/ 菊石 ***see* Ammonoidea** 107
Ammonoidea /ˌæməˈnɔɪdɪə/ 菊石亞綱 107
ammonoids /ˌæməˈnɔɪdz/ 菊石類 107
amorphous /əˈmɔrfəs/ 非晶質；無定形的 44
Amphibia /æmˈfɪbɪə/ 兩棲綱 109
amphibians /æmˈfɪbɪənz/ 兩棲類 ***see* Amphibia** 109
amphiboles /æmfɪˌbolz/ 角閃石類 57
amphibolite facies /æmˈfɪbəˌlaɪt ˈfeʃɪiz/ 角閃岩相 92
amygdale /əˈmɪgdel/ 杏仁孔 73
amygdaloidal /əˈmɪgdəˌlɔɪdl/ 杏仁狀的 73
amygdule /əˈmɪgdjul/ 小杏仁體 ***see* amygdaloidal** 73
anaerobic /ænˌeəˈrobɪk/ 厭氧的 81, 101

analysis, chemical /ə'næləsɪs, 'kɛmɪkḷ/ 化學分析 17
anatase /'ænə,tes/ 鋭鈦礦 49
anatexis /,ænə'teksɪs/ 深熔作用 93
ancestral /æn'sɛstrəl/ 祖先的 102
andalusite /,ændəlu'saɪt/ 紅柱石；赤柱石 59
andesine /ændɪzin/ 中長石 56
andesite /'ændɪ,zaɪt/ 安山岩 76
andradite /æn'drɑ,daɪt/ 鈣鐵石榴石 58
Angiospermae /'ændʒɪo,spɝm/ 被子植物門 110
angiosperms /'ændʒɪo,spɝmz/ 被子植物類 ***see* Angiospermae** 110
angle, interfacial /'æŋgḷ, ,ɪntɚ'feʃəl/ 晶面角 40
angle of repose /'æŋgḷ ɑv rɪ'poz/ 休止角；安息角 21
angular /'æŋgjəlɚ/ 稜角狀的 83
angular unconformity /'æŋgjəlɚ ,ʌnkən'fɔrmətɪ/ 角度不整合 ***see* unconformity** 118
anhedral /æn'hɛdrəl/ 他形的 45
anhydrite /æn'haɪdraɪt/ 硬石膏 52
anion /'æn,aɪən/ 陰離子 15
anisotropic /,ænaɪsə'trɑpɪk/ 各向異性的；有方向性的 47
Annelida /ə'nɛlɪdə/ 環節動物門 105
annelids /'ænḷɪdz/ 環節蟲類 ***see* Annelida** 105
anorthite /æn'ɔr,θaɪt/ 鈣長石 57
anorthosite /æn'ɔrθə,saɪt/ 斜長岩 79
antecedent (drainage) /,æntə'sidṇt/ 先成（水系） 27
anthophyllite /,ænθə'fɪlaɪt/ 直閃石 ***see* amphiboles** 57
Anthozoa /,ænθə'zoə/ 珊瑚蟲綱 ***see* Coelentera** 105
anthracite /'ænθrə,saɪt/ 無烟煤 89
anticline /'æntɪ,klaɪn/ 背斜 124
antidinorium /'æntɪklaɪ'nɔrɪəm/ 複背斜 124
antiform /'æntɪfɔrm/ 背形 124
apatite /'ænə,taɪt/ 磷灰石 49
aphanitic /'æfə,naɪtɪk/ 非顯晶的；隱晶的 72
apophysis /ə'pɑfəsɪz/ 岩枝 66
Appalachian /,æpə'letʃɪən/ 阿帕拉契期的 116
apparent age /ə'pærənt edʒ/ 視年齡；表觀年齡 121
apparent dip /ə'pærənt dɪp/ 視傾斜 123
aquilfer /'ækwəfɚ/ 含水層 146
Arachnida /ə'ræknɪdə/ 蜘蛛綱 108
arachnids /ə'ræknɪdz/ 蜘蛛類 ***see* Arachnida** 108
aragonite /ə'rægə,naɪt/ 文石；霰石 51
arch /ɑrtʃ/ 海蝕穹 38
Archaean /ɑr'kiən/ 太古宙（的） 114
Archaeocyatha /,ɑrkɪ'ɑsaɪ'æθə/ 古杯動物門 104
arenaceous /,ærɪ'neʃəs/ 砂質的 ***see* arenite** 85
arenite /'ærənaɪt/ 砂屑岩 85
arête /æ'ret/ 刃脊；冰蝕脊 31
argillaceous /,ɑrdʒɪ'leʃəs/ 泥質的 85
argillite /'ɑrdʒɪ,laɪt/ 泥質岩 85
arkose /'ɑrkos/ 長石砂岩 87
arroyo /ə'rɔɪo/ 旱谷 25
arsenopyrite /,ɑrsṇo'paɪraɪt/ 砷黃鐵礦；毒砂 50
artesian /ɑr'tiʒən/ 自流的；承壓的 146
Arthropoda /ɑr'θrɑpədə/ 節肢動物門 108
arthropods /'ɑrθrə,pɑdz/ 節肢動物 ***see* Arthropoda** 108
asbestos /æs'bɛstəs/ 石棉 61
ash, volcanic /æʃ, vɑl'kænɪk/ 火山灰 69
asphalt /'æsfɔlt/ 地瀝青 89
assemblage /ə'sɛmblɪdʒ/ 組合；集合 100
assimilation /ə,sɪmḷ'eʃən/ 同化作用 62
asthenosphere /,æsθə'nosfɪr/ 軟流圈 9
astrobleme /,æstrə,blim/ 古隕擊坑 131
asymmetrical fold /,esɪ'mɛtrɪk fold/ 不對稱褶皺 126
atmosphere /'ætməs,fɪr/ 大氣圈 9
atoll /'ætɑl/ 環礁 35
atom /'ætəm/ 原子 152
atomic number /ə'tɑmɪk 'nʌmbɚ/ 原子序數 152
atomic weight /ə'tɑmɪk wet/ 原子量 152
augen gneiss /'ɔrgən naɪs/ 眼球狀片麻岩 97
augite /'ɔdʒaɪt/ 普通輝石 57
aulacogen /'ɔlə'kodʒɪn/ 塹溝 133
aureole, metamorphic /'ɔrɪ,ol, ,mɛtə'mɔrfɪk/ 接觸，變質帶 92
authigenic /,ɔθɪ'dʒɛnɪk/ 自生的 84
average /'ævərɪdʒ/ 平均值 152
Aves /'eviz/ 鳥綱 109
axial plane (crystallography) /'æksɪəl plen/ 軸面（結晶學 ***see* axis** 41
axial plane (of fold) /'æksɪəl plen/ 軸面（褶皺的） 123
axial ratio /'æksɪəl 'reʃo/ 軸率 41
axis, crystallographic /'æksɪs, ,krɪstḷ'ɑgrəfɪk/ 晶軸 41

axis of symmetry /ˈæksɪs ɑv ˈsɪmɪtrɪ/ 對稱軸 42
azurite /ˈæʒəˌraɪt/ 藍銅礦 51

B horizon /bi həˈraɪzn̩/ B 層 23
back-arc upwelling /bæk-ɑrk ʌpˈwɛlɪŋ/ 弧後上湧 138
backshore /bækʃɔr/ 濱後；後濱 37
balance /ˈbæləns/ 平衡 152
banding, in igneous rocks /bændɪŋ, ɪn ˈɪgnɪəs rɑks/ 火成岩中的帶狀構造 63
bar /bɑr/ 沙壩；沙洲 38
barchan dune /bɑrˈkɑndun/ 新月沙丘；彎月沙丘 22
barites /ˈbɛraɪtiz/ 重晶石 52
barrier reef /ˈbærɪɚ rif/ 堡礁；堤礁 38
barrier island /ˈbærɪɚ ˈaɪlənd/ 堤礁島 38
baryte /bəˈraɪt/ 重晶石 52
basalt /bəˈsɔlt/ 玄武岩 77
base /bes/ 鹼 152
base level /bes ˈlɛvl̩/ 基準面 24
basement /ˈbesmənt/ 基底 114
basement complex /ˈbesmənt ˈkɑmplɛks/ 基底雜岩 114
basic (rocks) /ˈbesɪk/ 基性的(岩) 74
basin, structural /ˈbesn̩, ˈstrʌktʃərəl/ 構造盆地 125
batholith /ˈbæθəˌlɪθ/ 岩基 64
bathyal /ˈbæθɪrl/ 半深海的 34
bathylith /ˈbæθɪlɪθ/ 岩基 ***see* batholith** 64
bauxite /ˈbɔksaɪt/ 鋁土礦 23
beach /bitʃ/ 海灘 37
  raised /rezd/ 上升海灘 39
bed /bɛd/ 層 80, 113
bedding /ˈbɛdɪŋ/ 層理 80
  convolute /ˈkɑnvəˌlut/ 旋捲層理 82
  cross /krɔs/ 交錯層理 82
  current /ˈkɝənt/ 水流層理；波狀層理 82
bedding, dune /ˈbɛdɪŋ, djun/ 沙丘層理 82
  graded /gredɪd/ 遞變層理；粒級層理 82
bedding-plane /ˈbɛdɪŋ-plen/ 層面 80
beheaded (stream) /bɪˈhɛdɪd/ 斷頭河 25
Belemnoidea /ˈbɛləmˌnɔɪdɪə/ 箭石目 107
belemnites /ˈbɛləmˌnaɪts/ 箭石 107
Benioff zone /ˈbɛniɑf zon/ 畢烏夫帶 137
benthonic /ˈbɛnˈθɑnɪk/ 底棲的 ***see* benthos** 100
benthos /ˈbɛnθɑs/ 底棲生物 100
Bergschrund /ˈbɛrgʃrund/ 冰後隙；壁前大縫 28
berm /bɝm/ 後濱階地 37
beryl /ˈbɛrəl/ 綠柱石；綠玉 60
beta quartz /ˈbetə kwɔrts/ β 石英；高温石英 55
biocoenose /baɪəsɪˈnos/ 生物群落 101
biofacies /ˌbaɪoˈfeʃɪiz/ 生物相 117
biogenic /ˌbaɪoˈdʒɛnɪk/ 生物成因的 101
bioherm /ˈbaɪoˌhɝm/ 生物礁 101
biolith /ˈbaɪəlɪθ/ 生物岩 101
bioseries /ˌbaɪəˈsɪrɪz/ 生物系列 103
biosphere /ˈbaɪəˌsfɪr/ 生物圈 98
biostratigraphical /ˈbaɪoˌstrætiˈgræfɪkl/ 生物地層的 117
giostrome /ˈbaɪəstrom/ 生物層 101
biotite /ˈbaɪəˌtaɪt/ 黑雲母 55
birds /bɝdz/ 鳥類 ***see* Aves** 109
birefrimgence /ˌbaɪrɪˈfrɪndʒəns/ 雙析射 47
bitumen /bɪˈtjumən/ 瀝青 89
bitumenous coal /bɪˈtjumənəs kol/ 烟煤 89
black-band ironstone /ˈblækˌbænd ˈaɪɚnˌston/ 黑泥鐵礦 88
block (structural) /blɑk/ 地塊(構造的) 133
block, volcanic /blɑk/ 火山塊 69
blue john /blu dʒɑn/ 螢石 52
bomb, volcanic /bɑm, vɑlˈkænɪk/ 火山彈 69
borehole /ˈborˌhol/ 鑽孔 144
boss /bɔs/ 岩瘤 65
botryoidal /ˌbɑtrɪˈɔɪdəl/ 葡萄狀的 45
bottomset beds /ˈbɔtəmsɛt bɛdz/ 底積層 82
boulder /ˈboldɚ/ 漂礫 87
boulder clay /ˈboldɚ kle/ 泥礫；漂礫土 29
boundary /ˈbaʊndərɪ/ 邊界 152
box fold /bɑks fold/ 箱狀褶皺 126
Brachiopoda /ˈbrækɪəˌpɑdə/ 腕足動物門 105
brachiopods /ˈbrækɪəˌpɑdz/ 腕足類 ***see* Brachiopoda** 105
breccia /ˈbrɛtʃɪə/ 角礫岩 87
  volcanic /vɑlˈkænɪk/ 火山 69
Bryozoa /ˌbraɪəˈzoə/ 苔蘚動物門 106
bryozoans /ˌbraɪəˈzoənz/ 苔蘚蟲 ***see* Bryozoa** 106
burst, evolutionary /bɝst, ˌɛvəˈluʃənˌɛrɪ/ 演化突變 103

butte /bjut/ 孤山 33
bytownite /ˈbaɪtaʊnaɪt/ 培長石；倍長石 56

C horizon /si həˈraɪzn̩/ C 層 23
Cainozoic /ˌkaɪnəˈzoɪk/ 新生代 115
calc tufa /kælk ˈtjufə/ 鈣華，石灰華 21
calcareous /kælˈkɛrɪəs/ 鈣質的 86
calcareous tufa /kælˈkɛrɪəs ˈtjufə/ 石灰華，鈣華 21
calc-alkaline (rocks) /kɑlˈælkəˌlaɪn/ 鈣鹼性（岩） 75
calcite /ˈkælsaɪt/ 方解石 51
caldera /kælˈdirə/ 破火山口 68
Caledonian /ˌkælɪˈdonɪən/ 加里東期的 116
Caledonides /kælɪˈdonaɪdɪz/ 加里東造山帶 116
caliche /kəˈlitʃi/ 鈣質層 23
Cambrian /ˈkæmbrɪən/ 寒武紀（的） 114
cancrinite /ˈkæŋkrɪnaɪt/ 鈣霞石 58
cannel coal /ˈkænl̩ kol/ 燭煤 89
canyon /ˈkænjən/ 峽谷 25
  submarine /ˌsʌbməˈrin/ 海底峽谷 35
cap rock /kæp rɑk/ 蓋層 144
capture, (river and stream) /ˈkæptʃɚ/ 襲奪河；掠奪河 25
carbon minerals /ˈkɑrbɑn ˈmɪnərəlz/ 含碳礦物 48
carbon-14 dating /ˈkɑrbɑnˌforˈtin ˈdetɪŋ/ 碳 14 年齡測定法 120
carbonaceous (sediments) /ˌkɑrbəˈneʃəs/ 碳質的（沉積） 89
carbonate /ˈkɑrbənɪt/ 碳酸鹽 16
Carboniferous /ˌkɑrbəˈnɪfərəs/ 石炭紀（的） 114
cartilage /ˈkɑrtl̩ɪdʒ/ 軟骨 152
cassiterite /kəˈsɪtəˌraɪt/ 錫石 48
cast /kæst/ 鑄型 98
cataclasis /ˈkætəˌklæsɪs/ 碎裂作用 94
cataclastic /ˌkæəˈklæstɪk/ 碎裂的 94
Catastrophism /kəˈtæstrəˌfɪzəm/ 災變說 112
catchment area /ˈkætʃmənt ˈɛrɪə/ 集水區；集水面積 146
cation /ˈkætˌaɪən/ 陽離子 15
cauldron subsidence /ˈkɔldrən səbˈsaɪdn̩s/ 頂蓋沉陷；火山口沉陷 67
cave, sea /kev/, /si/ 海蝕洞 37
Cca horizon /si həˈraɪzn̩/ C 鈣層 23
celestite /ˈsɛlɪsˌtaɪt/ 天青石 52
celestine /ˈsɛlɪsˌtin/ 天青石 52
cell /sɛl/ 細胞 153
cell, unit /sɛl, ˈjunɪt/ 晶胞 40
cement /səˈmɛnt/ 膠結物 84
cementation /ˌsimənˈteʃən/ 膠結作用 84
Cenozoic /ˌsinəˈzoɪk/ 新生代的 ***see*** **Cainozoic** 115
central eruption /ˈsɛntrəl ɪˈrʌpʃən/ 中心噴發 68
centre of symmetry /ˈsɛntɚ ɑv ˈsɪmɪtrɪ/ 對稱中心 42
cephalopoda /ˈsɛfələˌpɑdə/ 頭足綱 107
cephalopods /ˈsɛfələˌpɑdz/ 頭足類 ***see*** **Cephalopoda** 107
chain structure (silicates) /tʃen ˈstrʌktʃɚ/ 鏈狀結構（矽酸鹽） 53
chalcedony /kælˈsɛdn̩ɪ/ 石髓；玉髓 55
chalcocite /ˈkælkəˌsaɪt/ 輝銅礦 50
chalcophile /ˌkælkəˈfaɪl/ 親銅的；親硫的 18
chalcopyrite /ˌkælkəˈpaɪraɪt/ 黃銅礦 50
chalk /tʃɔk/ 白堊 86
channel /ˈʧænl̩/ 水道 153
charge /tʃɑrdʒ/ 電荷 153
chemical composition /ˈkɛmɪkl̩ ˌkɑmpəˈzɪʃən/ 化學成分 15
chemical compound /ˈkɛmɪkl̩ ˈkɑmpaʊnd/ 化合物 15
chemical equilibrium /ˈkɛmɪkl̩ ˌikwəˈlɪbrɪəm/ 化學平衡 17
chemical precipitate /ˈkɛmɪkl̩ prɪˈsɪpətɪt/ 化學沉澱作用 ***see*** **precipitation** /si prɪˌsɪpəˈteʃən/ 見沉澱作用 18
chemical reaction /ˈkɛmɪkl̩ rɪˈækʃən/ 化學反應 17
chemical weathering /ˈkɛmɪkl̩ ˈwɛðərɪŋ/ 化學風化 20
chemistry /ˈkɛmɪstrɪ/ 化學 153
chert /tʃɝt/ 燧石 86
chevron fold /ˈʃɛvrən fold/ 尖棱褶皺 126
chilled margin /tʃɪld ˈmɑrdʒɪn/ 冷凝邊 66
chilled zone /tʃɪld zon/ 冷凝帶 ***see*** **chilled margin** 66
chlorite /ˈkloraɪt/ 綠泥石 61
chondrite /ˈkɑnˌdraɪt/ 球粒隕石 149
chondrule /ˈkɑndrul/ 隕石球粒 ***see*** **chondrite** 149
chordata /kɔrˈdetə/ 脊索動物門 109
chordates /ˈkɔrdets/ 脊索動物 ***see*** **chordata** 109
chromite /ˈkromaɪt/ 鉻鐵礦 49
chron /kron/ 時 113

chronostratigraphical (unit) /ˌkrɔnəstrætɪˈgræfɪkl̩/ 年代地層的(單位) 113
cinder cone /ˈsɪndɚ kon/ 火山渣錐 68
cinnabar /ˈsɪnəˌbɑr/ 辰砂；銀朱 50
cirque /sɝk/ 冰斗 31
cladistics /klæˈdɪstɪks/ 親緣分枝法 103
cladogenesis /ˌklædəˈdʒɛnəsɪs/ 分枝演化 103
class (taxonomic) /klæs/ 綱(分類系統的) 99
classify /ˈklæsəˌfaɪ/ 分類 153
clast /ˈklæst/ 碎屑 85
clastic (sediments) /ˈklæstɪk/ 碎屑狀的(沉積) 85
clay /kle/ 黏土 88
  red /rɛd/ 紅黏土 36
clay-ironstone /ˈkle ˈaɪɚnˌston/ 泥鐵礦 88
clay minerals /kle ˈmɪnərəlz/ 黏土礦物 61
cleavage (rock) /ˈklivɪdʒ/ 劈理(岩石) 95
  false /fɔls/ 假劈理 95
  flow /flo/ 流狀劈理 95
  fracture /ˈfræktʃɚ/ 破劈理 95
  shear /ʃɪr/ 剪劈理 95
  slaty /ˈsletɪ/ 板狀劈理 95
cleavage (mineral) /ˈklivɪdʒ/ 解理；分裂性(指礦物) 44
cleavage-plane /ˈklivɪdʒ-plen/ 解理面；分裂平面 44
cleave /kliv/ 劈開 ***see*** **cleavage** 44
cliff /klɪf/ 海崖；懸崖 37
clinometer /klaɪˈnɑmətɚ/ 測斜儀 148
clinopyroxenes /ˌklaɪnəˈpaɪrɑkˌsin/ 單斜輝石類 57
clinozoisite /ˌklaɪnəˈzɔɪsaɪt/ 斜黝簾石 60
close /kloz/ 閉合 ***see*** **closure** 123
closed system /klozd ˈsɪstəm/ 封閉系統 19
closure /ˈkloʒɚ/ 閉合 123
coal /kol/ 煤 89
  bituminous /bɪˈtjumənəs/ 烟煤 89
  cannel /ˈkænl̩/ 燭煤 89
  sub-bituminous /sʌb ˌbɪˈtjumənəs/ 次烟煤 89
  rank of /ræŋk ɑv/ 煤級 89
coal measures /kol ˈmɛʒɚs/ 煤系 89
coal seam /kol sim/ 煤層 89
coarse /kors/ 粗的 153
coast /kost/ 海岸 37
  drowned /draʊnɪd/ 淹沒海岸 39
  emergent /ɪˈmɝdʒənt/ 上升海岸 39
  primary /ˈpraɪˌmɛrɪ/ 原生海岸 39
  ria /ˈriɑ/ 里亞式海岸 39
  secondary /ˈsɛkənˌdɛrɪ/ 次生海岸 39
  submerged /səbˈmɝdʒd/ 下沉海岸 39
coastline /ˈkostˌlaɪn/ 海岸線 37
cobble /ˈkɑbl̩/ 中礫 87
Coelenterata /siˌlɛntəˈrætə/ 腔腸動物門 105
coesite /kəʊˈzaɪt/ 柯石英 55
columnar (habit) /kəˈlʌmnɚ/ 圓柱狀的(習狀) 45
columnar jointing /kəˈlʌmnɚ dʒɔɪntɪŋ/ 柱狀節理 67
columnar structure /kəˈlʌmnɚ ˈstrʌktʃɚ/ 柱狀構造 ***see*** **columnar jointing** 67
combine /kəmˈbaɪn/ 使結合 153
community, fossil /kəˈmjunətɪ, ˈfɑsl/ 化石群落 100
compaction /kəmˈpækʃən/ 壓實 84
compass, prismatic /ˈkʌmpəs, prɪzˈmætɪk/ 稜鏡羅盤 148
competent bed /ˈkɑmpətənt bɛd/ 硬岩層；強岩層 126
complex /kəmˈplɛks/ 複雜的 153
component (chemical) /kəmˈponənt/ 化學組元 19
compose /kəmˈpoz/ 組成 153
composite cone /kəmˈpɑzɪt kon/ 複合火山錐 68
compositional zoning /ˌkɑmpəˈzɪʃənl̩ ˈzonɪŋ/ 成分分帶 63
composition, chemical /ˌkɑmpəˈzɪʃən, ˈkɛmɪkl̩/ 化學成分 15
compound, chemical /ˈkɑmpaʊnd, ˈkɛmɪkl̩/ 化合物 15
concave /kɑnˈkev/ 凹的 153
concentric /kənˈsɛntrɪk/ 同心的 153
concentric fold /kənˈsɛntrɪk fold/ 同心褶皺 124
concertina fold /ˌkɑnsɚtinə fold/ 手風琴形褶皺 126
concordant (age) /kɑnˌkɔrdnt/ 一致的(年齡) 121
concordant (intrusion) /kɑnˈkɔrdnt/ 整合侵入 65
concretion /kɑnˈkriʃən/ 固結作用 84
conductor /kənˈdʌktɚ/ 導體 153
conduit (volcanic) /ˈkɑndɪt/ 火山道 68
cone /kon/ 錐體 153
cone, volcanic /kon, vɑlˈkænɪk/ 火山錐 68
  composite /kəmˈpɑzɪt/ 複合火山錐 68
  cinder /ˈsɪndɚ/ 火山渣錐 68
  spatter /ˈspætɚ/ 熔岩滴錐；寄生熔岩錐 69
cone-sheet /kon-ʃit/ 錐狀岩席 67

confining pressure /kən'faınıŋ 'prɛʃɚ/ 圍壓 93
conformable /kən'fɔrməbl/ 整合的 118
conformity /kən'fɔrmətı/ 整合 118
conglomerate /kən'glɑmərıt/ 礫岩 87
connate water /'kɑnet 'wɔtɚ/ 原生水 146
consequent (stream) /'kɑnsə,kwɛnt/ 順向（河）；順斜（河） 27
conservative margin /kən'sɝvətıv 'mɑrdʒın/ 保守性邊緣 135
consolidation /kən,sɑlə'deʃən/ 固結 84
constructive margin /kən'strʌktıv 'mɑrdʒın/ 增生性邊緣 135
contact /'kɑntækt/ 接觸面 148
contact goniometer /'kɑntækt ,gonı'ɑmıtɚ/ 接觸測角儀 ***see*** **goniometer** 40
contact metamorphism /'kɑntækt, ,mɛtə'mɔrfızm̩/ 接觸變質作用 90
contact /'kɑntækt zon/ 接觸帶 92
contemporaneous /kən,tɛmpə'renıəs/ 同時的 118
continent /'kɑntənənt/ 大陸 153
continental accretion /,kɑntə'nɛntl̩ ə'kriʃən/ 大陸增長 140
continental drift /,kɑntə'nɛntl̩ drıft/ 大陸漂移 134
continental rise /,kɑntə'nɛntl̩ raız/ 大陸隆 34
continental shelf /,kɑntə'nɛntl̩ ʃɛlf/ 大陸架 34
continental slope /,kɑntə'nɛntl̩ slop/ 大陸坡 34
contour /'kɑntʊr/ 等值線 153
contract /kən'trækt/ 收縮 154
convection /kən'vɛkʃən/ 對流 154
convection current (in mantle) /kən'vɛkʃən 'kɝənt/ 地幔對流 142
convergence /kən'vɝdʒəns/ 趨同 103
zone of /zon əv/ 會聚帶 137
convex /kɑn'vɛks/ 凸的 154
convolute bedding /'kɑnvə,lut 'bɛdıŋ/ 旋捲層理 82
coral reef /'kɑrəl rif/ 珊瑚礁 38
corals /'kɑrəlz/ 珊瑚 105
cordierite /'kɔrdıə,raıt/ 堇青石 60
cordillera /kɔr'dılərə/ 山系；雁列山脈 133
core /kor/ 地核 9
corrasion /kə'reʒən/ 刻蝕；流蝕 24
correlation /,kɔrə'leʃən/ 對比 117
correspond to /,kɔrə'spɑnd tu/ 相當於 154
corrie /'kɔrı/ 冰斗 31
corrosion /kə'roʒən/ 侵蝕；熔蝕；腐蝕 20
corundum /kə'rʌndəm/ 剛玉 49
cosmic abundances /'kɑzmık ə'bʌndənsız/ 宇宙豐度 ***see*** **abundances of elements** 18
country-rock /'kʌntrı rɑk/ 原岩；圍岩 65
crater /'kretð/ 月坑 150
craton /'kretɔn/ 克拉通 133
creep, soil /krip, sɔıl/ 土滑 143
crescent marks /'krɛsn̩t marks/ 新月形痕 83
crest (fold) /krɛst/ 背斜脊 124
Cretaceous /krı'teʃəs/ 白堊紀（的） 115
crevasse /krə'væs/ 冰裂隙 28
Crimoidea /kraı'nɔıdıə/ 海百合綱 106
crinoids /'kraınɔıdz/ 海百合 ***see*** **Crinoidea** 106
cristobalite /krı'stobə,laıt/ 方英石 55
cross bedding /krɔs 'bɛdıŋ/ 交錯層理 82
cross-cutting /krɔs 'kʌtıŋ/ 橫切的 65
cross-section /krɔs 'sɛkʃən/ 橫切面 148
crude /krud/ 原油 144
crude oil /krud ɔıl/ 原油 144
crust /krʌst/ 地殼 9
Crustacea /krʌs'teʃıə/ 甲殼綱 108
crustaceans /krʌs'teʃənz/ 甲殼類 ***see*** **Crustacea** 108
cryptocrystalline /,krıpto'krıstəlın/ 隱晶結構的 44
cryptocrystalline (texture) /,krıpto'krıstəlın/ 隱晶的（結構） 72
cryptoexplosion structure /,krıptoık'splozən 'strʌktʃɚ/ 隱爆構造 131
crystal /'krıstl̩/ 晶體 40
zoned /zond/ 環帶狀晶體 46
crystal form /'krıstl̩ fɔrm/ 晶形 ***see*** **form** 40
crystal lattice /'krıstl̩ 'lætıs/ 晶體點陣；晶格 40
crystal nucleation /'krıstl̩ ,njuklı'eʃən/ 晶體成核作用 ***see*** **nucleation of crystals** 19
crystal settling /krıstl̩ 'sɛtıŋ/ 晶體沉降 63
crystal system /krıstl̩ 'sıstəm/ 晶系 43

crystal, twin /krɪstḷ, twɪn/ 雙晶 41
crystalline /ˈkrɪstḷɪn/ 晶質的 40, 44
crystalline texture /ˈkrɪstḷɪn ˈtɛkstʃɚ/ 晶質結構的 72
crystallinity /ˈkrɪstḷɪnətɪ/ 結晶度 44
crystallize /ˈkrɪstḷˌaɪz/ 使結晶 40
crystallized /ˈkrɪstḷˌaɪzd/ 結晶的 44
crystalloblastic /ˌkrɪstələˈblæstɪk/ 變晶質的 94
crystallographic axis /ˌkrɪstḷˈɑgrəfɪk ˈæksɪs/ 結晶軸 41
crystallography /ˌkrɪstḷˈɑgrəfɪ/ 結晶學 40
crystallographic /ˌkrɪstḷˈɑgrəfɪk/ 結晶學的 40
  axis /ˈæksɪs/ 晶軸 41
cube /kjub/ 立方體 41
cubic system /ˈkjubɪk ˈsɪstəm/ 等軸晶系；立方晶系 43
cuesta /ˈkwɛstə/ 單面山 33
cumulate /ˈkjumjəlɪt/ 堆積岩 63
cupola /ˈkjupələ/ 岩鍾 64
cuprite /ˈkjupraɪt/ 赤銅礦 48
curie point /ˈkjurɪ pɔɪnt/ 居里點 14
current, turbidity /ˈkɝənt, tɝˈbɪdətɪ/ 濁流 36
  density /ˈdɛnsətɪ/ 密度流 ***see*** **turbidity current** 36
current bedding /ˈkɝənt ˈbɛdɪŋ/ 水流層理；波狀層理 82
current ripples /ˈkɝənt ˈrɪpḷz/ 流痕 ***see*** **ripple mark** 83
cwm /kum/ 冰斗 31
cycle of erosion /ˈsaɪkḷ əv ɪˈroʒən/ 侵蝕循環 32
cyclic sequence /ˈsaɪklɪk ˈsikwəns/ 旋迴層序 112
cylindrical fold /səˈlɪndrɪkəl fold/ 圓柱狀褶皺 126
cylosilicates /saɪˈkloˈsɪlɪkɪts/ 環狀矽酸鹽類 53
cyclothem /ˈsaɪkləθɛm/ 韻律層 112

dacite /ˈdesaɪt/ 英安岩 76
daughter element /ˈdɔtɚ ˈɛləmənt/ 子元素 19
decke /dɛdə/ 推覆體 130
décollement /ˌdeˈkɔləmənt/ 滑脱 130
decompose /ˌdikəmˈpoz/ 分解 154
decussate (texture) /dɪˈkʌset/ 交錯的（結構） 72
dedolomitization /ˌdiˈdɑləməˌtaɪzeʃen/
  去白雲石化作用 86
deformation /ˌdifɔrˈmeʃən/ 形變 122
deformational fabric /ˌdifɔrˈmeʃənəl ˈfæbrɪk/ 變形組構
  see **tectonite** 97
degradation /ˌdɛgrəˈdeʃən/ 陵夷作用 32
delta /ˈdɛltə/ 三角洲 26
deltaic /dɛlˈteɪk/ 三角洲的 26
deltaic (environment) /dɛlˈteɪk/ (/ɪnˈvaɪrənmənt/)
  三角洲（環境） 81
dendritic /dɛnˈdrɪtɪk/ 樹枝狀 45
dendrochronology /ˌdɛndrokrəˈnɑlədʒɪ/ 年輪測年學 121
density /ˈdɛnsətɪ/ 密度 154
density current /ˈdɛnsətɪ ˈkɝənt/ 密度流 36
denudation /ˈdɪˈnjuˌdeʃən/ 剝蝕作用 32
deposit /dɪˈpɑzɪt/ 沉積；礦床 80
deposition /ˌdɛpəˈzɪən/ 沉積作用 80
depositional environment /ˌdɛpəˈzɪʃənḷ ɪnˈvaɪrənmənt/
  沉積環境 81
derived fossil /dəˈraɪvd ˈfɑsḷ/ 移積化石；轉生化石 111
desert varnish /ˈdɛzɚt ˈvɑrnɪʃ/ 沙漠岩漆 22
destructive margin /dɪˈstrʌktɪv ˈmɑrdʒɪn/
  破壞性邊緣 135
detrital /dɪˈtraɪtḷ/ 碎屑的 85
devitrification /diˌvɪtrəfɪˈkeʃṇ/ 去玻作用 63
Devonian /dəˈvonɪən/ 泥盆紀（的） 114
dextral fault /ˈdɛkstrəl fɔlt/ 右行平移斷層 129
diabase /ˈdaɪəˌbes/ 輝綠岩 78
diachronous /daɪəˈkrɑnəs/ 穿時的；歷時性的 117
diagenesis /ˌdaɪəˈdʒɛnəsɪs/ 成岩作用 84
diagram /ˈdaɪəˌgræm/ 圖解；圖表 155
diameter /daɪˈæmətɚ/ 直徑 155
diamond /ˈdaɪəmənd/ 金剛石；金剛鑽 48
diapir /ˈdaɪəpɪr/ 底闢構造 131
diastem /ˌdaɪəˈstim/ 沉積停頓 118
diastrophism /daɪˈæstrəˌfɪzəm/ 地殼變動 132
Diatomaceae /daɪəˈtɑməsiəɪ/ 矽藻族 110
diatoms /ˈdaɪətəmz/ 矽藻 ***see*** **Diatomacae** 110
diatreme /ˈdaɪətrim/ 火山爆發口（道） 69
dichroism /ˈdaɪkroˌɪzəm/ 二色性 47
differential erosion /ˌdɪfəˈrɛnʃəl ɪˈroʒən/
  差異侵蝕 21
differential weathering /ˌdɪfəˈrɛnʃəl ˈwɛðərɪŋ/
  差異風化 21
differentiation, magmatic /ˌdɪfəˌrɛnʃɪˈeʃən ˌmægmətɪk/
  岩漿分異 ***see*** **magmatic differentiation** 62
diffusion /dɪˈfjuʒən/ 擴散 63

dike /daɪk/ 岩牆 ***see* dyke** 67
dimension /dəˈmɛnʃən/ 維 155
dimorphism /daɪˈmɔrfɪzəm/ 同質二像 46
dinosaurs /ˈdaɪnəˌsɔrz/ 恐龍類 109
diopside /daɪˈɑpsaɪd/ 透輝石 57
diorite /ˈdaɪəˌraɪt/ 閃長石 76
dip /dɪp/ 傾角 123
true /tru/ 真傾角 123
apparent /əˈpærənt/ 視傾角 123
dip-slip fault /dɪp-slɪp fɔlt/ 傾向滑動斷層 129
dip-slope /dɪp-slop/ 傾向坡 33
directed pressure /dəˈrɛktɪd ˈprɛʃɚ/ 定向壓力 93
directed stress /dəˈrɛktɪd strɛs/ 定向應力 ***see* stress** 122
disconformity /ˌdɪskənˈfɝmətɪ/ 假整合 118
discontinuity /ˌdɪskɑntəˈnuətɪ/ 間斷面 10
discordant (age) /dɪsˈkɔrdn̩t/ (/edʒ/) 不一致的（年齡） 121
discordant (intrusion) /dɪsˈkɔrdn̩t/ (/ɪnˈtruʒən/) 不整合（侵入） 65
disharmonic fold /ˌdɪshɑrˈmɑnɪk/ 不協調褶皺 126
displacement (fault) /dɪsˈplesmənt/ (/fɔlt/) 位移（斷層） 128
dissolve /dɪˈzɑlv/ 溶解 155
distillation /ˌdɪstl̩ˈeʃən/ 蒸餾 155
distributary /dɪˈstrɪbjʊˌtɛrɪ/ 汊流 26
distribute /dɪˈstrɪbjʊt/ 分佈 155
divergence /dəˈvɝdʒəns/ 趨異 102
zone of /zon ɑv/ 離散帶 136
diversification /dəˌvɝsəfəˈkeʃən/ 異化 102
diversity (species) /dəˈvɝsətɪ/ (/ˈspiʃɪz/) 種的變異度 102
division /dəˈvɪʒən/ 劃分 155
dolerite /ˈdɑləˌraɪt/ 粗玄岩 78
dolomite /ˈdɑləˌmaɪt/ 白雲岩 51
dolomitization /ˈdɑləməˌtaɪzeʃen/ 白雲石化作用 86
dome /dom/ 穹窿 125
piercement /pɪrsmənt/ 刺穿穹丘 131
salt /sɔlt/ 鹽丘 131
downthrow /ˈdaʊnˌθro/ 下落的 128
downwards /ˈdaʊnwɚdz/ 向下 155
downwarp /daʊnwɔrp/ 坳陷（區） 125
drag fold /dræg fold/ 牽引褶皺 126
drainage, antecedent /ˈdrenɪdʒ, ˌæntəˈsidn̩t/ 先成水系 27
drainage, inconsequent /l̩ˈdrenɪdʒ, ɪnˈkɑnsəˌkwɛnt/ 非順向水系 27
superimposed /ˌsupərɪmˈpozd/ 叠置水系 27
drainage-pattern /ˈdrenɪdʒ-ˈpætɚn/ 水系類型 27
dendritic /dɛnˈdrɪtɪk/ 樹枝狀水系 27
radial /ˈredɪəl/ 放射狀水系 27
rectangular /rɛkˈtæŋgjəlɚ/ 長方水系 27
system /ˈsɪstəm/ 水系 25
trellis /ˈtrɛlɪs/ 格狀 27
drift, continental /drɪft, ˌkɑntəˈnɛntl̩/ 大陸漂移 134
drift (glacial) /drɪft/ (/ˈgleʃəl/) 冰磧 31
drill /drɪl/ 鑽 155
drowned coast /draʊnd kost/ 淹沒海岸 39
drumlin /ˈdrʌmlɪn/ 鼓丘；冰磧丘 30
dry valley /draɪ ˈvælɪ/ 乾谷 25
dune /djun/ 沙丘 22
barchan /bɑrˈkɑn/ 新月形沙丘；彎月沙丘 22
longitudinal /ˌlɑndʒəˈtjudn̩l/ 縱向沙丘 22
parabolic /ˌpærəˈbɑlɪk/ 抛物線沙丘 22
transverse /trænsˈvɝs/ 橫向沙丘 22
whaleback /ˈhwelˌbæk/ 鯨背沙丘 22
dune bedding /djun ˈbɛdɪŋ/ 沙丘層理 82
dunite /ˈdjunaɪt/ 純橄欖岩 79
duricrust /ˈdjurɪkrʌst/ 鈣質殼；硬殼 23
dust, volcanic /dʌst, vɑlˈkænɪk/ 火山塵 69
dyke /daɪk/ 岩牆 67
neptunean /nɛpˈtjuniən/ 水成岩牆 84
sandstone /ˈsændˌston/ 砂岩岩牆 84
dyke-swarm /ˈdaɪkˌswɔrm/ 岩牆群 67
dynamic metamorphism /daɪˈnæmɪk ˌmɛtəˈmɔrfɪzm̩/ 動力變質作用 91
dynamothermal metamorphism /ˌdaɪnəmɑˈθɝml̩ ˌmɛtəˈmɔrfɪzm̩/ 動熱變質作用 91

Earth sciences /ɝθ ˈsaɪənsɪz/ 地球科學 9
earth flow /ɝθ flo/ 土流 143
Earth, structure of /ɝθ, ˈstrʌktʃɚ ɑv/ 地球結構 9
earthquake /ˈɝθˌkwek/ 地震 12
Echinodermata /ɪˌkaɪnəˈdɝmətə/ 棘皮動物門 106

echinoderms /ɛˈkaɪnəˌdɝmz/ 棘皮類動物
see **Echinodermata** 106
Echinoidea /ɛkɪˈnɔɪdɪə/ 海胆綱 106
echinoids /ɪˈkaɪnɔɪdz/ 海胆 *see* **Echinoidea** 106
eclogite /ˈɛklədʒaɪt/ 榴輝岩 97
eclogite facies /ˈɛklədʒaɪt feʃɪiz/ 榴輝岩相 92
effluent (stream) /ˈɛflʊənt/ 潛水補給河；側流河 25
ejectamenta (volcanic) /ɪˌdʒɛktəˈmɛntə/ (/vɑlˈkænɪk/)
噴出物(火山) 69
electron /ɪˈlɛktrɑn/ 電子 155
element /ˈɛləmənt/ 元素 15
elements, abundances of /ˈɛləmənts, əˈbʌndənsɪz ɑv/
元素豐度 18
embayment /ɛmˈbemənt/ 海灣 39
emergent coast /ɪˈmɝdʒənt kost/ 上升海岸 39
emplacement /ɪmˈplesmənt/ 侵位 64
en echelon /ɛn ˈʒʃəˌlɑn/ 雁列狀 155
end-member /ɛnd ˈmɛmbɚ/ 端員 46
end-moraine /ɛnd moˈren/ 尾磧 29
energy /ˈɛnɚdʒɪ/ 能 155
engineering geology /ˌɛndʒəˈnɪrɪŋ dʒiˈɑlədʒɪ/
工程地質學 138
englacial moraine /ɛnˈgleʃəl moˈren/ 內磧；內冰磧 29
enrichment /ɪnˈrɪtʃmənt/ 富集作用 62
enstatite /ˈɛnstətaɪt/ 頑火輝石 57
environment, depositional /ɪnˈvaɪrənmənt, ˌdɛpəˈzɪʃənl̩/
沉積環境 81
marine /məˈrin/ 海洋環境 81
sedimentary /ˌsɛdəˈmɛntərɪ/ 沉陸環境 81
Eocene /ˈiəˌsin/ 始新世(的) 115
eolian weathering /ɪˈolɪən ˈwɛðərɪŋ/ 風力風化 22
eon /ˈiən/ 宙 *see* **aeon** 113
epeirogenic movements /ˌɛpaɪˈrɑdʒənɪk ˈmuvmənts/
造陸運動 125
epibole /əˈpɪbəlɪ/ 極盛帶 117
epicentre /ˈɛpɪˌsɛntɚ/ 震中 12
epicontinental sea /ˌɛpɪˌkɑntəˈnɛntl̩/ 陸緣海 34
epidotes /ˈɛpɪˌdots/ 綠簾石類 60
epifauna /ˌɛpɪˈfɔnə/ 體外寄生動物群 101
epitaxis /ˈɛpɪˌtæksɪs/ 取向附生 46
epizoon /ˌɛpɪˈzoɑn/ 體外寄生動物 101
epoch /ˈɛpək/ 世 113
equant /iˈkwɔnt/ 等量綱的 45
equigranular (texture) /ˈikwəgrænjəlɚ/ 等粒狀(結構) 73
equilibrium /ˌikwəˈlɪbrɪəm/ 均衡 155
equilibrium, chemical /ˌikwəˈlɪbrɪəm, ˈkɛmɪkl̩/
化學平衡 17
equivalent /ɪˈkwɪvələnt/ 等價物 156
era /ˈɪrə/ 代 113
erode /ɪˈrod/ 侵蝕 20
erosion /ɪˈroʒən/ 侵蝕 20
differential /ˌdɪfəˈrɛnʃəl/ 差異侵蝕 21
erosional cycle /ɪˈroʒənl̩ ˈsaɪkl̩/ 侵蝕旋迴 32
erratic, glacial /əˈrætik, ˈgleʃəl/ 冰川漂礫；冰川漂石 31
eruption (volcanic) /ɪˈrʌpʃən/ 噴發(火山) 68
central /ˈsɛntrəl/ 中心噴發 68
fissure /ˈfɪʃɚ/ 裂隙噴發 68
phreatic /friˈætɪk/ 蒸氣 70
escarpment /ɛˈskɑrpmənt/ 陡崖 33
esker /ˈɛskɚ/ 蛇形丘；蛇狀丘 30
essential mineral /əˈsɛnʃəl ˈmɪnərəl/ 主要礦物 75
estuarine (environment) /ˈɛstʃʊəˌrɪn/ (/ɪnˈvaɪrənmənt/)
河口灣的(環境) 81
eugeosyncline /juˌdʒioˈsɪnklaɪn/ 優地槽 132
euhedral /juˈhidrəl/ 自形的 45
Eurypterida /juˈrɪptərɪdə/ 板足鱟亞綱 108
eurypterids /juˈrɪptərɪdz/ 板足鱟 *see* **Eurypterida** 108
eustatic (movements) /ˈjustətɪk/ (/ˈmuvmənts/)
海面升降(運動) 36
eutectic point /juˈtɛktɪk pɔɪnt/ 共晶點；低共熔點 71
evaporation /ɪˌvæpəˈreʃən/ 蒸發 156
evaporite /ɪˌvæpəˈraɪt/ 蒸發岩 85
evolution /ˌɛvəˈluʃən/ 演化；進化 102
explosive /ɪkˈsplosɪv/ 突發式演化 103
evolutionary burst /ˌɛvəˈluʃənˌɛrɪ bɝst/ 演化突變 103
exfoliation /ɛksˌfolɪˈeʃən/ 葉狀剝落 20
expand /ɪkˈspænd/ 膨脹 156
explosive evolution /ɪkˈsplosɪv ˌɛvəˈluʃən/
突發式演化 103
exposure /ɪkˈspoʒɚ/ 露頭點 122
exsolution /ˈɛksˈsəˈluʃən/ 出溶作用 71
external /ɪkˈstɝnl̩/ 外部的 156

extinction /ɪk'stɪŋkʃən/ 滅絕 102
extrusive /ɪk'strusɪv/ 噴出的 69

face (crystal) /fes/ 晶面(結晶) 40
face (structural attitude) /fes/ 朝向(構造產狀) 127
facies /'feʃɪɪz/ 相 117
  metamorphic /ˌmɛtə'mɔrfɪk/ 變質相 92
  fossil /'fɑsl̩/ 指相化石 111
facies fauna /'feʃɪɪz 'fɔnə/ 指相動物群
  ***see* facies fossil** 111
false cleavage /fɔls 'klivɪdʒ/ 假劈理 95
family /'fæməlɪ/ 科 99
fan, outwash /fæn, 'aʊtˌwɑʃ/ 冰水扇形地；外洗扇 30
fan fold /fæn fold/ 扇形褶皺 124
fault /fɔlt/ 斷層 128
  dextral /'dɛkstrəl/ 右行平移斷層 129
  dip-slip /'dɪp ˌslɪp/ 傾向滑動斷層 129
  listric /'lɪstrɪk/ 鏟狀斷層 130
  normal /'nɔrml̩/ 正斷層 128
  oblique-slip /ə'blik, slɪp/ 斜向滑動斷層 129
  reverse /rɪ'vɝs/ 逆斷層 128
  rotational /ro'teʃənl̩/ 旋轉斷層 129
  sinistral /'sɪnɪstrəl/ 左行平移斷層 129
  strike-slip /'straɪk, slɪp/ 平移斷層；走向滑動斷層 129
  tear /tɪr/ 捩斷層 129
  thrust /θrʌst/ 逆衝斷層 129
  transcurrent /træns'kɝənt/ 橫推斷層 129
  wrench /rɛntʃ/ 扭斷層；走向斷層 129
fault block /fɔlt blɑk/ 斷塊 129
fault breccia /fɔlt 'brɛtʃɪə/ 斷層角礫岩 129
fault gouge /fɔlt gaʊdʒ/ 斷層泥 129
fault plane /fɔlt plen/ 斷層面 128
fault zone /fɔlt zon/ 斷層帶 128
fault scarp /fɔlt skɑrp/ 斷層崖 129
fauna /'fɔnə/ 動物群 98
fayalite /'feəlaɪt/ 鐵橄欖石 58
feldspars /'fɛldˌspɑrz/ 長石類 56
feldspathoids /'fɛldˌspæθɔɪd/ 副長石類；似長石類 58
felsic (minerals) /'fɛlsɪk/ 長英質的(礦物) 75
felsite /'fɛlsaɪt/ 霏細岩；長英岩 77
felspars /'fɛlˌspɑrz/ 長石類 ***see* feldspars** 56
felspathoids /fɛl'spæθɔɪdz/ 似長石類
  ***see* feldspathoids** 58
fenster /'fɛnstə/ 蝕窗 130
ferromagnesian minerals /ˌfɛromæg'niʒən 'mɪnərəlz/
  鎂鐵礦物類 55
ferrosilite /ˌfɛro'sɪlaɪt/ 鐵輝石 57
ferruginous (sediments) /fɛ'rudʒənəs/
  鐵質的(沉積物) 88
fibrous (habit) /'faɪbrəs/ 纖維狀(習性) 45
field work /fild wɝk/ 野外工作 148
field geology /fild dʒi'ɑlədʒɪ/ 野外地質學 148
fine /faɪn/ 細的 156
fiord /fjord/ 峽灣 31
fire curtain /faɪr 'kɝtn̩/ 火簾 70
fire fountain /faɪr 'faʊntn̩/ 火噴泉 70
fissile /'fɪsl̩/ 易剝裂的 80
fissure /'fɪʃɚ/ 裂隙 21
fissure eruption /'fɪʃɚ ɪ'rʌpʃən/ 裂隙噴發 68
fish /fɪʃ/ 魚類 ***see* Pisces** 109
flexure /'flɛkʃɚ/ 撓曲 133
flint /flɪnt/ 打火石 86
flood-plain /'flʌd ˌplen/ 氾濫平原 26
flora /'florə/ 植物群；植物區系 110
flow cast /flo kæst/ 流動底模 83
flow cleavage /flo 'klivɪdʒ/ 流狀劈理 95
flow fold /flo fold/ 流狀褶皺 126
fluorite /'fluəˌraɪt/ 螢石 52
fluorspar /'fluɚˌspɑr/ 氟石 52
fluvial (environment) /'fluvɪəl/ 河成的(環境) 81
fluvio-glacial deposits /'fluvɪə-'gleʃəl dɪ'pɑzɪts/
  冰水沉積 30
focus (earthquake) /'fokəs/ 震源 12
fold /fold/ 褶皺 123
  asymmetrical /ˌesɪ'mɛtrɪkl/ 不對稱褶皺 126
  box /bɑks/ 箱狀褶皺 126
  chevron /'ʃɛvrən/ 尖棱褶皺 126
  concentric /kən'sɛntrɪk/ 同心褶皺 124
  concertina /ˌkɑnsɚ'tinə/ 手風琴形褶皺 126
  cylindrical /sə'lɪndrɪkəl/ 圓柱狀褶皺 126
  disharmonic /ˌdɪshɑr'mɑnɪk/ 不協調褶皺 126
  drag /dræg/ 拖曳褶皺 126

flow /flo/ 流狀褶皺 126
isoclinal /ˌaɪsəˈklaɪnl̩/ 等斜 127
overturned /ˌovɚˈtɝnd/ 倒轉 127
parallel /ˈpærəˌlɛl/ 平行褶皺 124
recumbent /rɪˈkʌmbənt/ 平臥 127
shear /ʃɪr/ 剪切褶皺 126
similar /ˈsɪmələ˞/ 相似褶皺 124
slip /slɪp/ 滑褶皺 126
symmetrical /sɪˈmɛtrɪkl̩/ 對稱褶皺 126
fold, zig-zag /fold, ˈzɪgzæg/ 鋸齒狀褶皺 126
fold-axis /ˈfold ˈæksɪs/ 褶皺軸 123
foliated /ˈfolɪɪtɪd/ 葉片狀的 45
foliation /ˌfolɪˈeʃən/ 葉理 95
footwall /ˈfʊtˌwɔl/ 下盤 128
Foraminifera /fəˌræməˈnɪfərə/ 有孔蟲目 104
forams /foˈremz/ 有孔蟲 104
force /fors/ 力 156
foreland /ˈforlənd/ 前陸 133
foreset beds /forˈsɛt bɛdz/ 前積層 82
fore-shock /forˈʃɑk/ 前震 13
foreshore /ˈforˌʃor/ 前濱 37
form (crystallographic) /fɔrm/ (結晶學的)晶形；格式 40
formation /fɔrˈmeʃən/ 組 113
formula (chemical) /ˈfɔrmjələ/ 化學式 156
forsterite /ˈfɔrtəˌrait/ 鎂橄欖石 58
fossil /ˈfɑsl̩/ 化石 98
derived /dəˈraɪvd/ 移積化石；轉生化石 111
fossilize /ˈfɑsl̩ˌaɪz/ 化石化 98
fossiliferous /ˌfɑsəˈlɪfərəs/ 含化石的 98
fossil community /ˈfɑsl̩ kəˈmjunətɪ/ 化石群落 100
fossil record /ˈfɑsl̩ rɛkɚd/ 化石記錄 98
fraction /ˈfrækʃən/ 分餾部分 62
fracture (mineral) /ˈfræktʃɚ/ 斷口(礦物) 44
fracture (rock) /ˈfræktʃɚ/ 破裂(岩石) 122
fracture cleavage /ˈfræktʃɚ ˈklivɪdʒ/ 破劈理 95
fragment /ˈfrgmənt/ 碎屑 156
frequency /ˈfrikwənsɪ/ 頻率 156
frost heave /frɔst hiv/ 凍脹 30
frost wedging /frɔst wɛdʒɪŋ/ 冰楔作用 30
fullers' earth /ˈfʊlɚz ɝθ/ 漂白土 61
fumarole /ˈfjuməˌrol/ 噴氣孔 70

gabbro /ˈgæbro/ 輝長岩 76
gal /gæl/ 伽 11
galena /gəˈlinə/ 方鉛礦 50
gangue /gæŋ/ 脈石 145
garnets /ˈgɑrnɪts/ 石榴石類 58
gas /gæs/ 氣體 156
gas, natural /gæs, ˈnætʃərəl/ 天然氣 144
Gastropoda /ˈgæstrəˌpodə/ 腹足綱 107
gastropods /ˈgæstrəˌpodz/ 腹足動物 *see* **Gastropoda** 107
geanticline /dʒiˈæntɪklaɪn/ 地背斜 132
gene /dʒin/ 基因 156
genotype /ˈdʒɛnoˌtaɪp/ 屬模 ***see*** **type** 99
genus /ˈdʒinəs/ 屬 99
geochemical cycle /ˌdʒioˈkɛmɪkl̩ ˈsaɪkl̩/ 地球化學旋迴 18
geochemistry /ˌdʒioˈkɛmɪstrɪ/ 地球化學 15
geochronology /ˌdʒiokrəˈnɑlədʒɪ/ 地質紀年學；地質年代學 120
geography /dʒiˈɑgrəfɪ/ 地理 156
geological column /ˌdʒiəˈlɑdʒɪkl̩ ˈkɑləm/ 地質柱狀圖 113
geological survey /ˌdʒiəˈlɑdʒɪkl̩ ˈsɝve/ 地質調查 148
geology /dʒiˈɑlədʒɪ/ 地質學 9
geomagnetism /ˌdʒioˈmægnəˌtɪzəm/ 地磁 14
geomorphology /ˌdʒiəmɔrˈfɑlədʒɪ/ 地貌學；地形學 32
geophysics /ˌdʒioˈfɪzɪks/ 地球物理學 11
geosyncline /ˌdʒioˈsɪnklaɪn/ 地槽 132
geotechnics /ˌdʒioˈtɛknɪks/ 土力學 143
geothermal gradient /ˌdʒioˈθɝməl ˈgredɪənt/ 地熱增温率；地温梯度 93
geothermal heat flow /ˌdʒioˈθɝməl hit flo/ 地熱流 93
geyser /ˈgaɪzɚ/ 間歇噴泉 70
ghost stratigraphy /gost strəˈtɪgrəfɪ/ 殘跡地層 92
glacial erratic /ˈgleʃəl əˈrætik/ 冰川漂礫；冰川漂石 31
glacial lake /ˈgleʃəl lek/ 冰河湖 28
glacial period /ˈgleʃəl ˈpɪrɪəd/ 冰期 28
glacial striae /ˈgleʃəl ˈstraɪˌi/ 冰川擦痕 29
glacial striations /ˈgleʃəl straɪˈeʃənz/ 冰川條痕 29
glaciation /ˌgleʃɪˈeʃən/ 冰川作用 28
glacier /ˈgleʃɚ/ 冰川；冰河 28

alpine /ˈælpaɪn/ 阿爾卑斯型冰川 28
lake /lek/ 冰川湖 28
mountain /ˈmaʊntn̩/ 高山冰川 28
piedmont /ˈpidmɑnt/ 山麓冰川 28
valley /ˈvælɪ/ 山谷冰川 28
glaciofluvial deposits /ˌglesɪˈɑˈfluvɪəl dɪˈpɑzɪts/ 冰水沉積 30
glaciology /ˌglesɪˈɑlədʒɪ/ 冰川學 28
glass /glæs/ 火山玻璃 77
glassy /ˈglæsɪ/ 玻璃質的 72
glauconite /ˈglɔkənaɪt/ 海綠石 55
glaucophane /ˈglɔkəfen/ 藍閃石 57
glide /glaɪd/ 滑移 143
Globigerina ooze /gloˌbɪdʒəˈraɪnə uz/ 抱球蟲軟泥 36
Gnathostoma /næˈθostomə/ 有頜超綱 109
gneiss /naɪs/ 片麻岩 97
gold, native /gold/, /ˈnetɪv/ 自然金 48
Gondwanaland /gɑndˈwɑnəˌlænd/ 岡瓦納古陸 139
goniometer /ˌgonɪˈɑmɪtɚ/ 測角儀 40
contact /ˈkɑntækt/ 接觸測角儀 40
reflecting /rɪˈflɛktɪŋ/ 反射測角儀 40
graben /ˈgrɑbən/ 地塹 133
gradation /greˈdeʃən/ 均夷作用 32
grade, metamorphic /gred, ˌmɛtəˈmɔrfɪk/ 變質等級 ***see*** **metamorphic grade** 91
graded bedding /gredɪd ˈbɛdɪŋ/ 遞變層理；粒級層理 82
gradient (stream) /ˈgredɪənt/ (/strim/) 河流坡降 24
grain /gren/ 晶粒；顆粒 72
grain boundary /gren ˈbaʊndərɪ/ 晶粒間界 72
granite /ˈgrænɪt/ 花崗岩 76
granitization /ˌgrænɪtɪˈzeʃən/ 花崗岩化 63
granoblastic /ˌgrænəˈblæstɪk/ 花崗變晶狀 94
granodiorite /ˌgrænəˈdaɪəraɪt/ 花崗閃長岩 76
granulite /ˈgrænjulaɪt/ 麻粒岩 97
granulite facies /ˈgrænjulaɪt ˈfeʃɪiz/ 麻粒岩相 92
graph /græf/ 曲線圖 157
graphite /ˈgræfaɪt/ 石墨 48
graptolites /ˈgræptəlaɪts/ 筆石 ***see*** **Graptolithina** 109
Graptolithina /ˈgrɑptolɪˌθɪnə/ 筆石綱 109
gravel /ˈgrævl̩/ 礫石 87
gravimeter /grəˈvɪmətɚ/ 重力儀，重力計 11
gravitational acceleration /ˌgrævəˈteʃənl̩ ækˌsɛləˈreʃən/ 重力加速度 11
gravity /ˈgrævətɪ/ 重力 11
specific /spɪˈsɪfɪk/ 比重 44
gravity anomaly /ˈgrævətɪ əˈnɑməlɪ/ 重力異常 11
gravity meter /ˈgrævətɪ ˈmitɚ/ 重力儀，重力計 11
gravity separation /ˈgrævətɪ ˌsɛpəˈreʃən/ 重力分離 63
gravity tectonics /ˈgrævətɪ tɛkˈtɑnɪks/ 重力構造 130
graywacke /ˈgreˌwæk/ 硬砂岩 ***see*** **greywacke** 87
greensand /ˈgrinˌsænd/ 綠砂岩 87
greenschist facies /ˈgrinʃɪst ˈfeʃɪiz/ 綠色片岩相 92
greywacke /ˈgreˌwæk/ 雜砂岩 87
grit /grɪt/ 粗砂岩 87
grossularite /ˈgrɑsjuləˌraɪt/ 鈣鋁岩榴石 58
groundmass /ˈgraʊndˌmæs/ 基質 73
groundwater /ˈgraʊndˈwɔtɚ/ 地下水 146
group (stratigraphical) /grup/ 群（地層學的） 113
Gutenberg discontinuity /ˈgutn̩ˌbɝg ˌdɪskɑntəˈnuətɪ/ 古登堡間斷面 10
guyot /ˈgaɪət/ 海底平頂山 35
Gymnospermae /ˈdʒɪmnəˌspɝməɪ/ 裸子植物亞門 110
gymnosperms /ˈdʒɪmnəˌspɝmz/ 裸子植物類 ***see*** **Gymnospermae** 110
gypsum /ˈdʒɪpsəm/ 石膏 52
habit /ˈhæbɪt/ 習性 45
habitat /ˈhæbəˌtæt/ 生境 100
hade /hed/ 伸角；斷層餘角 128
haematite /ˈhɛməˌtaɪt/ 赤鐵礦 48
half-life /ˈhæfˌlaɪf/ 半衰期 19
halide /ˈhælaɪd/ 鹵素 16
halite /ˈhælaɪt/ 石鹽 52
hand specimen /hænd ˈspɛsəmən/ 手標本 147
hanging valley /ˈhæŋɪŋ ˈvælɪ/ 懸谷 31
hanging wall /ˈhæŋɪŋ wɔl/ 上盤 128
hardness /ˈhɑrdnɪs/ 硬度 44
haüyne /ɔˈwin/ 盤方石 58
heat flow /hit flo/ 熱流 ***see*** **geothermal heat flow** 93
heave (fault) /hiv/ (/fɔlt/) 平錯（斷層；水平斷距 128
heave, frost /hiv/ (/frɔst/) 凍脹 30
hematite /ˈhɛməˌtaɪt/ 赤鐵礦 ***see*** **haematite** 48
hemera /ˌhɛmərə/ 極盛時期 117

Hercynian /hɚˈsɪnɪən/ 海西期的 116
hexagonal system /hɛkˈsægənəl ˈsɪstəm/ 六方晶系 43
highlands, lunar /ˈhaɪləndz, ˈlunɚ/ 月球高地 150
high quartz /haɪ kwɔrts/ 高温石英 ***see* beta quartz** 55
hills, abyssal /hɪl, əˈbɪsl̩/ 深海丘陵 34
hinge (fold) /hɪndʒ/ (/fold/) 褶皺轉折端；(褶皺)樞紐 123
historical geology /hɪsˈtɔrɪkl̩ dʒiˈɑlədʒɪ/ 歷史地質學 112
Holocene /ˈhɑləˌsin/ 全新世(的) 115
holocrystalline /ˌhɑləˈkrɪstl̩ɪn/ 全晶質的 72
holotype /ˈhɑləˌtaɪp/ 全型 ***see* type** 99
homoeomorphy /ˌhomɪəˈmɔrfɪ/ 異種同態 103
horizon (stratigraphical) /həˈraɪzn̩/ 層位(地層學的) 112
horizontal /ˌhɑrəˈzɑntl̩/ 水平的 157
hornblende /ˈhɔrnˌblɛnd/ 普通角閃石 57
hornblendite /ˈhɔrnˌblɛndaɪt/ 角閃石岩 79
hornfels /ˈhɔrnfɛls/ 角頁岩；角岩 96
horst /hɔrst/ 地壘 133
hot spot /hɑt spɑt/ 熱點 142
hyaline /ˈhaɪəlɪn/ 玻璃狀的 72
hybrid /ˈhaɪbrɪd/ 混染岩 62
hybridization /ˌhaɪbrɪdəˈzeʃən/ 混染作用 62
hydration /haɪˈdreʃən/ 水合作用 157
hydrocarbon minerals /ˌhaɪdroˈkɑrbən ˈmɪnərəlz/ 碳氫礦物 89
hydrogeology /ˌhaɪdrodʒiˈɑlədʒɪ/ 水文地質學 146
hydrology /haɪˈdrɑlədʒɪ/ 水文學 146
hydrolysis /haɪˈdrɑləsɪs/ 水解作用 18
hydrosphere /ˈhaɪdrəˌsfɪr/ 水圈 34
hydrostatic pressure /ˌhaɪdrəˈstætɪk ˈprɛʃɚ/ 靜水壓力 93
hydrostatic stress /ˌhaɪdrəˈstætɪk strɛs/ 靜水應力 ***see* stress** 122
hydrothermal /ˌhaɪdrəˈθɝml̩/ 熱液的 63
hydrothermal deposit /ˌhaɪdrəˈθɝml̩ dɪˈpɑzɪt/ 熱液礦床 145
hydrous /ˈhaɪdrəs/ 含水的 157
hydroxyl, group /haɪˈdrɑksɪl, grup/ 羥基 16
Hydrozoa /ˌhaɪdrəˈzoə/ 水螅綱 ***see* Coelentera** 105
hypabyssal /ˌhɪpəˈbɪsəl/ 半深成的；淺成的 64
hypidiomorphic /haɪˌpɪdɪəˈmɔrfɪk/ 半自形的 72
Iapetus sea /aɪˈæpətəs si/ 亞皮特斯海 141
ice age /aɪs edʒ/ 冰河時代 28
ice-dammed lake /ˈaɪsˌdæmd lek/ 冰塞湖 28
ice sheet /aɪs ʃit/ 冰蓋；冰野 28
idioblastic /ˈɪdɪəˌblæstɪk/ 自形變晶的 94
idiomorphic /ˌɪdɪəˈmɔrfɪk/ 自形的 72
idocrase /ˈɪdoˌkres/ 符山石 60
igneous /ˈɪgnɪəs/ 火成的 62
igneous intrusion /ˈɪgnɪəs ɪnˈtruʒən/ 火成侵入體 64
igneous (rocks) /ˈɪgnɪəs/ 火成岩 62
ignimbrite /ˌɪgˈnɪmbraɪt/ 熔結凝灰岩 70
illite 伊萊石；水白雲母 61
ilmenite /ˈɪlmənaɪt/ 鈦鐵礦 49
imbricate structure /ˈɪmbrɪkɪt ˈstrʌktʃɚ/ 叠瓦構造 130
impermeable /ɪmˈpɝmɪəbl̩/ 不透水的 146
impregnation /ˌɪmprɛgˈneʃən/ 浸染 84
incised meander /ɪnˈsaɪzd mɪˈændɚ/ 深切曲流；刻蝕曲流 26
inclusion /ɪnˈkluʒən/ 包體 46
incompetent bed /ɪnˈkɑmpətənt bɛd/ 軟岩層；弱岩層 126
incongruent melting /ɪnˈkɑŋgruənt ˈmɛltɪŋ/ 異元熔融；不一致熔融 71
inconsequent (drainage) /ɪnˈkɑnsəˌkwɛnt/ 非順向(水系)；非順斜(水系) 27
index fossil /ˈɪndɛks ˈfɑsl̩/ 標準化石 111
index mineral /ˈɪndɛks ˈmɪnərəl/ 指示礦物 91
indices (crystallographic) /ˈɪndəˌsiz/ 晶面指數 41
induration /ˌɪndjuˈreʃən/ 硬化 84
influent (stream) /ˈɪnfluənt/ 滲流 25
injection /ɪnˈdʒɛkʃən/ 貫入 64
inlier /ˈɪnˌlaɪr/ 內露層 118
inorganic /ˌɪnɔrˈgænɪk/ 無機的 17
inosilicates /ˌɪnəˈsɪlɪkɪt/ 鏈狀酸鹽類 54
Insecta /ɪnˈsektə/ 昆蟲綱 108
insects /ˈɪnsɛkts/ 昆蟲 ***see* Insecta** 108
inselberg /ˈɪnsəlbɝg/ 島山 33
in situ /ɪnˈsaɪtju/ 在原地 148
insolation /ˌɪnsoˈleʃən/ 曝曬 20
intense /ɪnˈtɛns/ 強烈的 157
intensity (earthquake) /ɪnˈtɛnsətɪ/ 烈度(地震) 12

interbedded /ˌɪntɚˈbɛdɪd/ 夾層的 119
intercalated /ɪnˈtɝkəˌletd/ 插入的 119
intercept (crystallographic) /ˈɪntɚˌsɛpt/ 截距(結晶學的) 41
interfacial angle /ˌɪntɚˈfeʃəl ˈæŋgl̩/ 晶面角；交互面夾角 40
interglacial period /ˌɪntɚˈgleʃəl ˈpɪrɪəd/ 間冰期 28
intergrowth /ˈɪntɚgroθ/ 連晶；共生混合體 73
intermediate (rocks) /ˌɪntɚˈmidɪɪt/ 中性的(岩) 74
interstratified /ˌɪntɚˈstrætəˌfaɪd/ 間層的；互層的 119
intraformational /ˌɪntrəfɔrˈmeʃənl/ 層內的 118
intrusion, igneous /ɪnˈtruʒən, ˈɪgnɪəs/ 火成侵入體 64
multiple /ˈmʌltəpl̩/ 多次侵入的；重復侵入的 65
intrusive /ɪnˈtrusɪv/ 侵入的 ***see*** **igneous intrusion** 64
inversion /ɪnˈvɝʃən/ 倒轉 ***see*** **inverted** 127
Invertebrata /ɪnˈvɝtəbrɪtə/ 無脊椎動物 104
invertebrates /ɪnˈvɝtəbrɪts/ 無脊椎動物 ***see*** **Invertebrata** 104
inverted /ɪnˈvɝtɪd/ 倒轉的 127
ion /ˈaɪən/ 離子 15
ionic radius /aɪˈɑnɪk ˈredɪəs/ 離子半徑 15
iridescence /ˌɪrəˈdɛsn̩s/ 暈色；彩虹色 47
iron (meteorite) /ˈaɪɚn/ (/ˈmitɪərˌaɪt/) 鐵隕石 149
ironstone /ˈaɪɚnˌston/ 富鐵岩石 88
island arc /ˈaɪlənd ɑrk/ 島弧 135
island, barrier /ˈaɪlənd, ˈbærɪɚ/ 堡礁島 38
isobath /ˈaɪsəˌbæθ/ 等深線 35
isochemical process /ˌaɪsəˈkɛmɪkl̩ ˈprɑsɛs/ 等化學過程 19
isochron /aɪˈsɑkrɔn/ 等時線 121
isoclinal fold /ˌaɪsəˈklaɪnl̩ fold/ 等斜褶皺 127
isograd /ˈaɪsəgræd/ 等變線 91
isomorphous series /ˌaɪsəˈmɔrfəs ˈsɪrɪz/ 同系列；同形系列 46
isopach /ˈaɪsəˌpæk/ 等厚線 122
isopachyte /ˌaɪsəˈpækaɪt/ 等厚線 122
isoseismal /ˌaɪsəˈsaɪzml/ 等震的 12
isoseismal line /ˌaɪsəˈsaɪzml̩ laɪn/ 等震線 12
isostasy /aɪˈsɑstəsɪ/ 地殼均衡說 11
isostatic adjustment /ˌaɪsəˈstætɪk əˈdʒʌstmənt/ 均衡調整 11
isostatic compensation /ˌaɪsəˈstætɪk ˌkɑmpənˈseʃən/ 均衡補償 11
isotherm /ˈaɪsəˌθɝm/ 等温線 93
isotope /ˈaɪsəˌtop/ 同位素 19
isotopic age /ˈaɪsəˌtopɪk edʒ/ 同位素年齡 121
isotropic /ˌaɪsəˈtrɑpɪk/ 各向同性的；無向性的 47

jet /dʒɛt/ 煤玉；黑玉 89
joint /dʒɔɪnt/ 節理 21
joint set /dʒɔɪnt sɛt/ 節理組 21
joint system /dʒɔɪnt ˈsɪstəm/ 節理系 21
Jurassic /dʒʊˈræsɪk/ 侏羅紀(的) 115
juvenile water /ˈdʒuvənl̩ ˈwɔtɚ/ 初生水；岩漿水 146

K horizon /ke-həˈraɪzn̩/ K 層 23
K-Ar dating /ke-ɑr detɪŋ/ 鉀氬年齡測定 120
kame /kem/ 冰礫阜；冠丘 30
kaolinite /ˈkeəlɪnˌaɪt/ 高嶺土 61
karst topography /kɑrst toˈpɑgrəfɪ/ 岩溶地形；喀斯特地形 33
Kimmerian /ˈkɪmərɪən/ 基米里期的 116
klippe /ˈklɪpə/ 飛來峯 130
knick-point /nɪk pɔɪnt/ 裂點；轉折點 24
kyanite /kaɪəˌnaɪt/ 藍晶石 59

laboratory /ˈlæbrəˌtorɪ/ 實驗室 147
labradorite /ˈlæbrədɔrˌaɪt/ 拉長石 56
laccolith /ˈlækəˌlɪθ/ 岩蓋 65
lacustrine (environment) /ləˈkʌstrɪn/ 湖泊的，湖成環境 81
lag /læg/ 滯後斷層 130
lagoon /ləˈgun/ 瀉湖；礁湖 38
lagoonal (environment) /ləˈgunl̩/ 瀉湖的；瀉湖環境 81
lahar /ˈlɑhɑr/ 火山泥流物 70
lake, glacial /lek, ˈgleʃəl/ 冰河湖 28
lake, glacier /lek, ˈgleʃɚ/ 冰川湖 28
ice-dammed /aɪs-dæmɪd/ 冰塞湖 28
marginal /ˈmɑrdʒɪnl̩/ 冰前湖；冰川邊緣湖 30
ox-bow /ˈɑksˌbo/ 牛軛湖 26
lamellae /ləˌmɛli/ 頁片 ***see*** **lamellar** 45
lamellar /ləˈmɛlɚ/ 頁片狀 45

Lamellibranchiata /lə͵mɛlə'bræŋkɪetə/ 瓣鰓綱 107
lamellibranchs /lə'mɛlɪbræŋks/ 瓣鰓類 ***see* Lamellibranchiata** 107
lamina /'læmənə/ 紋層 80
lamination /͵læmə'neʃən/ 紋理 80
lamprophyre /'læmprə͵faɪɚ/ 煌斑岩 78
landscape /'lænskep/ 景觀 157
landslide /'lænd͵slaɪd/ 滑坡 143
landslip /'lænd͵slɪp/ 地滑 143
lapilli /lə'pɪlaɪ/ 火山礫 69
Laramide /'lærəmaɪd/ 拉臘米期的 116
lateral moraine /'lætərəl mo'ren/ 側磧；側冰磧 29
laterite /'lætə͵raɪt/ 紅土 23
laterization /͵lætərə'zeʃən/ 磚紅壤化作用，紅土化作用 23
lattice, crystal /'lætɪs, 'krɪstḷ/ 晶體點陣；晶格 40
Laurasia /lɔ'reʒə/ 勞亞古陸 140
lava /'lɑvə/ 熔岩 70
lava-flow /'lɑvə flo/ 熔岩流 70
lava tube /'lɑvə tjub/ 熔岩管 70
lava tunnel /'lɑvə 'tʌnḷ/ 熔岩隧道 70
layer /'leɚ/ 層 157
layer lattice silicates /'leɚ 'lætɪs 'sɪlɪkɪts/ 層格矽酸鹽類 54
layering in igneous rocks /'leɚɪŋ ɪn 'ɪgnɪəs rɑk/ 火成岩中的層狀構造 63
lead-lead dating /lid-lid detɪŋ/ 鉛—鉛年齡測定 ***see* uranium-lead dating** 120
lead-uranium dating /lid-ju'renɪəm detɪŋ/ 鉛—鈾年齡測定 ***see* uranium-lead dating** 120
lens (rock) /lɛnz/ 透鏡體（岩石） 119
lens (glass) /lɛnz/ 放大鏡；玻璃透鏡 147
lenticular /lɛn'tɪkjəlɚ/ 扁豆狀的；透鏡狀的 119
lepidolite /lɪ'pɪdə͵laɪt/ 雲鋰母；鱗雲母 55
leucite /'lusaɪt/ 白榴石 58
leucocratic /͵lukə'krætɪk/ 淡色的 75
levée /'lɛvɪ/ 天然堤 26
lignite /'lɪgnaɪt/ 褐煤 89
limb (fold) /lɪm/ （褶皺）翼 123
limestone /'laɪm͵ston/ 石灰岩 86
limnic (environment) /'lɪmnɪk/ 湖沼的（環境） 81
limonite /'laɪmə͵naɪt/ 褐鐵礦 48
lineage /'lɪnɪɪdʒ/ 譜系；世系 102
lineament /'lɪnɪəmənt/ 線狀構造 131
lineation /͵lɪnɪ'eʃən/ 線理 95
liquidus /'lɪkwɪdəs/ 液相曲線 71
listric fault /'lɪstrɪk fɔlt/ 鏟狀斷層 130
lithifaction /͵lɪθɪ'fækʃən/ 石化作用 84
lithification /͵lɪθɪfə'keʃən/ 石化作用 84
lithofacies /'lɪθə͵feʃɪz/ 岩相 117
lithological /lɪ'θɑlədʒɪkḷ/ 岩性的 ***see* lithology** 85
lithology /lɪ'θɑlədʒɪ/ 岩性 85
lithophile /'lɪθə͵faɪl/ 親石的；親氧的 18
lithosphere /'lɪθə͵sfɪr/ 岩石圈 9
lithostatic pressure /͵lɪθə'stætɪk 'prɛʃɚ/ 地靜壓力 93
lithostratigraphical (unit) /͵lɪθəstrə'tɪgrəfɪkḷ/ 岩性地層的（單位） 113
lit-par-lit /'lit͵pɑr'lit/ 間層的 66
littoral /'lɪtərəl/ 海岸區的；潮汐區的 37
load (stream) /lod/ 河流泥沙；流水移運物 24
load cast /lod kæst/ 負荷印模 83
load pressure /lod 'prɛʃɚ/ 負荷壓力 93
lode /lod/ 礦脈 145
loess /'loɪs/ 黃土 22
longitudinal dune /͵lɑndʒə'tjudnḷ djun/ 縱向沙丘 22
longitudinal profile /͵lɑndʒə'tjudnḷ 'profaɪl/ 縱向剖面，河流縱斷 24
lopolith /'lɔpəlɪθ/ 岩盆 66
low quartz /lo kwɔrts/ 低温石英 ***see* alpha quartz** 55
lunar /'lunɚ/ 月球的 150
lunar highlands /'lunɚ 'haɪləndz/ 月球高地 150
lunar regolith /'lunɚ 'rɛgə͵lɪθ/ 月壤 150
lunar soil /'lunɚ sɔɪl/ 月土 150
lustre /'lʌstɚ/ 光澤 47
lustrous /'lʌstrəs/ 光亮的 ***see* lustre** 47
lutaceous /lu'teʃəs/ 泥質的 ***see* lutite** 85
lutite /'lutaɪt/ 泥屑岩 85
L-waves /'ɛl͵ wevz/ L 波 12

mafic (minerals) /'mæfɪk/ 鎂鐵質的（礦物） 75
magma /'mægmə/ 岩漿 62

magma chamber /ˈmægmə ˈtʃembɚ/ 岩漿房 64
magmatic differentiation /mægˈmætɪk ˌdɪfəˌrɛnʃɪˈeʃən/ 岩漿分異作用 62
magmatic stoping /mægˈmætɪk stapɪŋ/ 岩漿頂蝕作用 ***see* stoping** 65
magmatism /ˈmægnəˌtɪzəm/ 岩漿作用 62
magnesite /ˈmægnəˌsaɪt/ 菱鎂礦 51
magnet /ˈmægnɪt/ 磁鐵 157
magnetic anomaly /mægˈnɛtɪk əˈnɑməlɪ/ 磁異常 14
magnetic field /mægˈnɛtɪk fild/ 磁場 157
magnetic pole /mægˈnɛtɪk pol/ 地磁極 14
magnetism, terrestrial /ˈmægnəˌtɪzəm, təˈrɛstrɪəl/ 大地磁場 14
magnetite /ˈmægnəˌtaɪt/ 磁鐵礦 49
magnetization, remanent /ˌmægnətɪˈzeʃən, ˈrɛmənənt/ 剩餘磁化 14
magnetometer /ˌmægnəˈtɑmətɚ/ 磁力儀 14
magnitude (earthquake) /ˈmægnəˌtjud/ 震級(地震) 12
malachite /ˈmæləˌkaɪt/ 孔雀石 51
Mammalia /mæˈmelɪə/ 哺乳綱 109
mammals /ˈmæml̩z/ 浦乳類 109
mantle /ˈmæntl̩/ 地幔 9
mantle rock /ˈmæntl̩ rɑk/ 覆蓋層 23
mantled gneiss dome /ˈmæntl̩d naɪs dom/ 覆蓋的片麻岩穹丘 131
marble /ˈmɑrbl̩/ 大理岩 96
mare (lunar) /ˈmɛrɪ/ 月海 150
mare ridge /ˈmɛrɪ rɪdʒ/ 月海脊 150
margin /ˈmɑrdʒɪn/ 邊緣 157
  plate /plet/ 板塊邊緣 135
  conservative /kənˈsɝvətɪv/ 保守性 135
  constructive /kənˈstrʌktɪv/ 增生性 135
  destructive /dɪˈstrʌktɪv/ 破壞性 135
marginal lake /ˈmɑrdʒɪnl̩ lek/ 冰前湖；冰川邊緣湖 30
marginal plateau /ˈmɑrdʒɪnl̩ plæˈto/ 陸緣高原 34
maria (lunar) /ˈmɛrɪə/ 月海 ***see* mare** 150
marine /məˈrin/ 海洋的；海成的 34, 157
marine environments /məˈrin ɪnˈvaɪrənmənts/ 海洋環境 81
marine swamp /məˈrin swamp/ 沿海沼澤 38
marine terrace /məˈrin ˈtɛrɪs/ 海蝕階地 39
marl /mɑrl/ 泥灰岩 88
mascon /ˈmæskɔn/ 質量瘤 150
massive (habit) /ˈmæsɪv/ 塊狀的(習性) 45
matrix /ˈmetrɪks/ 填質 73
mature (stream) /məˈtjur/ 壯年(河流) 24
M-discontinuity /ɛm-dɪskɑntəˈnuətɪ/ M 不連續面 ***see* Mohorovičić discontinuity** 10
meander /mɪˈændɚ/ 曲流；河曲 26
  incised /ɪnˈsaɪzd/ 深切河曲；刻蝕曲流 26
mechanical weathering /məˈkænɪkl̩ ˈwɛðərɪŋ/ 機械風化 20
medial moraine /ˈmidɪəl moˈren/ 中磧；中冰磧 29
medium /ˈmidɪəm/ 中等的 157
melanocratic /ˌmɛlənəˈkrætɪk/ 暗色的 75
melting, incongruent /ˈmɛltɪŋ, ɪnˈkɑŋgruənt/ 異元熔融；不一致熔融 71
member (stratigraphical) /ˈmɛmbɚ/ 段(地層學的) 113
mesa /ˈmesə/ 方山、平頂山 33
mesosphere /ˈmɛzəˌsfɪə/ 中圈 9
Mesozoic /ˌmɛsəˈzo·ɪk/ 中生代(的) 115
meta-(prefix) /ˌmɛtə/ (詞頭)表示"變質" 90
metal /ˈmɛtl̩/ 金屬 17
metallogenetic province /məˌtælədʒəˈnɛtɪk ˈprɑvɪns/ 成礦區 145
metamorphic aureole /ˌmɛtəˈmɔrfɪk ˈɔrɪˌol/ 接觸變質帶 92
metamorphic facies /ˌmɛtəˈmɔrfɪk ˈfeʃɪiz/ 變質相 92
metamorphic grade /ˌmɛtəˈmɔrfɪk gred/ 變質等級 91
metamorphic rocks /ˌmɛtəˈmɔrfɪk rɑks/ 變質岩 90
metamorphic zones /ˌmɛtəˈmɔrfɪk zonz/ 變質帶 91
metamorphism /ˌmɛtəˈmɔrfɪzm̩/ 變質作用 90
  contact /ˈkɑntækt/ 接觸變質作用 90
  dynamic /daɪˈnæmɪk/ 動力變質作用 90
metamorphism, dynamothermal /ˌmɛtəˈmɔrfɪzm̩, ˌdaɪnəməˈθɝml̩/ 動熱變質作用 91
  regional /ˈridʒnəl/ 區域變質作用 90
  thermal /ˈθɝml̩/ 熱力變質作用 90
metaquartzite /ˌmɛtəˈkwɔrtsaɪt/ 變質石英岩 96
metasomatism /ˌmɛtəˈsomətɪzm̩/ 交代作用 91
Metazoa /ˌmɛtəˈzoə/ 後生動物 104
meteoric water /ˌmitɪˈɔrɪk ˈwɔtɚ/ 大氣降水 146

meteorite /ˈmitɪər,aɪt/ 隕石 149
stony /ˈstonɪ/ 石隕石 149
method /ˈmɛθəd/ 方法 157
micas /ˈmaɪkəz/ 雲母族 55
micrite /ˈmaɪkrraɪt/ 泥晶石灰岩 86
microcline /ˈmaɪkrə,klaɪn/ 微斜長石 56
microcrystalline (texture) /,maɪkroˈkrɪstəlɪn/ 微晶的(結構) 72
microfauna /,maɪkroˈfɔnə/ 微動物群 99
microfossil /,maɪkroˈfɑsəl/ 微體化石 99
microgranite /ˈmaɪkrə,grænɪk/ 微花崗岩 76
micropalaeontology /,maɪkro,pelɪɑnˈtɑlədʒɪ/ 微體古生物學 99
microplankton /,maɪkroˈplæŋktən/ 小型浮游生物 100
microplate /,maɪkroˈplet/ 微板塊 134
microscope /ˈmaɪkrə,skop/ 顯微鏡 147
petrological /piˈtrɑlədʒɪkl̩/ 岩石顯微鏡 147
microseism /ˈmaɪkrəsazm̩/ 微震 13
mid-oceanic ridge /ˈmɪd,oʃɪˈænɪk rɪdʒ/ 洋中脊 35
mid-oceanic ridge basalt /ˈmɪd,oʃɪˈænɪk rɪdʒ ˈbæsɔlt/ 洋中脊玄武岩 ***see*** **MORB** 77
migmatite /ˈmɪgmə,taɪt/ 混合岩 97
migration (of oil) /maɪˈgreʃən/ 運移(石油) 144
milligal /ˈmɪlɪgæl/ 毫伽 11
mine /maɪn/ 礦井；礦坑 145
mineral /ˈmɪnərəl/ 礦物 44
accessory /ækˈsɛsərɪ/ 副礦物；附生礦物 75
essential /əˈsɛnʃəl/ 主要礦物 75
mineral deposits /ˈmɪnərəl dɪˈpɑzɪts/ 礦床 145
mineralization /,mɪnərələˈʒeʃən/ 礦化 145
mineralogy /,mɪnəˈælədʒɪ/ 礦物學 44
minor intrusions /ˈmaɪnə ɪnˈtruʒənz/ 小侵入體 67
Miocene /ˈmaɪə,sin/ 中新世(的) 115
miogeosyncline /,maɪə,dʒiəˈsɪŋklaɪn/ 冒地槽 132
misfit stream /ˈmɪs,fɪt strim/ 不稱河 25
Mississippian /,mɪsəˈsɪpɪən/ 密西西比紀(的) 114
mobile belt /ˈmobil bɛlt/ 活動帶 132
Moho, Mohorovičić discontinuity /ˈmoho,mohoroˈvitʃitʃ ,dɪskɑntəˈnuətɪ/ 莫霍界面，莫霍洛維奇間斷面 10
Mohs' scale /mos skel/ 莫氏硬度計 44
molecular structure /məˈlɛkjələ ˈstrʌktʃə/ 分子結構 15
molecule /ˈmɑlə,kjul/ 分子 15
Mollusca /məˈlʌskə/ 軟體動物門 107
molluscs /ˈmɑləsks/ 軟體動物 ***see*** **Mollusca** 107
molybdenite /məˈlɪbdɪ,naɪt/ 輝鉬礦 50
monadnock /məˈnædnɑk/ 殘丘 33
monocline /ˈmɑnə,klaɪn/ 單斜褶皺 124
monoclinic system /,mɑnəˈklɪnɪk ˈsɪstəm/ 單斜晶系 43
monomineralic /,mɔnə,mɪnərælɪk/ 單礦物的 75
montmorillonite /,mɑntməˈrɪlənaɪt/ 蒙脱石 61
monzonite /ˈmɑnzə,naɪt/ 二長岩 78
moraine /moˈren/ 冰磧 29
englacial /ɛnˈgleʃəl/ 內磧；內冰磧 29
lateral /ˈlætərəl/ 側磧；側冰磧 29
medial /ˈmidɪəl/ 中磧；中冰磧 29
terminal /ˈtɝmənl̩/ 終磧；端冰磧 29
MORB (洋中脊玄武岩) 77
mould /mold/ 印模 98
mountain glacier /ˈmaʊntn̩ ˈgleʃə/ 高山冰川 28
mud cracks /mʌd kræks/ 泥裂 83
mud flow /mʌd flo/ 泥流 143
mudstone /ˈmʌd,ston/ 泥岩 88
multiple intrusion /ˈmʌltəpl̩ ɪnˈtruʒən/ 多次侵入體；重復侵入體 65
muscovite /ˈmʌskə,vaɪt/ 白雲母 55
mutation /mjuˈteʃən/ 突變 102
mylonite /ˈmaɪlə,naɪt/ 糜棱岩 97

nappe /næp/ 納布；推覆體 130
native element /ˈnetɪv ˈɛləmənt/ 自然元素 15
native gold /ˈnetɪv gold/ 自然金 48
native sulphur /ˈnetɪv ˈsʌlfə/ 自然硫 48
natural gas /ˈnætʃərəl gæs/ 天然氣 144
natural selection /ˈnætʃərəl səˈlɛkʃən/ 自然選擇；天擇 102
Nautiloidea /ˈnɔtə,lɔɪdɪə/ 鸚鵡螺亞綱 107
nautiloids /ˈnɔtə,lɔɪdz/ 鸚鵡螺 ***see*** **Nautiloidea** 107
neck, volcanic /nɛk, vɑlˈkænɪk/ 火山頸 68
nekton /ˈnɛktɑn/ 自泳生物 100
Neogene /ˈniodʒin/ 晚第三紀(的) 115
nepheline /ˈnɛfəlɪn/ 霞石 58
Neptunean dyke /nɛpˈtjuniən daɪk/ 水成岩牆 84

neritic /nɪ'rɪtɪk/ 淺海的 34
neritic zone /nɪ'rɪtɪk zon/ 淺海帶 34
nesosilicates /nisə'sɪlɪkɪt/ 孤島狀矽酸鹽類 53
net slip /nɛt slɪp/ 總滑距 128
neutralization /ˌnjutrələ'zeʃən/ 中和作用 17
NiFe 鎳鐵(體) 17
nodule /'nɑdʒul/ 結核 84
non-depositional unconformity /ˌnɑn-dɛˌpə'zɪʃənḷ ˌʌnkən'fɔrmətɪ/ 停積不整合 ***see* unconformity** 118
non-metal /nɑn'mɛtḷ/ 非金屬 17
non-sequence /nɑn'sikwəns/ 小間斷 118
normal /'nɔrmḷ/ 法線 40
normal fault /'nɔrmḷ fɔlt/ 正斷層 128
nosean /nəzɪən/ 黝方石 58
notch, wave-cut /nɑtʃ, wev-kʌt/ 海蝕龕 37
nucleation of crystals /ˌnjuklɪ'eʃən əv 'krɪstḷz/ 晶體的晶核作用 19
nucleus, atomic /'njuklɪəs, ə'tɑmɪk/ 原子核 157
nucleus, crystal /'njuklɪəs, 'krɪstḷ/ 晶核 19
nuclei /'njuklɪˌaɪ/ 晶核 19
nuée ardente /nu'e ɑr'dontɪ/ 熾熱火山雲 70

oblique-slip fault /ə'blik slɪp fɔlt/ 斜向滑動斷層 129
obsequent (stream) /'ɑbsəkwənt/ 逆向(河); 反斜(河) 27
obsidian /əb'sɪdɪən/ 黑曜岩 77
occur /ə'kɝ/ 出現 157
oceanic ridge /'oʃən rɪdʒ/ 洋脊 35
oceanic trench /ˌoʃɪ'ænɪk trɛntʃ/ 洋溝;海溝 35
oceanography /ˌoʃɪən'ɑgrəfɪ/ 海洋學 34
octahedron /ˌɑktə'hidrən/ 八面體 41
offlap /'ɔflæp/ 退覆 119
offset /'ɔfˌsɛt/ 水平斷錯 128
off-shore /'ɔf'ʃor/ 濱外的;離岸的 37
oghurd /'ogɚd/ 星狀沙丘 22
oil /ɔɪl/ 油 144
oil shale /ɔɪl ʃel/ 油頁岩 144
Oligocene /'ɑlɪgoˌsin/ 漸新世(的) 115
oligoclase /'ɑlɪgokles/ 奧長石;更長石;鈉灰長石 56
olivines /'ɑləˌvinz/ 橄欖石類 58
ontogeny /ɑn'tɑdʒənɪ/ 個體發育 102
oolite /'oəˌlaɪt/ 鮞狀岩 86
oolith /'oəlɪθ/ 鮞石 86
ooze /uz/ 軟泥;海泥 36
Globigerina /gləuˌbɪdʒə'raɪnə/ 抱球蟲軟泥 36
Radiolarian /ˌredɪo'lɛrɪən/ 放射蟲軟泥 36
opencast /'opən'kæst/ 露天開採的 145
open system /'opən 'sɪstəm/ 開放系統 19
ophiolite complex /'ɔfɪ'əlaɪt kəm'plɛks, 'kɑmplɛks/ 蛇綠岩 78
ophiolite assemblage /'ɔfɪəlaɪt ə'sɛmblɪdʒ/ 蛇綠岩組合 78
ophitic (texture) /ɑ'fɪtɪk/ 輝綠結構的 73
optical interference /'ɑptɪkḷ ˌɪntɚ'fɪrəns/ 光波干涉 158
optical properties, of minerals /'ɑptɪkḷ 'prɑpɚtɪz, ɔf'mɪnərəlz/ 礦物的光學性質 47
order (taxonomic) /'ɔrdɚ/ 目(分類學) 99
Ordovician /ˌɔrdə'vɪʃən/ 奧陶紀(的) 114
ore /ɔr/ 礦石 145
ore body /'ɔrˌbɑdɪ/ 礦體 145
organic /ɔr'gænɪk/ 有機的 17
organic (sediments) /ɔr'gænɪk/ 有機(沉積物) 85
organism /'ɔrgənˌɪzəm/ 有機體,生物 98
origin /'ɔrədʒɪn/ 起源 158
orogen /ɔ'rɑdʒən/ 造山帶 132
orogenesis /ˌɔrə'dʒɛnəsɪs/ 造山運動 ***see* orogeny** 132
orogenic belt /ə'rɑdʒənɪk bɛlt/ 造山帶 ***see* orogen** 132
orogenic period /ə'rɑdʒənɪk 'pɪrɪəd/ 造山期 116
orogeny /ɔ'rɑdʒənɪ/ 造山運動 132
orpiment /'ɔrpɪmənt/ 雌黃 50
orthite /'ɔrθaɪt/ 褐簾石 60
orthochemical /'ɔrθə'kɛmɪkḷ/ 正源化學的 85
orthoclase /'ɔrθəˌkles/ 正長石 56
orthogneiss /ˌɔrθə'dʒɛnəsɪs/ 正片麻岩;火成片麻岩 97
orthopyroxenes /ˌɔrθə'paɪərɔksin/ 斜方輝石類 57
orthoquartzite /ˌɔrθə'kwortsaɪt/ 正石英岩 87
orthorhombic system /ˌɔrθə'rɑmbɪk 'sɪstəm/ 斜方晶系;正交晶系 43
orthosilicates, structures /ˌɔrθə'sɪlɪkɪts 'strʌktʃɚ/ 正矽酸鹽,結構 ***see* nesosilicates** 53
oscillate /'ɑsḷˌet/ 擺動 158

oscillation ripples /ˌɑsəˈleʃən ˈrɪplz/ 對稱波痕，擺動波痕 ***see*** **ripple mark** 83
Ostracoda /ˌɑstrəˈkɑdə/ 介形亞綱 108
ostracods /ˌɑstrəkɑdz/ 介形類 ***see*** **Ostracoda** 108
outcrop /aʊtˌkrɑp/ 露頭 122
outlier /ˈaʊtˌlaɪɚ/ 外露層 118
outwash fan /ˈaʊtˌwɑʃ fæn/ 冰水扇形地；外洗扇 30
outwash plain /ˈaʊtˌwɑʃ plen/ 冰水沉積平原；外洗平原 31
overburden /ˈovɚˌbɝdn/ 剝離層 145
overlap /ˌovɚˌlæp/ 超覆 119
oversaturated /ˌovɚˈsætʃəˌretɪd/ 過飽和的 74
overstep /ˌovɚˈstɛp/ 跨覆 119
overthrust /ˌovɚˈθrʌst/ 逆掩斷層 129
overturned fold /ˌovɚˈtɝnd fold/ 倒轉褶皺 127
ox-bow lake /ˈɑksˌbo lek/ 牛軛湖 26
oxide /ˈɑksaɪd/ 氧化物 16

pahoehoe (lava) /pɑˈhoɪeˌhoˌɪ/ 繩狀熔岩 70
palaeobiogeography /ˌpelɪoˈbaɪoˌdʒɪˈɑgrəfɪ/ 古生物地理學 100
palaeobotany /ˌpelɪoˈbɑtənɪ/ 古植物學 110
Palaeocene /ˈpælɪrsin/ 古新世 115
palaeocurrent /ˌpelɪoˈkɝənt/ 古水流 82
palaeoecology /ˌpelɪoɪˈkɑlədʒɪ/ 古生態學 100
Palaeogene /ˌpelɪodʒin/ 早第三紀(的) 115
palaeogeography /ˌpelɪodʒɪˈɑgrəfɪ/ 古地理學 139
palaeomagnetism /ˌpelɪoˈmægnəˌtɪzəm/ 古地磁 14
palaeontology /ˌpelɪɑnˈtɑlədʒɪ/ 古生物學 98
stratigraphical /ˌstrætɪˈgræfɪkl/ 地層的 111
palaeopole location /ˌpelɪəpol loˈkeʃən/ 古地磁極位 14
Palaeozoic /ˌpelɪəˈzo·ɪk/ 古生代(的) 114
paleo-, *see* palaeo- /ˌpelɪo/ 古 -
palimpsest structure /ˈpælɪmpˌsɛst ˈstrʌktʃɚ/ 變餘構造 94
paludal (environment) /ˈpæljudl/ 沼澤(環境) 81
palynology /ˌpæləˈnɑlədʒɪ/ 孢粉學 110
Pangaea /pænˈdʒiə/ 泛古陸 139
panidiomorphic texture /pænˌɪdɪrˈmorfɪk ˈtɛkstʃɚ/ 全自形結構 72
parabolic dune /ˌpærəˈbɑlɪk dun/ 拋物線沙丘 22
paragneiss /ˈpærənaɪs/ 副片麻岩；水成片麻岩 97
paralic (environment) /ˈpɛrælɪk/ 近海的(環境) 81
parallel /ˈpærəˌlɛl/ 平行的 158
parallel fold /ˈpærəˌlɛl fold/ 平行褶皺 124
parameter (crystallographic) /pəˈræmətɚ/ (結晶學)參數；軸單位比 41
parent element /ˈpɛrənt ˈɛləmənt/ 母元素 19
particle /ˈpɑrtɪkl/ 粒子 158
peat /pit/ 泥炭 89
pebble /ˈpɛbl/ 小漂礫；卵石 87
pediment /ˈpɛdəmənt/ 麓原；山前侵蝕平原 33
pediplain /ˈpɛdɪplen/ 山麓侵蝕面平原 32
pedology /pɪˈdɑlədʒɪ/ 土壤學 23
pegmatite /ˈpɛgməˌtaɪt/ 偉晶岩 79
pelagic /pəˈlædʒɪk/ 遠洋的 100
Pelecypoda /ˌpɛlɪˈsɪpədə/ 斧足綱 107
pelecypods /ˌpɛlɪˈsɪpədz/ 斧足類 ***see*** **Pelecypoda** 107
penecontemporaneous /ˌpinɪkənˌtɛmpəˈrenɪəs/ 準同時的 118
peneplain /ˈpinəˌplen/ 準平原 32
peneplane /ˈpinəplen/ 準平原 32
Pennsylvanian /ˌpɛnsɪlˈvenɪən/ 賓夕法尼亞紀(的) 114
pericline /ˈpɛrəˌklaɪn/ 圍斜構造 125
periodotite /ˌpɛrɪˈdotaɪt/ 橄欖岩 79
periglacial /ˌpɛrəˈgleʃəl/ 冰緣的 31
period (geological) /ˈpɪrɪəd/ 紀(地質年代) 113
glacial /ˈgleʃəl/ 冰期 28
periodic table /ˌpɪrɪˈɑdɪk ˈtebl/ 週期表 18
perknite /ˈpɝknaɪt/ 輝閃岩類 79
permafrost /ˈpɝməˌfrɔst/ 永久凍土 30
permeability /ˌpɝmɪəˈbɪlətɪ/ 滲透率 146
Permian /ˈpɝmɪən/ 二疊紀(的) 114
perthite /ˈpɝθaɪt/ 條紋長石 56
petrogenesis /ˌpɛtrəˈdʒɛnəsɪs/ 岩石成因論；岩理學 62
petrographic province /piˈtrɑgrəfɪk ˈprɑvɪns/ 岩區 75
petrography /piˈtrɑgrəfɪ/ 岩相學、岩類學 62
petroleum /pəˈtrolɪəm/ 石油 144
petrology /piˈtrɑlədʒɪ/ 岩石學 62
petrological microscope /piˈtrɑlədʒɪkl ˈmaɪkrəˌskop/ 岩石顯微鏡 147
phacolith /ˈʃækəlɪθ/ 岩鞍 66

phaneritic (texture) /'fænəraɪtɪk/ 顯晶的(結構) 72
phanerocrystalline (texture) /ˌfænərə'krɪstəlaɪn/ 顯晶質的(結構) 72
Phanerozoic /ˌfænərə'zo·ɪk/ 顯生宇(的) 114
phase /fez/ 相 19
phenocryst /ˌfinəˌkrɪst/ 斑晶 73
phlogopite /'flɑgəpaɪt/ 金雲母 55
phonolite /'fonəˌlaɪt/ 響岩 77
phosphate /'fɑsfet/ 磷酸鹽 16
photic zone /'fotɪk zon/ 透光層 36
phreatic eruption /fri'ætɪk ɪ'rʌpʃən/ 蒸氣噴發 70
phyla /'faɪlə/ 門(分類系統) ***see* phylum** 99
phyllite /'fɪlaɪt/ 千枚岩；千層岩 96
phyllosilicates /ˌfɪlə'sɪlɪkɪts/ 層狀矽酸鹽類 54
phylogeny /faɪ'lɑdʒənɪ/ 系統發育 102
phylum /'faɪləm/ 門(分類學) 99
physics /'fɪzɪks/ 物理學 158
phytoplankton /ˌfaɪto'plæŋktən/ 浮游植物 100
piedmont glacier /'pidmɑnt 'gleʃɚ/ 山麓冰川 28
piedmontite /'pidmɑntaɪt/ 紅簾石 60
piercement dome /'pɪrsmənt dom/ 刺穿穹丘 131
pillow-lava /'pɪlo 'lævə/ 枕狀熔岩 70
pinacoid /'pɪnəkɔɪd/ 軸面 41
pingo /'pɪŋgo/ 冰核丘；平鍋 30
Pisces /'pɪsiz/ 魚綱 109
pisolite /'paɪsəlaɪt/ 豆石 86
pisolith /'paɪsəlaɪt/ 豆狀岩 86
pitch (of fold) /pɪtʃ/ 側伏角(褶皺的) 123
pitchblende /'pɪtʃˌblɛnd/ 瀝青鈾礦 48
pitchstone /'pɪtʃˌston/ 松脂岩；瀝青岩 77
placer deposit /'plesɚ dɪ'pɑzɪt/ 砂礦床 145
plagioclase feldspar /'pledʒɪəˌkles 'fɛldˌspɑr/ 斜長石 56
plain, abyssal /plen/, /ə'bɪsl̩/ 深海平原 34
plan /plæn/ 平面圖 158
plane /plen/ 平面 158
plane of symmetry /plen ɔv 'sɪmɪtrɪ/ 對稱面 42
plankton /'plæŋktən/ 浮游生物 100
plants /plænts/ 植物 ***see* palaeobotany** 110
plastic /'plæstɪk/ 塑性的 158
plate /plet/ 板塊 134
plate boundary /plet'baundərɪ/ 板塊邊界 135
plate margin /plet 'mɑrdʒɪn/ 板塊邊緣 135
plate tectonics /plet tɛk'tɑnɪks/ 板塊構造 134
plateau /plæ'to/ 高原 32
  marginal /'mɑrdʒɪnl̩/ 陸緣高地 34
  submarine /'sʌbməˌrin/ 海底高原 35
plateau basalt /plæ'to 'bæsɔlt/ 高原玄武岩 77
platform (structural) /'plætˌfɔrm/ (/'strʌktʃərəl/) 地臺(構造的) 133
platform, wave-cut /'plætˌfɔrm/ (/wev kʌt/) 浪蝕臺地；海蝕臺 37
Pleistocene /'plaɪstəˌsin/ 更新世 115
pleochroism /plɪ'ɑkroɪzəm/ 多色性 47
Pliocene /'plaɪəˌsin/ 上新世(的) 115
plug, volcanic /plʌgl, vɑl'kænɪk/ 火山栓 68
plume /plum/ 地幔羽 142
plunge /plʌndʒ/ 傾伏；傾伏角 123
pluton /'plutɑn/ 深成岩體 64
plutonic rocks /plu'tɑnɪk rɑks/ 深成岩類 64
pneumatolysis /ˌnjumə'tɑlɪsɪs/ 氣成 63
poikilitic (texture) /ˌpɔɪkɪ'lɪtɪk/ 嵌晶狀的(結構) 73
poikiloblastic (texture) /'pɔɪkəloˌblæstɪk/ 變嵌晶狀(結構) 94
polar wander /'polɚ 'wɑndɚ/ 地磁極游移 14, 141
polarity, reversed /po'lærətɪ, rɪ'vɝst/ 極性倒轉 14
polarized light /'poləˌraɪzd laɪt/ 偏振光，偏光 158
poles magnetic /polz mæg'nɛtɪk/ 地磁極 14
polymorphism /ˌpɑlɪ'mɔrfɪzm̩/ (同質)多晶形現象；同質多象 46
polyphyletic /ˌpɑlɪfaɪ'lɛtɪk/ 多系列演化的；多源演化的 103
Polyzoa /ˌpɑlɪ'zoə/ 苔蘚蟲 106
pore-fluid pressure /pɔr'fluɪd 'prɛʃɚ/ 孔隙流體壓力 93
pore space /pɔr spes/ 孔隙 84
Porifera /po'rɪfərə/ 多孔動物門 104
porosity /po'rɑsətɪ/ 孔隙度 84
porous /'porəs/ 多孔隙的 84
porphyritic (texture) /ˌpɔrfə'rɪtɪk/ 斑狀的(結構) 73
porphyroblast /pɔr'fɪrəblæst/ 變斑晶 94
porphyroblastic (texture) /pɔr'fɪrəblæstɪk/ 斑狀變晶的(結構) 94
porphyry /'pɔrfərɪ/ 斑岩 78

position /pə'zɪʃən/ 位置 158
postkinematic /post,kɪnə'mætɪk/ 造山運動後的 116
post-orogenic /post ɔ'rɑdʒənɪk/ 造山期後的 116
post-tectonic /post tɛk'tɑnɪk/ 構造後的 116
potash feldspar /'pɑt,æʃ 'fɛld,spɑr/ 鉀長石 56
potassium-argon dating /pə'tæsɪəm 'ɑrgɑn 'detɪŋ/ 鉀氬年齡測定法 120
Precambrian /pri 'kæmbrɪəm/ 前寒武紀(的) 114
precipitation, chemical /pri,sɪpə'teʃən, 'kɛmɪkl̩/ 化學沉澱作用 18
precipitation (atmospheric) /pri,sɪpə'teʃən/ (大氣)降水 146
Preferred orientation /prɪ'fɝd ,ɔrɪɛn'teʃən/ 優選方位 ***see*** **tectonite** 97
pressure /'prɛʃɚ/ 壓力 93
  confining /kən'faɪnɪŋ/ 圍壓 93
  directed /də'rɛktɪd/ 定向壓力 93
  hydrostatic /,haɪdrə'stætɪk/ 靜水壓力 93
  lithostatic /'lɪθə'stætɪk/ 地靜壓力 93
  load /lod/ 負荷壓力 93
Pore-fluid /pɔr'fluɪd/ 孔隙流體 93
primary coast /'praɪ,mɛrɪ kost/ 原生海岸 39
principal shock (earthquake) /'prɪnsəpl̩ ʃɑk/ 主震 13
prism /'prɪzəm/ 稜柱 41
  accretionary /ə'krɪʃən,ɛrɪ/ 加積稜柱體 138
prismatic (habit) /prɪz'mætɪk/ 柱狀(習性) 45
prismatic compass /prɪz'mætɪk 'kʌmpəs/ 稜鏡羅盤 148
process /'prosɛs/ 過程 158
profile, soil /'profaɪl, sɔɪl/ 土壤剖面 23
  stream /strim/ 河流剖面 24
prograde (metamorphism) /'progred/ 前進的(變質作用) 91
property /'prɑpɚtɪ/ 性質 158
proportion /prə'porʃən/ 比例 158
Proterozoic /,prɑtərə'zo·ɪk/ 元古宙的 114
proton /'protɑn/ 質子 159
Protozoa /,protə'zoə/ 原生動物門 104
provenance /'prɑvənəns/ 源岩區；源岩 117
psammite /'sæmaɪt/ 砂質岩 96
psephite /'sifaɪt/ 礫質岩 85
pseudomorph /'sjudəmɔrf/ 假象 46
Pteridophyta /'tɛrədo,faɪtə/ 真蕨植物門 110
Pteridospermae /'tɛrədo,spɝmi/ 種子蕨綱 110
ptygmatic (folding) /tɪg'mætɪk/ 腸狀的(褶皺) 95
pull-apart zone /pul ə'pɑrt zon/ 拉開帶 136
pumice /'pʌmɪs/ 浮岩 69
P-wave /pi wev/ P 波 12
pyramid /'pɪrəmɪd/ 稜錐 41
pyramidal (habit) /pɪ'ræmədl̩/ 錐狀(習性) 45
pyriboles /paɪ'rɪbolz/ 輝閃石類 57
pyrite /'paɪraɪt/ 黃鐵礦 50
pyroclastic (rocks) /,paɪro'klæstɪk/ 火成碎屑(岩) 69
pyrolusite /,paɪrə'lusaɪt/ 軟錳礦 49
pyrometamorphism /,paɪro,mɛtə'mɔrfɪzm/ 高熱變質作用 90
pyromorphite /,paɪrə'mɔrfaɪt/ 磷酸氯鉛礦 49
pyrope /'paɪrop/ 鎂鋁石榴石 58
pyroxenes /'paɪrɑk,sinz/ 輝石類 57
pyroxenite /paɪ'rɑksənaɪt/ 輝石岩 79

quantum evolution /'kwɑntəm ,ɛvə'luʃən/ 量子式演化 103
quaquaversal /,kwekwə'vɝsl̩/ 穹狀的 123
quarry /'kwɑrɪ/ 採石場 145
quartz /kwɔrts/ 石英 55
quartzite /'kwɔrtsaɪt/ 石英岩 87, 96
Quaternary /Kwə'tɝnərɪ/ 第四亞代(的) 115
quick clay /kwɪk kle/ 超靈敏黏土、不穩黏土 143

radiation (adaptive) /,redɪ'eʃən/ (/ə'dæptɪv/) 輻射(適應性) 102
radioactivity /'redɪ,oæk'tɪvətɪ/ 放射性 19
Radiolaria /,redɪo'lɛrɪə/ 放射蟲目 104
Radiolarian ooze /,redɪo'lɛrɪən uz/ 放射蟲軟泥 36
radiometric dating /,redɪ'ɑmətrɪk 'detɪŋ/ 放射性年齡測定法 120
raised beach /rezd bitʃ/ 上升海灘 39
range (of fossil) /rendʒ/ (/əv 'fɑsl̩/) (化石的)延續時限；延限 111
rank of coal /ræŋk əv kol/ 煤級 89
rare earth element /rɛr ɝθ 'ɛləmənt/ 稀土元素 15
ratio /'reʃo/ 比 159

Rayleigh wave /'relɪ wev/ 瑞利波 12
Rb-Sr dating /ru'b-'str 'detɪŋ/ 銣鍶年齡測定 120
reaction, chemical /rɪ'ækʃən, 'kɛmɪkḷ/ 化學反應 17
reaction rim /rɪ'ækʃən rɪm/ 反應邊 46
reaction series /rɪ'ækʃən 'sɪrɪz/ 反應系列 71
reactive /rɪ'æktɪv/ 活性的 17
realgar /rɪ'ælgɚ/ 雄黃 50
Recent /'risṇt/ 近代 115
recrystallization /ri'krɪstḷ,aɪzʃən/ 重結晶作用 93
recumbent fold /rɪ'kʌmbənt fold/ 平臥褶皺 127
red beds /rɛd bɛdz/ 紅層 88
red clay /rɛd kle/ 紅黏土 36
REE /re/ 稀土元素 15
reef /rif/ 礁 38
  barrier /'bærɪɚ/ 堡礁；堤礁 38
  coral /'kɑrəl/ 珊瑚礁 38
reef (mineral) /rif/ 含金石英脈（礦物） 145
refer /rɪ'fɝ/ 談及；涉及；查閱 159
reflecting goniometer /rɪ'flɛktɪŋ,gonɪ'amɪtɚ/ 反射測角儀 40
reflection seismology /rɪ'flɛkʃən saɪz'mɑlədʒɪ/ 反射地震學 13
refraction seismology /rɪ'frækʃən saɪz'mɑlədʒɪ/ 折射地震學 13
refractive index /rɪ'fræktɪv 'ɪndɛks/ 折射率 159
regional metamorphism /'ridʒnəl ,mɛtə'mɔrfɪzṃ/ 區域變質作用 90
regolith /'rɛgə,lɪθ/ 風化層；浮土 23
  lunar /'lunɚ/ 月壤 150
regression, marine /rɪ'grɛʃənl, lmə'rin/ 海退 119
rejuvenation(stream) /rɪ,dʒuvə'neʃən/ 河流回春作用 24
relict structure /'rɛlɪkt 'strʌktʃɚ/ 殘餘構造 94
relief /rɪ'lif/ 地勢 32
remanent magnetization /'rɛmənənt 'mægnə,taɪzʃən/ 剩餘磁化 14
replace /rɪ'ples/ 交代 159
replacement deposit /rɪ'plesmənt dɪ'pɑzɪt/ 交代礦床 145
repose, angle of /rɪ'poz, 'æŋgḷ ɔv/ 休止角、安息角 21
Reptilia /rɛp'tɪlɪə/ 爬蟲綱 109
reptiles /'rɛptḷz/ 爬蟲類 ***see*** **Reptilia** 109
reservoir (oil) /'rɛzɚ,vɔr/ 儲集層（石油） 144
residual /rɪ'zɪdʒuəl/ 殘留的 33
retrograde(metamorphism) /'rɛtrə,gred/ 退化的（變質作用） 91
reverse fault /rɪ'vɝs fɔlt/ 逆斷層 128
reversed polarity /rɪ'vɝst pu'lærətɪ/ 極性倒轉 14
rheomorphism /,rɪɑ'mɔrfɪzm/ 柔流變質作用；軟流變質作用 93
rhombohedral system /,rɑmbə'bɛdrəl 'sɪstəm/ 菱形晶系：三方晶系 43
rhyolite /'raɪə'laɪt/ 流紋岩 76
rhythmic sequence /'rɪðmɪk 'sikwəns/ 韻律層序 112
ria /'riɑ/ 里亞式海灣 39
ria coast /'ria kɔst/ 沉降海岸 39
Richter scale /'rɪktɚ skel/ 里氏震級表 ***see*** **magnitude** 12
ridge /rɪdʒ/ 脊 159
ridge, mare /rɪdʒ, mɛr/ 月海脊 150
  oceanic /,oʃɪ'ænɪk/ 洋脊 35
  mid-oceanic /'mɪd'oʃɪ'ænɪk/ 洋中脊 35
riebeckite /'rɪbɛkaɪt/ 鈉閃石 57
rift /rɪft/ 斷陷谷；裂陷 133
rift valley /rɪft 'vælɪ/ 裂谷 133
rigid /'rɪdʒɪd/ 剛性的 159
rill marks /rɪl mɑrks/ 流痕 83
rille /rɪl/ 月溪；月面溝紋 150
  normal /'nɔrmḷ/ 直月溪 150
  sinuous /'sɪnjuəs/ 曲月溪 150
ring-complex /rɪŋ 'kɑmplɛks/ 環狀雜岩 67
ring-dyke /rɪŋ daɪk/ 環狀岩牆 67
ring silicates /rɪŋ 'sɪlɪkɪts/ 環矽酸鹽 53
ripple mark /'rɪpḷ mɑrk/ 波痕 83
rise, continental /raɪz, kɑntə'nɛntḷ/ 大陸隆 34
river terrace /'rɪvɚ 'tɛrɪs/ 河成階地；河岸階地 25
rivers /'rɪvɚz/ 河流 24 - 27
roches moutonnees /roʃɪz mutɔ'nes/ 羊背石；羊背岩 29
rock /rɑk/ 岩石 62
rock fall /rɑk fɔl/ 岩崩 143
rock glacier /rɑk 'gleʃɚ/ 石冰川 21
rock mechanics /rɑk mə'kænɪks/ 岩石力學 143
rock salt /rɑk sɔlt/ 岩鹽 52

rock-stratigraphical (unit) /rɑk strəˈtɪgrəfɪkl̩/(ˈjunɪt)/ 岩石地層(單位) 113
rotational fault /roˈteʃənl̩ fɔlt/ 旋轉斷層 129
rounded /ˈraʊndɪd/ 圓形的 83
rubidium-strontium dating /ruˈbɪdɪəm ˈstrɑnʃɪəm detɪŋ/ 銣鍶年齡測定 120
rudaceous rocks /ruˈdeʃɪəs rɑks/ 礫狀岩
see **rudites** 85, 87
rudite /rʊdaɪt/ 礫屑岩 85
runoff /ˈrʌn,ɔf/ 經流 146
rutile /ˈrutil/ 金紅石 49

sabkha /ˈsɛbkə/ 鹼灘；鹽沼 38
salt /sɔlt/ 鹽 17
salt dome /sɔlt dom/ 鹽丘 131
sand /sænd/ 砂 87
sandstone /ˈsænd, ston/ 砂岩 87
sandstone dyke /ˈsænd,ston daɪk/ 砂岩岩牆 84
sanidine /ˈsænɪdin/ 透長石 56
saturation /,sætʃəˈreʃən/ 飽和 74
scheelite /ˈʃelaɪt/ 白鎢礦；重石礦 52
schiller /ˈʃɪlɚ/ 閃光 47
schist /ʃɪst/ 片岩 97
schistosity /ˈʃɪstosɪtɪ/ 片理 95
schlieren /ˈslɪrən/ 析離體；異離體 65
schuppen structure /ˈʃʊpən ˈstrʌktʃɚ/ 叠置構造 130
scoria /ˈskorɪə/ 火山渣 69
scoriaceous /ˈskorɪeʃəs/ 渣狀的 73
scour-and-fill /skaʊr ənd fɪl/ 沖淤作用 82
scree /skri/ 山麓碎石；岩屑堆 21
Scyphozoa /,saɪfəˈzoə/ 水母綱 *see* **Coelentera** 105
sea cave /si kev/ 海蝕洞 37
sea, epicontinental /sil, l,ɛpɪ,kɑntəˈnɛntl̩/ 陸緣海 34
sea-floor spreading /si flɔr sprɛdɪŋ/ 海底擴張 136
seamount /ˈsi,maʊnt/ 海山 35
secondary coast /ˈsɛkən,dɛrɪ kost/ 次生海岸 39
section (exposure) /ˈsɛkʃən/ 剖面(露頭) 148
section, thin /ˈsɛkʃən, θɪn/ 薄片 147
sedentary /ˈsɛdn̩,tɛrɪ/ 定居的 101
sediment /ˈsɛdəmənt/ 沉積物 80
sedimentary evironment /,sɛdəˈmɛntərɪ ɪnˈvaɪrənmənt/ 沉降環境 81
sedimentary structures /,sɛdəˈmɛntərɪ ˈstrʌktʃɚz/ 沉積構造 83
sedimentation /,sɛdəmənˈteʃən/ 沉積物形成作用 80
sedimentology /,sɛdəmənˈtalədʒɪ/ 沉積學 80
seif /sef/ 劍沙丘；直線沙丘 22
seismic /ˈsaɪzmɪk/ 地震的 12
seismograph /ˈsaɪzmə,græf/ 地震儀 12
seismology /saɪzˈmɑlədʒɪ/ 地震學 12
selenite /ˈsɛlə,naɪt/ 透石膏 52
sequence /ˈsikwəns/ 層序 112
cyclic /ˈsaɪklɪk/ 旋迴層序 112
rhythmic /ˈrɪðmɪk/ 韻律層序 112
series /ˈsɪrɪz/ 系列 159
series (stratigraphical) /ˈsɪrɪz/ 統(地層學的) 113
serpentine /ˈsɝpən,tin/ 蛇紋石 61
sessile /ˈsɛsl̩/ 固着的 101
shadow zone /ˈʃædo zon/ 陰影區；震影帶 12
shale /ʃel/ 頁岩 88
shear cleavage /ʃɪr ˈklivɪdʒ/ 剪劈理 95
shear fold /ʃɪr fold/ 剪切褶皺 126
shear stress /ʃɪr strɛs/ 剪應力 *see* **stress** 122
sheet silicates /ʃɪt ˈsɪlɪ,kets/ 片狀矽酸鹽類 54
shelf /ʃɛlf/ 陸架 34
shield /ʃild/ 地質 133
shield volcano /ʃild vɑlˈkeno/ 盾形火山 68
shingle /ˈʃɪngl̩/ 扁礫 37
shore /ʃɔr/ 海濱 37
shoreline /ʃɔr laɪn/ 海濱線 37
sial /ˈsaɪ,æl/ 矽鋁層；硅鋁層 10
siderite (mineral) /ˈsɪdə,raɪt/ 菱鐵礦(礦物) 51
siderite (meteorite) /ˈsɪdə,raɪt/ 鐵隕(隕石) 149
siderolite /ˈsɪdərə,laɪt/ 石鐵隕石 149
siderophile /ˈsɪdərəfaɪl/ 親鐵的 18
silica /ˈsɪlɪkə/ 矽石；矽氧：氧化矽 16
silicate /ˈsɪlɪkɪt/ 矽酸鹽；硅酸鹽 16
silicate structures /ˈsɪlɪkɪt ˈstrʌktʃɚ/ 矽酸鹽構造 53-4
siliceous /sɪˈlɪʃəs/ 矽質的 87
sill /sɪl/ 岩床 66
sillimanite /ˈsɪlɪmə,naɪt/ 矽線石 59

silt /sɪlt/ 粉砂 88
siltstone /ˈsɪltˌston/ 粉砂岩 88
silurian /səˈlʊrɪən/ 志留紀(的) 114
sima /ˈsaɪmə/ 矽鎂層；硅鎂層 10
similar fold /ˈsɪmələ˞ fold/ 相似褶皺 124
sinistral fault /ˈsɪnɪstrəl fɔlt/ 左行平移斷層 129
sink-hole /ˈsɪŋkˌhol/ 落水洞，溶岩 21
site investigation /saɪt ɪnˌvɛstəˈgeʃən/ 場地勘察 143
skarn /skɑn/ 夕卡岩 96
skeleton /ˈskɛlətn̩/ 骨骼 159
slate /slet/ 板岩 96
slaty cleavage /ˈsletɪˈ klivɪdʒ/ 板狀劈理 95
slickenside /ˈslɪkənˌsaɪd/ 斷層擦痕 129
slide /slaɪd/ 滑移斷層 130
slip /slɪp/ 滑距 128
slip fold /slɪp fold/ 滑褶皺 126
slope, continental /slop, ˌkɑntəˈnɛntl̩/ 大陸坡 34
slump /slʌmp/ 滑塌 143
smectite /ˈsmɛktaɪt/ 膠嶺石；蒙脱石 61
smithsonite /ˈsmɪθsənˌaɪt/ 菱鋅礦 51
soda feldspar /ˈsodə ˈfɛldˌspɑr/ 鈉長石 56
sodalite /ˈsodəˌlaɪt/ 方鈉石 58
sodium feldspar /ˈsodɪəm ˈfɛldˌspɑr/ 鹼性長石
see **feldspar** 56
soil /sɔɪl/ 土壤 23
soil creep /sɔɪl krip/ 土滑 143
soil horizon /sɔɪl həˈraɪzn̩/ 土壤層 23
soil mechanics /sɔɪl məˈkænɪks/ 土壤力學
see **rock mechanics** 143
sɔɪl ˈprofile/sɔɪl ˈprofaɪl/ 土壤剖面 23
soil, lunar /sɔɪl, lunə˞/ 月土 150
sole marks /sol mɑrks/ 底痕 83
solid solution /ˈsɑlɪd səˈluʃən/ 固溶體 46
solid solution series /ˈsɑlɪd səˈluʃən ˈsɪrɪz/
固溶體系列 46
solidus /ˈsɑlɪdəs/ 因相曲線 71
solifluction /ˌsɑlɪˈflʌkʃən/ 解凍泥流 30
solution /səˈluʃən/ 溶解 159
sorosilicates /ˌsɔrəˈsɪlɪkɪts/ 群島狀矽酸鹽類；儔矽
酸鹽類 53
sorting /sɔrtɪŋ/ 分選 80
sparite /ˈspæraɪt/ 亮晶石灰岩 86
spatter cone /ˈspætə˞ kon/ 熔岩滴錐：寄生熔岩錐 69
speciation /ˌspiʃɪˈeʃən/ 物種形成 103
species /ˈspiʃɪz/ 種 99
specific gravity /spɪˈsɪfɪk ˈgrævətɪ/ 比重 44
specimen /ˈspɛsəmən/ 標本 147
Spermatophyta /ˈspɝmətəˌfaɪtə/ 種子植物門 110
spermatophytes /ˈspɝmətəˌfaɪts/ 種子植物類
see **Spermatophyta** 110
spessartite /ˈspɛsə˞taɪt/ 錳鋁石榴石 58
sphalerite /ˈsfæləˌraɪt/ 閃鋅礦 50
sphene /sfin/ 榍石 59
spherulite /ˈsfɪrəˌlaɪt/ 球粒 73
spilite /ˈspaɪlit/ 細碧岩 78
spinels /spɪˈnɛlz/ 尖晶石族 49
spit /spɪt/ 沙嘴；岬 38
sponges /spʌndʒɪz/ 海綿 ***see*** **Porifera** 104
spreading, sea-floor /sprɛdɪŋ, siˌflor/ 海底擴張 136
spring /sprɪŋ/ 泉 146
spur, truncated /spɝl, trʌŋketɪd/ 削斷山嘴 31
stable /ˈstebl̩/ 穩定的 159
stack (sea) /stæk/(/si/) 海蝕柱 38
stage (stratigraphical) /stedʒ/ 階(地層學的) 113
stalactite /stəˈlæktaɪt/ 鐘孔石 21
stalagmite /stəˈlægmaɪt/ 石笋 21
states of matter /stets ɔv ˈmætə˞/ 物態 159
staurolite /ˈstɔrəˌlaɪt/ 十字石 59
stock /stɑk/ 岩株 65
stockwork /ˈstɑkˌwɝk/ 網脈 145
stony (meteorite) /ˈstonɪ/ 石隕石 149
stony-iron (meteorite) /ˈstonɪ aɪə˞n/ 石鐵隕石 149
stoping /stɑpɪŋ/ 頂蝕作用 65
strain /stren/ 應變 122
strata /ˈstretə/ 地層 ***see*** **stratum** 80
stratification /ˌstrætəfəˈkeʃən/ 層理 80
stratified /ˈstrætəˌfaɪd/ 分層的 80
stratigraphical palaeontology /ˌstrætɪˈgræfɪkl̩
ˌpelɪɑnˈtɑlədʒɪ/ 地層古生物學 111
stratigraphy /strəˈtɪgrəfɪ/ 地層學 112
stratovolcano /ˈstætəˌvɑlˈkeno/ 成層火山 68
stratum /ˈstretəm/ 地層 80

streak /strik/ 條痕 44
stream, beheaded /strim, bɪˈhɛdɪd/ 斷頭河 25
capture /ˈkæptʃɚ/ 襲奪河；掠奪河 25
consequent /ˈkɑnsəˌkwɛnt/ 順向河；順斜河 27
effluent /ˈɛfluənt/ 潛水補給河，側流河 25
influent /ˈɪnfluənt/ 滲流河 25
misfit /ˈmɪsˌfɪt/ 不稱河 25
obsequent /ˈɑbsəkwənt/ 逆向河；反斜河 27
subsequent /ˈsʌbsɪˌkwɛnt/ 後成河；順層河 27
stream gradient /strim ˈgredɪənt/ 河流坡降 24
stream profile /strim ˈprofaɪl/ 河流剖面 24
streams /strimz/ 江河 24–27
stress /strɛs/ 應力 122
directed /dəˈrɛktɪd/ 定向應力 122
hydrostatic /ˌhaɪdrəˈstætɪk/ 靜水應力 122
striae, glacial /ˈstraɪ·i, ˈgleʃəl/ 冰川擦痕 29
striations, glacial /straɪˈeʃənz, ˈgleʃəl/ 冰川條痕 29
strike /straɪk/ 走向 123
strike-slip fault /straɪk slɪp fɔlt/ 平移斷層；走向滑動斷層 129
stromatolites /ˈstroməmætəlaɪts/ 疊層石 101
structure (geological) /ˈstrʌktʃɚ/ 構造（地質的） 122
structure, molecular /ˈstrʌktʃɚ, məˈlɛkjəlɚ/ 分子結構 15
structures, sedimentary /ˈstrʌktʃɚz, ˌsɛdəˈmɛntərɪ/ 沉積構造 83
subangular /ˈsʌbˈæŋgjəlɚ/ 近稜角狀的 83
sub-bituminous coal /sʌb bɪˈtjumənəs kol/ 次烟煤 89
subduction zone /səbˈdʌkʃən zon/ 俯衝帶 137
sub-era /sʌb ˈɪrə/ 亞代 113
subfamily /sʌbˈfæməlɪ/ 亞科 99
subhedral /ˈsʌbˈhidrəl/ 半自形的 45
subjacent /sʌbˈdʒesn̩t/ 深成的；深淵的 65
submarine /ˌsʌbməˈrin/ 海底的 35
submarine canyon /ˌsʌbməˈrin ˈkænjən/ 海底峽谷 35
submarine plateau /ˌsʌbməˈrin plæˈto/ 海底高原 35
submerged coast /səbˈmɝdʒd kost/ 下沉海岸 39
subrounded /sʌbˈraundɪd/ 近圓形的 83
subsequent (stream) /ˈsʌbsɪˌkwɛnt/ 後成（河）；順層（河） 27
subsidence /səbˈsaɪdn̩s/ 沉陷 125
subsoil /ˈsʌbˌsɔɪl/ 底土，亞層土 23
subsurface /sʌbˈsɝfɪs/ 地下的 159
succession /səkˈsɛʃən/ 序列 112
sulphate /ˈsʌlfet/ 硫酸鹽 16
sulphide /ˈsʌlfaɪd/ 硫化物 16
sulphur, native /ˈsʌlfɚ, ˈnetɪv/ 自然硫 48
supercooling /ˌsupɚˈkulɪŋ/ 過冷却 19
superfamily /ˌsupɚ ˈfæməlɪ/ 總科；超科 99
superimposed (drainage) /ˌsupərɪmˈpozd/ 疊置（水系） 27
superposition (principle of) /ˌsupɚpəˈzɪʃən/ 疊覆（原理） 112
survey /ˈsɝve/ 測量 148
geological /ˌdʒiəˈlɑdʒɪkl̩/ 地質調查 148
suspension load /səˈspɛnʃən lod/ 懸移質 24
suture /ˈsutʃɚ/ 地縫合線 140
swallow-hole /ˈswɑlo hol/ 落水洞，溶岩 21
swamp, marine /swɑmp, məˈrɪn/ 沿海沼澤 38
swash marks /swaʃ marks/ 沖痕 83
s-wave /ɛs wev/ S 波 12
syenite /ˈsaɪəˌnaɪt/ 正長岩 76
syenodiorite /saɪəˌnəˈdaɪəˌraɪt/ 正長閃長岩 78
sylvite /ˈsɪlvaɪt/ 鉀鹽 52
symmetrical fold /sɪˈmɛtrɪkl̩ fold/ 對稱褶皺 126
symmetry /ˈsɪmɪtrɪ/ 對稱 159
(crystal) (/ˈkrɪstl̩/) 晶體對稱性 42
axis of /ˈæksɪs əv/ 對稱軸 42
centre of /ˈsɛntɚ əv/ 對稱中心 42
plane of /plen ɔv/ 對稱面 42
symmetry element /ˈsɪmɪtrɪ ˈɛləmənt/ 對稱要素 42
symplectic (texture) /sɪmˈplɛktɪk/ 後成合晶的結構 73
syncline /ˈsɪŋklaɪn/ 向斜 124
synclinorium /ˌsɪŋkləˈnɔrɪəm/ 複向斜 124
synform /ˈsɪnfɔrm/ 向形 124
synkinematic /sɪnˌkaɪnɪˈmætɪk/ 同造山運動的 116
synorogenic /sɪnˌorɑˈdʒənɪk/ 同造山期的 116
syntectonic /ˌsɪntɛkˈtɔnɪk/ 同構造的 116
syntexis /sɪnˈtɛksɪs/ 同熔作用 93
system open /ˈsɪstəm ˈopən/ 開放系統 19
closed /klozd/ 封閉系統 19
geological /ˌdʒiəˈlɑdʒɪkl̩/ 系（地層學的） 113

tabular (habit) /ˈtæbjələ˞/ 板狀（習性） 45
Taconic /ˈtækənɪk/ 塔康期的 116
talc /tælk/ 滑石 61
talus /ˈteləs/ 山麓堆積 21
taphonomy /tæpˈhɔnəmɪ/ 化石埋藏學 99
taphrogenesis /ˌtæfrəˈdʒɛnəsɪs/ 地裂作用 133
taxon /ˈtæksən/ 分類單位 99
taxonomy /tæksˈɑnəmɪ/ 分類學 99
tear fault /tɛr fɔlt/ 捩斷層 129
tectonics /tɛk ˈtɑnɪks/ 大地構造學；構造地質學 122
  plate /plet/ 板塊構造 134
tectonite /tɛkˈtɔnaɪt/ 構造岩 97
tectosilicates /tɛkˈtosɪlɪkɪts/ 架狀矽酸鹽類 54
tektite /ˈtɛkˌtaɪt/ 玻隕石 149
temperature /ˈtɛmprətʃə˞/ 温度 160
temperature-composition diagram /ˈtɛmprətʃə˞-kɑmpəˈzɪʃən ˈdaɪəˌgræm/ 温度組成圖解 71
tensile stress /ˈtɛnsl̩ strɛs/ 張應力 ***see* stress** 122
tension gash /ˈtɛnʃən gæʃ/ 張裂隙 21
terminal moraine /ˈtɝmənl̩ moˈren/ 終磧；終冰蹟 29
term /tɝm/ 術語 160
ternary diagram /ˈtɝnərɪ ˈdaɪəˌgræm/ 三元相圖 71
terra rossa /ˈtɛrɑ rɔsə/ 鈣質紅土 33
terrae (lunar)/ˈtɛri/ 月陸 150
terrace, marine /ˈtɛrɪs, məˈrin/ 海蝕階地 39
  river /ˈrɪvə˞/ 河成階地：河岸階地 25
terrestrial /təˈrɛstrɪəl/ 陸地的 81
terrestrial abundances (of elements) /təˈrɛstrɪəl əˈbʌndənsɪz/(/əv ˈɛləmənts/) 地球豐度（元素的）
  ***see* abundances of elements** 18
terrestrial magnetism /təˈrɛstrɪəl ˈmægnəˌtɪzəm/ 大地磁場 14
terrigenous (sediments) /tɛˈrɪdʒɪnəs/ 陸源的（沉積的） 85
Tertiary /ˈtɝʃɪˌɛrɪ/ 第三亞代（的） 115
Tethys /ˈtiθɪs/ 特提斯海 141
tetragonal system /tɛˈtrægənl̩ ˈsɪstəm/ 四方晶系 43
tetrahedron /ˌtɛtrəˈhidrən/ 四面體 41
texture (rock)/ˈtɛkstʃə˞/ 結構（岩石） 72
Thalweg /ˈtɑlvek/ 最深河谷底線 24
theory /ˈθiərɪ/ 學說；理論 160
thermal metamorphism /ˈθɝml̩ ˌmɛtəˈmɔrfɪzm̩/ 熱力變質作用 90
thin section /θɪn ˈsɛkʃən/ 薄片 147
tholeiite /θɔlɪaɪt/ 拉斑玄武岩 77
thorium-lead dating /ˈθorɪəm lid detɪŋ/ 釷一鉛年齡測定
  ***see* uranium-lead dating** /juˈrenɪəm/ 120
throw /θro/ 垂直斷距；落差 128
thrust /θrʌst/ 衝斷層 129
thrust fault /θrʌst fɔlt/ 逆衝斷層 129
thrust plane /θrʌst plen/ 逆衝斷層面 129
thrust sheet /θrʌst ʃit/ 逆衝斷層面 130
tide /taɪd/ 潮；潮汐 37
tide mark /ˈtaɪd ˌmɑrk/ 潮標 37
till /tɪl/ 冰磧物 29
tillite /ˈtɪlaɪt/ 冰磧岩 29
time, geological /taɪm, ˌdʒiəˈlɑdʒɪkl/ 地質時代 120
time plane /taɪm plen/ 同時面；時間面 112
tissue /ˈtɪʃu/ 組織 160
titanite /ˈtaɪtəˌnaɪt/ 石 59
tongue /tʌŋ/ 岩舌 66
topaz /ˈtopæz/ 黃晶；黃玉；黃石英 59
topset beds /tɔset bɛdz/ 頂積層 82
tourmaline /ˈturməlɪn/ 電氣石 60
trace element /tresˈɛləmənt/ 微量元素 15
trace fossil /tres ˈfɑsl̩/ 遺迹化石 98
trachyte /ˈtrekaɪt/ 粗面岩 77
traction load /ˈtrækʃən lod/ 底移質 24
transcurrent fault /trænsˈkɝənt fɔlt/ 橫推斷層 129
transform fault /trænsˈfɔrm fɔlt/ 轉換斷層 136
transgression, marine /trænsˈgrɛʃən, məˈrin/ 海進 119
transient (evolution) /ˈtrænʃənt/ 漸變階段（演化） 103
transport /trænsˈport/ 搬運 21
transportation /ˌtrænspə˞feʃən/ 搬運（作用） 21
transverse dune /trænsˈvɝs dun/ 橫向沙丘 22
trap, stratigraphical /træp, strəˈtɪgrəfɪkl̩/ 地層圈閉 144
  structural /ˈstrʌktʃərəl/ 構造圈閉 144
tree-ring dating /tri rɪŋ detɪŋ/ 樹木年輪年齡鑒定法 121
tremolite /ˈtrɛməˌlaɪt/ 透閃石 57
tremor /ˈtrɛmə˞/ 小震 13
trench, ocean /trɛntʃ, ˈoʃən/ 洋溝；海溝 35
trend (evolutionary) /trɛnd/ 趨勢（演化的） 103

triangular diagram /traɪˈæŋgjələ˞ ˈdaɪəˌgræm/ 三角形圖解 71
Trias /ˈtraɪəs/ 三疊系 115
Triassic /traɪˈæsɪk/ 三疊紀(的) 115
triclinic system/traɪˈklɪnɪk ˈsɪstəm/ 三斜晶系 43
tridymite /ˈtrɪdəˌmaɪt/ 鱗石英 55
trigonal system /ˈtrɪgənəl ˈsɪstəm/ 三方晶系 43
Trilobita /ˈtraɪləˌbaɪtə/ 三葉蟲綱 108
trilobites /ˈtraɪləˌbaɪts/ 三葉蟲 ***see*** **Trilobita** 108
triple junction /ˈtrɪpl̩ ˈdʒʌŋkʃən/ 三接合點;三重點 135
trough (fold) /trɔf/ 褶皺槽 124
trough faulting /trɔf fɔltɪŋ/ 槽形斷層作用 133
true dip /tru dɪp/ 真傾斜 ***see*** **dip** 123
truncated spur /ˈtrʌŋketɪd spɝ/ 削斷山嘴 31
tsunami /tsuˈnami/ 海嘯 36
tufa /ˈtufə/ 石灰華 ***see*** **calcareous tufə** 21
tuff /tʌf/ 凝灰岩 69
turbidite /ˈtɝbɪdaɪt/ 濁積岩 ***see*** **turbidity current** 36
turbidity current /tɝbɪdətɪ ˈkɝənt/ 濁流 36
twin crystal /twɪn ˈkrɪstl̩/ 孿晶;雙晶 41
type-area /taɪpˈerɪə/ 標準地區 118
type locality /taɪp loˈkælətɪ/ 標準地點 118
type (specimen) /taɪp/ 模式(標本)99

ultrabasic (rocks) /ˈʌltrəˈbesɪk/ 超基性的(岩) 75
ultramafic (rocks) /ˈʌltrəˈmæfɪk/ 超鎂鐵質的(岩) 75
unconformity /ˌʌnkənˈfɔrmətɪ/ 不整合 118
unconsolidated /ˌʌnkənˈsɑləˌdetɪd/ 未固結的 84
undersaturated /ˌʌndə˞ˈsætʃəˌretɪd/ 不飽和的 74
Uniformitarianism /ˌjunəˌfɔrməˈtɛrɪənɪzm̩/ 均變說;天律不變說 112
uninverted /ˌʌnɪnˈvɝtɪd/ 未倒轉的 127
unit /ˈjunɪt/ 單元;單位 160
unit cell /ˈjunɪt sɛl/ 晶胞 40
unit form /ˈjunɪt fɔrm/ 單元晶形 41
unmetamorphosed /ˈʌnˌmɛtəˈmɔrfozd/ 未變質的 90
unreactive /ˌʌnrɪˈæktɪv/ 惰性的 ***see*** **reactive** 17
unstratified /ʌnˈstrætəfaɪd / 不成層的 80
unweathered /ˈʌnˈwɛðə˞d/ 未風化的 20
U-Pb dating /juˈrenɪəm-lɛd ˈdetɪŋ/ U-Pb 年齡測定 ***see*** **uranium-lead dating** 120
uplift /ˈʌpˌlɪft/ 上升 125
upthrow /ˈʌpˈθro/ 上升的 128
upward /ˈʌpwə˞d/ 向上 160
upwarp /ˈʌpˈwɔrp/ 陸起(區) 125
uraninite /juˈrænəˌnaɪt/ 晶質鈾礦 48
uranium-lead dating /juˈrenɪəm lɛd detɪŋ/ 鈾鉛年齡測定 120
U-shaped valley /juʃept ˈvælɪ/ U 形谷 31
uvarovite /uˈvɑrəfaɪt/ 鈣鉻石榴石 58

valley glacier /ˈvælɪ ˈgleʃə˞/ 山谷冰川 28
valley, dry /ˈvælɪ, draɪ/ 乾谷 25
hanging /ˈhæŋɪŋ/ 懸谷 31
U-shaped /ˈjuˌʃept/ U 形谷 31
variation /ˌvɛrɪˈeʃən/ 變化 160
Variscan /ˈvɑrɪskən/ 華里西期的 116
varnish, desert /ˈvɑrnɪʃ, ˈdɛzə˞t/ 沙漠岩漆 22
varve-count /ˈvɑrv-kaunt/ 紋泥計齡法 121
vary /ˈvɛrɪ/ 改變 160
vascular plants /ˈvæskjələ˞ plænts/ 維管植物 110
vein /ven/ 礦脈;脈體 145
velocity /vəˈlɑsətɪ/ 速度 160
vent, volcanic /vɛnt, vɑlˈkænɪk/ 火山噴出口;火山通道 68
ventifact /ˈvɛntɪˌfækt/ 風棱石 22
Vermes /ˈvɝmiz/ 蠕形動物;蠕蟲總門 105
vermiculite /vɝmɪkjəˌlaɪt/ 蛭石 61
Vertebrata /ˌvɝtəˈbretə/ 脊椎動物亞門 109
vertebrates /ˈvɝtəˌbrets/ 脊椎動物 ***see*** **Vertebrata** 109
vertical /ˈvɝtɪkl̩/ 垂直的 160
vesicle /ˈvɛsɪkl̩/ 氣孔 73
vesicular /vəˈsɪkjulə˞/ 多孔的 73
vesuvianite /vɪˈsuvɪənˌaɪt/ 維蘇威石 60
vibrate /ˈvaɪbret/ 振動 160
volatile (constituent) /ˈvɑlətl̩/ 揮發(組份) 18
volcanic /vɑlˈkænɪk/ 火山的 ***see*** **volcano** 68
volcanic, agglomerate /vɑlˈkænɪk, əˈglɑmərɪt/ 火山集塊岩 69
volcanic ash /vɑlˈkænɪk æʃ/ 火山灰塵 69
volcanic block /vɑlˈkænɪk blɑk/ 火山塊 69
volcanic bomb /vɑlˈkænɪk bɑm/ 火山彈 69

volcanic breccia /vɑlˈkænɪk ˈbrɛtʃɪə/ 火山角礫岩 69
volcanic cone /vɑlˈkænɪk kon/ 火山錐 68
volcanic conduit /vɑlˈkænɪk ˈkɑndɪt/ 火山道 68
volcanic dust /vɑlˈkænɪk dʌst/ 火山塵 69
volcanic eruption /vɑlˈkænɪk ɪˈrʌpʃən/ 火山噴發 68
volcanic neck /vɑlˈkænɪk nɛk/ 火山頸 68
volcanic plug /vɑlˈkænɪk plʌg/ 火山栓 68
volcanic vent /vɑlˈkænɪk vɛnt/ 火山口 68
volcanism /ˈvɑlkənɪzm̩/ 火山活動
see **volcano** 68
volcano /vɑlˈkeno/ 火山 68
shield /ʃild/ 盾形火山 68
volume /ˈvɑljəm/ 體積 161
V-shaped valley /vi ʃept ˈvælɪ/ V 形河谷 25
vulcanism /ˈvʌlkənɪzəm/ 火山作用 *see* **volcano** 68

wadi /ˈwɑdɪ/ 間歇乾谷；旱谷 25
wall rock /wɔl rɑk/ 圍岩 145
warping /wɔrpɪŋ/ 翹曲 125
washout /ˈwɑʃˈaʊt/ 沖蝕溝 82
waterfall /ˈwɔtɚˌfɔl/ 瀑布 25
water, connate /ˈwɔtɚ, ˈkɑnet/ 原生水 146
juvenile /ˈdʒuvənl̩/ 初生水；岩漿水 146
meteoric /ˌmitɪˈɔrɪk/ 大氣降水 146
watershed /ˈwɔtɚˌʃɛd/ 分水界；流域 25
water table /ˈwɔtɚ ˈtebl̩/ 地下水位 146
wave base /wev bes/ 波蝕低面 36
wave-cut platform /wev kʌt ˈplætˌfɔrm/ 浪蝕臺地；海蝕臺 37
wave-cut notch /wev kʌt natʃ/ 海蝕龕 37
wave motion /wev ˈmoʃən/ 波動 161
wavelength /wev lɛŋkθ/ 波長 161
way-up /we ʌp/ 層位頂面向上 127
weathering /ˈwɛðərɪŋ/ 風化 20
chemical /ˈkɛmɪkl̩ / 化學風化 20
mechanical /məˈkænɪkl̩/ 機械風化 20
aeolian /iˈolɪən/ 風力風化 22
differential /ˌdɪfəˈrɛnʃəl / 差異風化 21
eolian /ɪˈolɪən/ 風力風化 *see* **aeolian** 22
Weichert-Gutenberg discontinuity/ˈwaɪkɚt-ˈgutn̩bæg ˌdɪskɑntəˈnuətɪ/ 魏徹特-古登堡間斷面 10
well-logging /wɛl ˈlɔgɪŋ/ 測井 144
well-rounded /ˈwɛl ˈraʊndɪd/ 渾圓的 83
whaleback dune /ˈhwelˌ bæk dun/ 鯨背沙丘 22
white mica /hwaɪt ˈmaɪkə/ 白雲母 ***see* muscovite** 55
window /ˈwɪndo/ 構造窗 130
witherite /ˈwɪðəraɪt/ 毒重石 51
wolframite /ˈwʊlfrəmˌaɪt/ 黑鎢礦；鎢錳鐵礦 52
worms /wɝmz/ 蠕蟲類 *see also* Annelida 105
wrench fault /rɛntʃ fɔlt/ 扭斷層 129
wrinkle ridge /ˈrɪŋkl̩ rɪdʒ/ 月面皺脊 150

xenoblastic /ˈzənɑbæstɪk/ 他形變晶的 94
xenocryst /ˈzɛnəˌkrɪst / 捕擄晶 73
xenolith /ˈzɛnəlɪθ/ 捕擄體 65

young (*v*) /jʌŋ/ 面向 127

zeolites /ˈziəˌlaɪts/ 沸石類 61
zeugen /ˈzjugən/ 風蝕桌狀石；風蝕柱 21
zig-zag fold /ˈzɪgzæg fold/ 鋸齒狀褶皺 126
zinc blende /zɪŋk blɛnd/ 閃鋅礦 50
zircon /ˈzɝkɑn/ 鋯石 59
zoisite /ˈzɔɪsaɪt/ 黝簾石 60
zone /zon/ 帶 161
zone (crystallographic) /zon/ 晶帶 40
(metamorphic) (/ˌmɛtəˈmɔrfɪk/) (變質)帶 91
neritic /nɪˈrɪtɪk/ 淺海 34
(stratigraphical) (/strəˈtɪgrəfɪkl̩/) (地層學的)帶 117
zone fossil/zon ˈfɑsl̩/ 分帶化石 111
zoned crystal /zond ˈkrɪsl̩/ 環帶狀晶體 46
zone of convergence /zon əv kənˈvɝdʒəns/ 會聚帶 137
zone of divergence /zon əv dəˈvɝdʒəns/ 離散帶 136
zoning, compositional /ˈzonɪŋ, ˌkɑmpaˌzɪʃənl̩/ 成分分帶 63